AF831212

In

Network O
Challenges
communica
readers wit
attaining se

This boo
maximizin
protocols a
flaws, and c

This boo
knowledge
mization in
predictive
integration
attaining pe

This boo
and cutting
whether it
book unlea
various sec

Network Optimization in Intelligent Internet of Things Applications

Principles and Challenges

Edited by
Payal Khurana Batra, Pawan Singh Mehra,
and Sudeep Tanwar

CRC Press
Taylor & Francis Group
Boca Raton London New York

CRC Press is an imprint of the
Taylor & Francis Group, an **informa** business

Designed cover image: ShutterStock Images

First edition published 2025
by CRC Press
2385 NW Executive Center Drive, Suite 320, Boca Raton FL 33431

and by CRC Press
4 Park Square, Milton Park, Abingdon, Oxon, OX14 4RN

CRC Press is an imprint of Taylor & Francis Group, LLC

British Library Cataloguing-in-Publication Data
A catalogue record for this book is available from the British Library

ISBN: 9781032521534 (hbk)
ISBN: 9781032521763 (pbk)
ISBN: 9781003405535 (ebk)

DOI: 10.1201/9781003405535

Typeset in Times
by Newgen Publishing UK

Contents

Preface

The phrase Internet of Things (IoT) was first used in 1999 by Kevin Ashton. In IoT, billions of tiny, resource-constrained devices are connected to share data among them. A large amount of data generated by the various IoT devices creates an overhead on the network, leading to many network-related issues such as congestion, energy conservation heterogeneity, scalability, reliability, quality of service (QoS), and security. Network optimization is critical in ensuring efficient and seamless connectivity among interconnected devices. This book comprehensively describes the essential aspects of IoT and various network optimizations from beginner- to advanced-level perspectives. It describes the fundamentals of IoT and its network optimization wherein discussions are carried out for system architecture, protocols, tools and technologies, applications, and various network optimization methods. The impact of artificial intelligence and intelligent computing is provided for depicting the enhancement of existing systems. Numerous network optimization-related issues are covered to guarantee the smooth operation of IoT ecosystems, enable real-time communication, and unleash the full potential of networked devices across a range of industries. This book is compiled to address maximal issues related to IoT network optimization with some existing improved solutions.

Preface

About the Editors

Payal Khurana Batra, Phd, is an assistant professor in the Department of Computer Science Engineering and Information Technology at Jaypee Institute of Information Technology, Noida, India. She earned her B.Tech. in Computer Science and Engineering from Kurukshetra University, Kurukshetra, India, in 2004 and her M.E. in Information Technology Degree from Panjab University, Chandigarh, India, in 2009. She earned her PhD from Jaypee Institute of Information Technology, Noida, India, in 2017. She has more than ten years of teaching experience. She has patents to her name in the domain of IoT. She has several publications in international conferences and journals indexed in SCI and Scopus. She is an active researcher in WSNs, IoT, soft computing techniques, and machine learning.

Pawan Singh Mehra, PhD, (Senior Member, IEEE) is an assistant professor at Delhi Technological University, New Delhi. He earned his PhD in Computer Engineering from Jamia Millia Islamia and M.Tech. (Hons) in Computer Science and Engineering from the Center for Development of Advanced Computing (CDAC). He completed a Bachelor of Engineering from RJIT, Gwalior. He has about 14 years of Teaching and Research Experience. He has authored more than 50 publications in international journals and conferences indexed in SCIE, Scopus, ESCI, and Web of Science. He is an active reviewer in international journals indexed in IEEE/ACM Transactions, SCIE, and Scopus. He was awarded the Research Excellence Award in 2022, 2023, and 2024. He is a Senior Member of IEEE, ACM, Fellow of IETE, and Life member of CSI and ISTE. He has published two books on the internet of things and edited a book for GATE. His research interests include wireless sensor networks, internet of things, quantum computing, and cryptography, blockchain, cryptography and network security, and artificial intelligence.

Sudeep Tanwar (Senior Member, IEEE) is a professor in the Computer Science and Engineering Department at the Institute of Technology, Nirma University, India. Tanwar is a visiting Professor at Jan Wyzykowski University in Polkowice, Poland, and the University of Pitesti in Pitesti, Romania. Tanwar's research interests include blockchain technology, wireless sensor networks, fog computing, smart grids, and IoT. He has authored 2 books and edited 13 books and more than 200 technical papers, including top journals and top conferences, such as IEEE TNSE, TVT, TII, WCM, Networks, ICC, GLOBECOM, and INFOCOM. Tanwar initiated the research field of blockchain technology adoption in various verticals in 2017. His h-index is 44. Tanwar actively serves his research communities in multiple roles. He is currently serving in the editorial boards of *Physical Communication, Computer Communications, International Journal of Communication System*, and *Security and Privacy*. He was awarded the best research papers from IEEE GLOBECOM 2018, IEEE ICC 2019, and Springer ICRIC-2019. He has served at many international conferences as a member of the organizing committee, such as publication chair for

FTNCT-2020, ICCIC 2020, WiMob 2019, member of advisory board for ICACCT-2021, ICACI 2020, workshop co-chair for CIS 2021, and general chair for IC4S 2019, 2020, ICCSDF 2020. Tanwar is a final voting member for the IEEE ComSoc Tactile Internet Committee in 2020. He is a member of IEEE, CSI, IAENG, ISTE, CSTA, and the Technical Committee on Tactile Internet of IEEE Communication Society. He is leading the ST research lab, where group members are working on the latest cutting-edge technologies.

Contributors

Reshu Agarwal
Amity University
INDIA

Harsh Aggarwal
Tecnia Institute of Advanced Studies, GGSIPU
INDIA

Puneet Kumar Aggarwal
ABES Engineering College
INDIA

Sangeeta Arora
KIET Group of Institutions
INDIA

Saumya Bansal
Bharati Vidyapeeth's Institute of Computer Applications and Management
INDIA

Gaurav Bathla
Chandigarh University
INDIA

Payal Khurana Batra
Jaypee Institute of Information Technology
INDIA

Arpita Jadhav Bhatt
Jaypee Institute of Information Technology
INDIA

Diksha Chawla
Delhi Technological University
INDIA

Yogita Chhabra
Jaypee Institute of Information Technology
INDIA

Deepakshi Dabas
Delhi Technological University
INDIA

Aryan Kumar Dubey
Shambhunath Institute of Engineering and Technology
INDIA

Pankaj Kumar Dubey
Kamla Nehru Institute of Technology
INDIA

Madhuri Gupta
Bennett University
INDIA

Umesh Gupta
Bennett University
INDIA

Meenal Jain
Jaypee Institute of Information Technology
INDIA

Parita Jain
KIET Group of Institutions
INDIA

Aatif Jamshed
ABES Engineering College
INDIA

Himanshu Khajuria
Amity University
INDIA

Zuha Khan
Amity University
INDIA

Mohit Lalit
Chandigarh University
INDIA

Pawan Singh Mehra
Delhi Technological University
INDIA

Kajol Mittal
Jaypee Institute of Information Technology
INDIA

Biswa Prakash Nayak
Amity University
INDIA

Deepika Pantola
Bennett University
INDIA

Anant Raghav
Tecnia Institute of Advanced Studies, GGSIPU
INDIA

Shreya Sahgal
Amity University
INDIA

Anita Sahoo
Jaypee Institute of Information Technology
INDIA

B. Santhosh
Dayananda Sagar College of Engineering
INDIA

Aryaman Sharma
Delhi Technological University
INDIA

Dolly Sharma
Amity University
INDIA

Jatin Sharma
Delhi Technological University
INDIA

Swati Sharma
KIET Group of Institutions
INDIA

Vikrant Shokeen
Maharaja Surajmal Institute of Technology
INDIA

Bindeshwar Singh
Kamla Nehru Institute of Technology
INDIA

Chintan Singh
Amity University
INDIA

Deependra Singh
Kamla Nehru Institute of Technology
INDIA

Mandeep Singh
Raj Kumar Goel Institute of Technology
INDIA

Pratibha Singh
Krishna Engineering College
INDIA

Surender Singh
Head Career Program, Code Quotient Pvt. Ltd
INDIA

Sanya Verma
Tecnia Institute of Advanced Studies, GGSIPU
INDIA

K. Viswanath
Siddaganga Institute of Technology, Tumkur
INDIA

Yash
Tecnia Institute of Advanced Studies, GGSIPU
INDIA

Part I

Introduction to IoT, Network Objectives, and Optimization Techniques

1 IoT Networks
Issues, Challenges, and Objectives

Yogita Chhabra and Arpita Jadhav Bhatt

1.1 INTRODUCTION

"IoT is a network of physical objects with embedded technology that allows them to communicate, perceive, and respond to internal or external parameters or circumstances" (Gartner) [1–6]. IoT consists of tangible objects, including devices, cars, buildings, and machines with connectivity, sensors, and software installed to enable data collection, sharing, and analysis [7]. IoT is a vast network of unconnected devices that can communicate with one another and with individuals via the internet and also characterized as a group of networked devices that can communicate and share data over the internet without a human being being present [1,3]. In addition to more complex systems such as transportation and utilities, these devices can comprise everyday objects such as sensors and appliances.

This network of interconnected devices enables real-time data collection and processing, delivers insightful information, and facilitates more efficient and effective decision-making. Challenges include security and data management [2,8]. By 2030, IoT insights estimate that there will likely be more than 50 billion connected devices globally, reflecting IoT technology's rapid growth and proliferation [9].

IoT has been a concept for decades, but in recent years, it has grown quickly and has been widely adopted because of technological improvements, affordable sensors, and connectivity. As a result, the things will essentially explode along with the number of connected things. The number of internet-connected devices is likely to surpass 50 billion in the not-too-distant future [9]. As anticipated, IoT is expected to connect many devices, reaching billions and potentially even trillions of interconnected things. IoT has become pervasive today, finding applications across diverse fields such as manufacturing, healthcare, agriculture, and transportation. It is a transformative technology driving innovation and efficiency in numerous industries [4].

One of the major advantages of IoT is its capacity to open up new avenues for automation and optimization, which boosts productivity, lowers costs, and produces better results. For instance, IoT in healthcare can provide remote patient monitoring, enabling medical practitioners to deliver more individualized and prompt care [10]. IoT has the potential to streamline supply chains and production processes in the

DOI: 10.1201/9781003405535-2

industrial industry, cutting waste and increasing efficiency [11] . Real-time vehicle tracking using IoT can enhance safety and ease traffic in the transportation sector. IoT holds immense potential to revolutionize various aspects of our lives. It can improve the efficiency and productivity of businesses, enhance the quality of life of individuals, and help us better understand and manage the world around us [12]. For example, in the smart home, the IoT can enable devices such as thermostats, lights, and appliances to communicate with each other and the homeowner's smartphone, allowing for remote control and automation of the home environment.

In industry, IoT can be used for real-time monitoring of supply chains, predictive maintenance of machinery, and optimization of manufacturing processes. In agriculture, precision farming, crop growth monitoring, and irrigation and fertilization optimization can all be done using the Internet of Things.

Certainly, addressing risks and problems in the development of IoT is essential for its safe and sustainable growth. Factors such as security, privacy, and interoperability play a significant role in mitigating potential challenges. Striking a balance between the benefits and drawbacks of IoT is crucial as the technology continues to evolve; by addressing these concerns, we can maximize the benefits while minimizing the associated risks.

The IoT can change virtually every area of our lives and create new opportunities for innovation, growth, social benefit, and domains, leading to increased efficiency, reduced costs, and improved outcomes for individuals and organizations alike [13].

The rest of the chapter is organized as follows. Section 1.2 details the IoT architecture. Section 1.3 details the applications of IoT in various fields with some case studies. Objectives, issues, and challenges are presented in Section 1.4. In Section 1.5, recent tools and technologies used in the field of IoT are discussed. The future scope and conclusion are described in Section 1.6.

1.2 IOT NETWORK ARCHITECTURE

IoT network architecture refers to the overall structure and design of the network infrastructure that enables seamless communication and connectivity among IoT devices. It includes all components – hardware, software, and protocols – employed in IoT contexts to support data processing, data exchanging, and decision-making. Figure 1.1 depicts the architecture of IoT networks, which comprises several linked layers that allow for effective and secure transmission of information and interaction. The application layer, data processing layer, network layer, and sensor layer are a few examples of these layers.

Application Layer: The uppermost layer of an IoT architecture is called the application layer, and it acts as a communication channel between an IoT system and users and other applications. It includes a range of IoT device data-using apps, services, and analytics. Functionalities, including monitoring, controlling, data analytics, visualization, and decision-making, are made possible by this layer. A few examples of application layer uses include smart home automation,

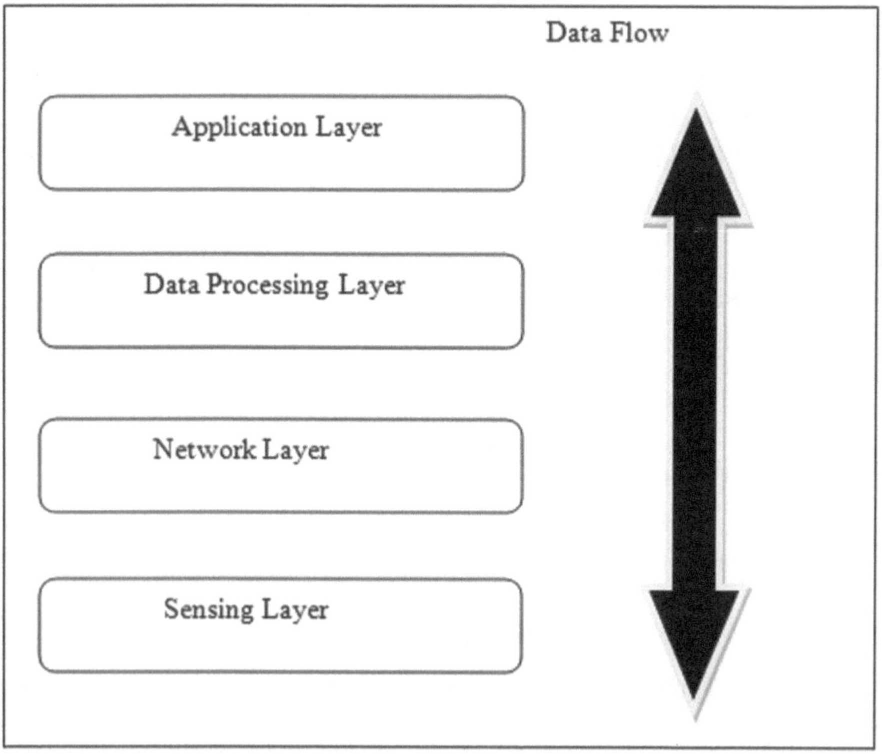

FIGURE 1.1 IoT network's layered architecture.

industrial process optimization, asset tracking, healthcare monitoring, and environmental monitoring.

Data Processing Layer (Middleware Layer): An interface between the application layer and the network layer is provided by the middleware layer, also known as the data processing layer. It provides the software framework to facilitate data management, communication, and interoperability between multiple devices and applications. Data normalization, device recognition (discovery), data filtering, and protocol translation are tasks handled by the data processing layer before the relevant data are transferred to the application layer.

Network Layer: The network layer's job is to enable and create connectivity among IoT devices and to make it easier for them to communicate effectively. It uses various technologies and communication protocols, including cellular networks, Bluetooth, Wi-Fi, ZigBee, and even direct device-to-device communication (for instance, through ad hoc networks). Between devices and gateways, the network layer enables secure and reliable data transmission.

Sensing Layer (Perception Layer): The Sensing Layer, the lowest layer of IoT architecture, consists of physical objects or "things" that are integrated with actuators and sensors. These devices gather environmental data or execute actions based on instructions they receive. Sensors capture a wide range of data, including temperature, humidity, pressure, and motion. Actuators enable devices to perform physical actions based on commands received from higher layers.

The architecture of IoT networks has evolved significantly over the years to address various challenges and enable the efficient functioning of interconnected devices. The key developments include the shift from centralized architecture to distributed architecture, the establishment of interoperability and standardization, enhancements in privacy and security measures, the integration of analytics and edge computing, the adoption of hybrid cloud-edge architecture, and the integration of AI and machine learning [14]. These advancements have facilitated scalability, improved data processing, optimized resource usage, and enhanced IoT networks' overall performance and capabilities.

1.3 APPLICATIONS OF IOT NETWORKS AND DEVICES

The section details the application of IoT networks and devices in numerous areas and some case studies.

1.3.1 Applications of IoT Network

The Internet of Things is a smart device-connected technology with expanding use cases in several sectors. A consistent standard for IoT ecosystems is now required because of their increasing use across numerous industries [15, 16]. Here are some common applications which are detailed in Figure 1.2.

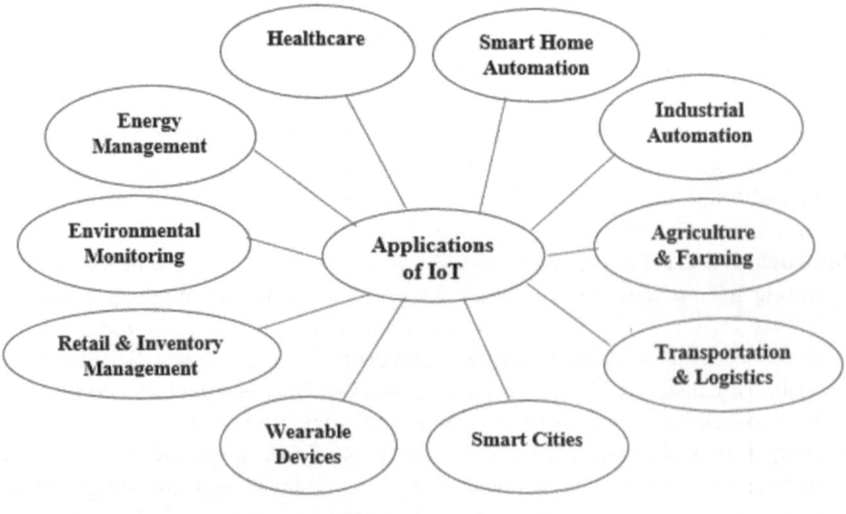

FIGURE 1.2 Applications of IoT network.

1.3.1.1 Smart Home Automation

IoT networks enable the interconnection and automation of various devices in a home, such as security cameras, lighting systems, thermostats, and appliances. Users can monitor and control these devices remotely using smartphones or voice assistants.

1.3.1.2 Industrial Automation

IoT networks enable the automation and optimization of industrial processes by connecting and controlling sensors, actuators, and machines. This improves safety, efficiency, and productivity in industrial and manufacturing settings.

1.3.1.3 Agriculture and Farming

IoT networks facilitate precision agriculture by deploying sensors to monitor humidity, temperature, soil moisture, and crop health. This information assists farmers in enhancing their irrigation techniques, optimizing fertilizer usage, and controlling pests, leading to increased crop production and reduced resource inefficiency.

1.3.1.4 Transportation and Logistics

IoT networks are employed for fleet management, logistics tracking, and supply chain optimization. Global Positioning System (GPS) and sensor devices in vehicles can monitor real-time location and vehicle performance, optimize routes, and enhance driver safety.

1.3.1.5 Smart Cities

IoT networks are essential for building smart cities. They support smart street lighting, waste management, parking systems, traffic management, and public safety initiatives, enhancing the overall city infrastructure optimization [5].

1.3.1.6 Wearable Devices

IoT is the backbone of wearable technology such as smartwatches, fitness trackers, and medical devices. These devices collect and analyze user data, provide personalized insights, and enable seamless integration with other IoT systems.

1.3.1.7 Healthcare

IoT networks are used in healthcare for connected medical devices, wearable health trackers, and remote patient monitoring. This enables healthcare providers to gather real-time information, monitor patients' vital signs, provide timely interventions, and improve overall patient care.

1.3.1.8 Energy Management

IoT networks are used for monitoring and controlling energy consumption in homes, industries, and buildings. Energy management systems, connected appliances, and smart meters help users to track energy usage, optimize efficiency, and reduce cost.

1.3.1.9 Retail and Inventory Management

IoT networks facilitate inventory tracking and management in retail stores and warehouses. Connected sensors can monitor stock levels, track inventory movements, and enable automated reordering to streamline operations and reduce stock-outs.

1.3.1.10 Environmental Monitoring

IoT networks can be used to analyze and monitor environmental conditions such as water quality, air quality, weather patterns, and noise levels. These data can be utilized for environmental (climate) research, pollution control, and urban planning.

We can conclude that based on mentioned applications, IoT applications can be found in virtually every aspect of society and daily life.

1.3.2 Case Studies: Application of IoT Devices and Their Social Impacts

1.3.2.1 Agriculture: Precision Farming for Improved Crop Management

This case study illustrates the integration of IoT sensors, weather data, and analytics into agricultural practices for efficient crop management. It exemplifies the utilization of IoT devices to monitor soil moisture levels, analyze real-time weather data, and implement automated irrigation systems. As a result, it leads to improved crop yields and decreased water consumption.

1.3.2.2 Industrial Manufacturing: Predictive Maintenance for Enhanced Efficiency

This case study exemplifies the utilization of IoT technologies for predictive maintenance in the manufacturing industry. In order to capture real-time data, it demonstrates the use of IoT sensors on equipment. Anomalies and potential equipment breakdowns can be identified by data analysis, enabling preventative maintenance procedures. By using such procedures, companies may reduce downtime, increase production efficiency, and minimize maintenance expenses, all of which enhance operational efficiency.

1.3.2.3 Healthcare: Remote Patient Monitoring for Personalized Care

This case study focuses on IoT-based remote patient monitoring devices that enable medical personnel to monitor patients' health metrics and vital signs remotely. It discusses the use of wearable technology, including networked sensors and smartwatches, to collect real-time data and send it to healthcare professionals. The study emphasizes how IoT enables the early diagnosis of medical concerns, enhances patient outcomes, and minimizes hospital readmissions.

1.3.2.4 Smart Cities: Transforming Urban Parking with Smart Parking Systems

This case study demonstrates the use of smart parking systems powered by the Internet of Things in metropolitan areas. It investigates the utilization of IoT mobile applications and sensors to deliver real-time updates on parking availability, assisting drivers in locating vacant parking spaces more effectively. The study exemplifies how IoT optimizes the utilization of parking spaces, alleviates traffic congestion, and improves the overall parking experience for residents in cities.

1.3.2.5 Energy Management: Optimizing Building Efficiency through IoT Solutions

This case study centres on the utilization of IoT in energy management specifically for buildings. It explores the incorporation of IoT devices, smart meters, and energy management platforms to enable real-time control and monitoring of energy consumption. The study highlights the implementation of IoT-driven energy optimization techniques such as demand response and smart grid integration, resulting in substantial energy cost reductions and the advancement of sustainable practices. By harnessing IoT in energy management, buildings can enhance efficiency, decrease environmental impact, and elevate overall performance [17]. This section included the application of IoT devices. The following section details the objectives, issues, and challenges of IoT devices.

1.4 OBJECTIVES, ISSUES, CHALLENGES, AND IMPORTANCE OF IOT NETWORKS

1.4.1 Objectives of IoT Network

IoT networks have evolved over the years to address challenges and achieve key objectives. These networks aim to establish seamless connectivity among devices, collect and analyze data, automate processes, optimize resource usage, ensure security and privacy, promote interoperability and standardization, enable real-time responsiveness, and support scalability and flexibility [18–20. By pursuing these objectives, IoT networks facilitate transformative applications and advancements across various industries and domains [21].

1.4.2 Issues

The combination of IoT architecture and IoT devices brings forth several key issues that need to be addressed. Scalability is a crucial issue because as the IoT ecosystem develops, it is harder to accommodate more devices and handle the growing amount of data flow. Another important concern is security which involves the requirement to protect sensitive information and guarantee the integrity of IoT devices and their communications against unauthorized access and potential cyberattacks. A barrier to enabling seamless integration and communication across various IoT devices, platforms, and protocols is interoperability. The development of common standards and protocols that provide interoperability and easy communication among IoT devices depends on standardization initiatives. Another crucial factor is power efficiency, which aims to reduce the energy consumption of IoT devices, especially those using scarce power sources. Designing IoT infrastructure and devices must also take into account adequate data handling, dependability, and continuous connectivity. The potential of IoT may be fully realized by overcoming these problems and allowing safe, effective, and scalable IoT infrastructures.

1.4.3 Challenges of IoT Network

The difficulties of IoT networks can have a variety of effects, including impeding the growth and effectiveness of IoT networks, raising the risk of data branches and cyberattacks, making it more difficult to integrate and communicate among devices, making it more difficult to handle and analyze large volumes of data, and affecting the overall dependability and performance of IoT systems. For IoT networks to achieve their full potential and have a successful deployment and operations, these issues must be resolved.

1.4.4 Importance of IoT Network

IoT networks have the potential to transform the way we work and live by providing new possibilities for innovation and development while improving efficiency, productivity, and quality of life.

1.4.4.1 Increased Efficiency

IoT networks may provide companies and organizations with real-time data and insights that aid in the optimization of their processes and operations, resulting in increased output and efficiency.

1.4.4.2 Cost Saving

By automating and improving operations, IoT networks may help organizations minimize costs in areas such as energy consumption, maintenance, and waste reduction.

1.4.4.3 Improved Decision-Making

IoT networks having access to real-time data may offer insights to support more rapid and informed decision-making, which results in better outcomes.

1.4.4.4 Enhanced Customer Experiences

Utilizing sensors, beacons, and other connected devices, IoT networks may help organizations offer customers experiences that are more personalized and responsive.

1.4.4.5 New Business Models

By offering new services or products that take advantage of IoT capabilities, IoT networks can open up new opportunities for businesses to generate income.

1.4.4.6 Improved Safety and Security

Real-time monitoring and alarms made possible by IoT networks can contribute to better security and safety in sectors such as healthcare, transportation, and public safety.

1.4.5 Overview of the Current State of Development

IoT is currently seeing tremendous expansion, rising adoption, and technological improvements. Healthcare, transportation, manufacturing, energy, and smart cities are just a few of the sectors and domains where IoT has already become a reality [12, 22]. It is now possible to analyze and collect enormous amounts of data in real time because of the increasing number of connected sensors and devices, high-speed internet, and cloud computing [23]. The development of low-cost and low-power sensors, wireless communication protocols, edge computing, and data analytics has all advanced significantly in terms of technology [24]. As a result, more advanced and intelligent IoT systems have been made possible, allowing them to manage and monitor a variety of physical processes and systems on their own [21]. IoT must overcome a few obstacles before it can realize its full potential. The requirement for qualified people to develop, install, and manage IoT systems is one of them, as are concerns with standardization and interoperability, data ownership and governance, security, and privacy. IoT network issues, objectives, and challenges are listed in Table 1.1.

The advancements in IoT tools and technologies are described in the next section.

TABLE 1.1
Comparison Table for IoT Networks, Focusing on Issues, Objectives, and Challenges

Issues	Objectives	Challenges
Data security	• Enhance IoT devices and data security with robust measures. • Maintain the accuracy, privacy, and trustworthiness of data.	• Vulnerabilities in IoT devices and networks. • Encryption and secure transmission of data.
Network coverage	• Construct infrastructure to provide better communication in difficult locations. • Increase network coverage to cover distant or underserved areas.	• Limited network coverage area of some places. • Bridge connectivity gaps in rural or challenging areas.
Latency	• Improve network resource utilization and bandwidth allocation to minimize latency. • Reduce latency for seamless real-time communication and time-sensitive applications.	• Reduce data transmission delays and improve response times. • Overcome network congestion and bandwidth limitations.
Scalability	• Create scalable IoT networks and optimize resource allocation for efficient management of network bandwidth.	• Optimize network bandwidth to minimize congestion and efficiently handle a large number of connected devices.

(continued)

TABLE 1.1 (Continued)
Comparison Table for IoT Networks, Focusing on Issues, Objectives, and Challenges

Issues	Objectives	Challenges
Interoperability	• Facilitate smooth integration of IoT ecosystems through cross-platform communication and collaboration. • Set unified standards and protocols for seamless device and system interoperability.	• Lack of standardization and compatibility challenges among diverse devices and protocols. • Enable seamless integration and data sharing among diverse IoT platforms.
Data management	• Design streamlined data management approaches for effectively handling and analyzing extensive volumes of data produced by IoT devices. • Ensure efficient data storage and processing.	• Deriving practical insights from extensive datasets. • Handle and process large data volumes effectively.
Power efficiency	• Maximize power efficiency and extend battery life in IoT devices.	• Manage energy usage and enhance battery longevity in IoT devices.
Reliability	• Failure recovery and fault tolerance. • Stable and reliable network connectivity.	• Ensure stable and resilient IoT network connectivity. • Implement fault tolerance and recovery mechanisms.
Privacy concerns	• Enforce privacy-by-design and transparent data practices.	• Ensure data privacy and obtain user consent.
Cost	• Explore cost-effective solutions and technologies for IoT network deployment and maintenance.	• Mitigate the issue of costly deployment and maintenance expenses.
Device compatibility	• Define protocols and standards to enable device compatibility and seamless interoperability.	• Encourage compatibility among devices from diverse manufacturers and platforms.

1.5 ADVANCEMENT OF TOOLS AND TECHNOLOGIES IN IOT

IoT technology refers to the group of techniques, programs, and protocols that allow physical objects, sensors, and systems to connect and communicate with one another through the internet [25, 26]. It includes a wide range of tools and capabilities that cooperate to make it possible for linked devices to share data and operate one another with no interruptions.

1.5.1 Recently Used Tools in IoT

There are many different types of IoT tools, and each one performs a particular function in the management and development of IoT applications [26]. These tools are used for a variety of tasks related to managing the overall system and developing IoT applications.

Several IoT tools that are widely used by developers and organizations are listed below.

1.5.1.1 Arduino

The open-source electronic platform Arduino offers both software and hardware tools for creating IoT projects and prototypes. Because it offers a variety of microcontroller boards, sensors, and libraries, it is well-linked by both professionals and enthusiasts.

1.5.1.2 Raspberry Pi

It is a credit card-sized computer that serves as a versatile IoT platform. It may be used to create IoT devices and home automation systems and supports a number of different operating systems, and it also contains general-purpose input/output (GPIO) pins for hardware interface.

1.5.1.3 AWS IoT Core

Amazon Web Services (AWS) offers AWS IoT Core, a fully managed solution, to manage, connect, and securely exchange data between IoT devices and the cloud. It includes capabilities such as message routing, device provisioning, and connecting with other AWS.

1.5.1.4 Google Cloud IoT Core

A safe and scalable IoT platform is offered by Google Cloud as a Google Cloud IoT Core. It enables users to safely manage, connect, and utilize data from millions of IoT devices that are dispersed around the world. For data storage, processing, and visualization, it connects with other Google Cloud services.

1.5.1.5 Microsoft Azure IoT Hub

A cloud-based service called Microsoft Azure IoT Hub enables two-way communication between IoT devices and Azure cloud. Along with other Azure services, it offers data consumption, device administration, and real-time monitoring capabilities.

1.5.1.6 Particle

A range of hardware development kits, cloud services, and developer tools are available through the IoT platform Particle. In order to develop and scale IoT applications, it offers device management, over-the-air firmware upgrades, and data storage capabilities.

1.5.1.7 IBM Watson IoT Platform

An enterprise-grade IBM Watson IoT platform enables organizations to securely manage, evaluate, and connect IoT data. Using IBM Watson AI technology, it provides device management, analytics, and cognitive capabilities.

1.5.1.8 Ubidots

A cloud-based IoT platform called Ubidots makes it easier to gather, analyze, and visualize IoT data. It supports interaction with famous hardware platforms and communication protocols and includes drag-and-drop capabilities for creating IoT dashboards,

1.5.1.9 ThingsBoard

ThingsBoard is an open-source IoT platform that provides data visualization, collection, processing, and device management functionalities. It supports various connectivity protocols and offers customizable dashboards and rule engines for IoT applications.

1.5.1.10 Losant

Losant is an enterprise IoT platform that helps organizations build, manage, and scale IoT solutions. It provides tools for device management, data visualization, workflow automation, and integration with other enterprise systems.

1.5.2 Technology Used in the Field of IoT

In these sections, we describe the key components and technologies used in IoT networks.

Table 1.2 details the key components and description of technologies used in the field of IoT [27].

In the field of IoT, numerous technologies have emerged to enable and enhance connectivity, data management, and device control. Table 1.3 mentions some recent technologies in the field of IoT.

TABLE 1.2
Key Components and Technologies Used in IoT Network

Key Components	Technology Description
Connectivity technologies	Bluetooth, cellular networks, Wi-Fi, Zigbee, Z-Wave, Ethernet, NFC, and LPWAN (LoRaWAN, NB-IoT)
Security	Identity and access management (IAM), encryption algorithms (AES, RSA), public key infrastructure (PKI), and secure communication protocols (TLS/SSL)
Sensors and actuators	Sensors for light, temperature, humidity, proximity, motion, accelerometers, pressure, GPS, and actuators
Cloud computing and edge computing	Cloud platforms (Google Cloud IoT, AWS IoT, Azure IoT) and edge computing frameworks (AWS Greengrass, Azure IoT Edge)
Protocols	DDS, MQTT, HTTP, WebSocket, AMQP, OPC UA, and CoAP
Blockchain	Distributed ledger technology for enhanced security and trust in IoT transactions
Artificial Intelligence of Things (AIoT)	Intelligent decision-making and automation through IoT system integration with AI and machine learning algorithms
Machine learning and data analytics	Apache Kafka, Apache Spark, Hadoop, TensorFlow, and PyTorch
Application development	IoT platforms (Google Cloud IoT Core, Azure IoT Hub, AWS IoT Core), SDKs and APIs, and programming languages (C, C++, Python, Java)

TABLE 1.3
Technologies Used in IoT Networks with Description

Technology	Description
Wi-Fi	Provides high-speed connectivity across short distances.
Bluetooth	Enables short-range communication between devices.
Zigbee	A low-data-rate, low-power wireless protocol suitable for IoT devices.
Z-Wave	A wireless protocol optimized for home automation and smart devices.
Contiki	A freely available operating system development specifically for IoT devices with limited resources.
LoRaWAN	For IoT applications, long-range, low-power wireless protocol.
NB-IoT (Narrowband IoT)	A cellular standard for low-power, wide-area IoT connectivity.
Sigfox	A global network dedicated to low-power IoT device communication.
AWS (Amazon Web Services)	A comprehensive cloud platform for IoT device management, data storage, and analytics.
Google Cloud IoT	Provides tools and services for managing, analyzing, and visualizing IoT data.
Microsoft Azure IoT	Provides a range of services, such as device provisioning, data integration, and analytics, for managing and developing IoT solutions.
Edge gateway	Serves as an intermediary between the cloud and IoT devices, processing and filtering data at the edge.
Edge analytics	Running analytics algorithms directly on IoT devices or at the edge to improve real-time decision-making and reduce latency.
Fog computing	Distributing computing resources and intelligence at the edge of the network, closer to the IoT devices.
RFID (radio-Frequency identification)	Tracking and identification of tags on objects using electromagnetic fields.
NFC (Near-Field Communication)	Enable short-range communications between devices, commonly used for contactless payment and data transfer.
Thread	A mesh networking protocol developed for low-power IoT devices.
RIOT	An alternative open-source operating system for low-power Internet of Things devices.
FreeRTOS	A popular real-time operating system for microcontrollers often used in IoT applications.
TinyOS	An operating system that is lightweight and designed for wireless sensor networks.

(*continued*)

TABLE 1.3 (Continued)
Technologies Used in IoT Networks with Description

Technology	Description
OAuth	A protocol for secure authorization and authentication of IoT devices and applications.
CoAP (Constrained Application Protocol)	A lightweight protocol for resource-constrained IoT devices to enable efficient communication over the internet.
OPCUA (Open Platform Communications Unified Architecture)	A standard for industrial IoT to enable secure and interoperability communication between systems and devices.
PKI (Public Key Infrastructure)	A security framework that uses cryptographic techniques to provide secure device authentication, data encryption, and integrity.
MQTT (Message Queuing Telemetry Transport)	A messaging protocol that is frequently used in the IoT and guarantees effective communication between servers and devices.
Blockchain	A distributed ledger technology allows transparent and secure transactions, enhancing security and trust in IoT systems.
Machine Learning	Utilizing algorithms to analyze and interpret IoT data for predictive maintenance, anomaly detection, and optimization.
Stream Processing	Real-time analysis of data streams from IoT devices to enable immediate decision-making.
Data Visualization	Presenting IoT data in a visual format to gain insights and make informed decisions.

1.6 FUTURE SCOPE AND CONCLUSION

1.6.1 Future Scope

IoT network improvements in security, standardization, edge computing, 5G connection, sustainability, data analytics, and industry-specific applications all point to a bright future for IoT networks. With the help of these advancements, the IoT ecosystem will become more intelligent, connected, and secure, revolutionizing several industries and enhancing many facets of life and society. IoT networks' future has a lot of potential for efficiency, innovation, and enhanced connectivity across many different domains.

1.6.2 Conclusion

The Internet of Things networks present a variety of concerns, difficulties, and goals that must be addressed for the efficient operation and growth of IoT systems. The scalability of IoT networks, security concerns, interoperability issues, data management complexities, and reliability considerations pose significant hurdles. However, efforts in standardization, security enhancement, data analytics, and advancements in connectivity technologies offer potential solutions. By overcoming these challenges, IoT networks have the potential to revolutionize various sectors, including transportation, healthcare, smart cities, and industrial automation. The objectives of IoT networks include achieving seamless connectivity, ensuring data privacy and security, enabling efficient data management, promoting interoperability, and enhancing energy efficiency. Addressing IoT network issues, challenges, and objectives is crucial to unlock its full potential for societal benefit. Collaboration among stakeholders is vital to build robust and scalable networks that meet evolving needs.

REFERENCES

1. Jiang, P., Liang, H., Li, G., & Tan, X. (2017). The Internet of Things: A survey. *Information Systems Frontiers*, 19(2), 279–293.
2. Lee, I., & Lee, K. (2015). The Internet of things (IoT): Applications, investments, and enterprise challenges. *Business Horizons*, 58(4), 431–440.
3. Manyika, J., Chui, M., Bughin, J., Dobbs, R., Bisson, P., & Marrs, A. (2015). *Unlocking the potential of the internet of things*. McKinsey Global Institute.
4. Perera, C., Liu, C. H., Jayawardena, S., & Chen, M. (2014). A survey on the Internet of Things from industrial market perspective. *IEEE Access*, 2, 1660–1679.
5. Zanella, A., Bui, N., Castellani, A., Vangelista, L., & Zorzi, M. (2014). Internet of Things for smart cities. *IEEE Internet of Things Journal*, 1(1), 22–32.
6. "The Internet of Things (IoT)." www.gartner.com/en/information-technology/glossary/internet-of-things (accessed Feb. 25, 2022).
7. Chawla, D., & Mehra, P. S. (2023). QSMAH: A novel quantum-based secure cryptosystem using mutual authentication for healthcare in the internet of things. *Internet of Things*, 24, 100949. https://doi.org/10.1016/J.IOT.2023.100949.
8. International Telecommunication Union. ITU Internet Reports 2005: The Internet of Things, November 2005.
9. "Strategy Analytics: internet of Things now numbers 22 billion devices but where is the revenue?" Strategy Analytics online newsroom." www.businesswire.com/news/home/20190516005700/en/Strategy-Analytics-Internet-of-Things-Now-Numbers-22-Billion-Devices-But-Where-Is-The-Revenue (accessed Feb. 23, 2023).
10. "The Future of Telemedicine Technology with IoT and AI – Intel." Intel, www.intel.in/content/www/in/en/healthcare-it/telemedicine.html (accessed Feb. 22, 2023).
11. Watson, N. J. (2021). The role of IoT in advancing the circular economy: A review. *IEEE Internet of Things Journal*, 8(2), 804–819. https://doi.org/10.1109/JIOT.2020.2992857.
12. Atzori, L., Iera, A., & Morabito, G. (2010). The Internet of Things: A survey. *Computer Networks*, 54(15), 2787–2805.
13. Xia, F., Yang, J., Wang, C., & Zhao, K. (2021). Internet of things. *International Journal of Communication Systems*, 34(7), e4751. https://doi.org/10.1002/dac.4751.
14. Roman, R., Lopez, J., & Mambo, M. (2017). Mobile edge computing, Fog et al.: A survey and analysis of security threats and challenges. *Future Generation Computer Systems*, 78(Part 2), 680–698. https://doi.org/10.1016/j.future.2016.11.009.
15. Bandyopadhyay, D., & Sen, J. (2011). Internet of Things: Applications and challenges in technology and standardization. *Wireless Personal Communications*, 58(1), 49–69.
16. Khanna, A., & Kaur, S. (2020). Internet of Things (IoT), applications and challenges: A comprehensive review. *Wireless Personal Communications*, 114, 1687–1762.

17. Chen, M., Hao, Y., Zhang, M., Li, Y., & Leung, Y. (2016). Internet of Things for smart cities: Technologies, challenges and opportunities. *Wireless Networks*, 22(1), 1–15. https://doi.org/10.1007/s11276-015-0942-y.
18. Misra, S., Mukherjee, A., & Roy, A. (2021) *Introduction to IoT*. Cambridge University Press.
19. Ashton, K. (2009). That 'Internet of Things' thing. *RFID Journal*, 22(7), 97–114.
20. Miorandi, D., Sicari, S., De Pellegrini, F., & Chlamtac, I. (2012). Internet of things: Vision, applications and research challenges. *Ad Hoc Networks*, 10(7), 1497–1516.
21. Sharma, J., & Mehra, P. S. (2023). Secure communication in IOT-based UAV networks: A systematic survey. *Internet of Things*, 23, 100883. https://doi.org/10.1016/J.IOT.2023.100883.
22. Zhang, Y., Yang, Y., & Fang, Y. (2019). Security and privacy in the Internet of Things: A survey. *IEEE Internet of Things Journal*, 6(3), 4915–4934.
23. Al-Fuqaha, A., Guizani, M., Mohammadi, M., Aledhari, M., & Ayyash, M. (2015). Internet of Things: A survey on enabling technologies, protocols, and applications. *IEEE Communications Surveys & Tutorials*, 17(4), 2347–2376.
24. Li, S., Xu, L. D., & Zhao, S. (2015). The Internet of Things: A survey of topics and trends. *Information Systems Frontiers*, 17(2), 261–274.
25. Li, S., Xu, J., & Xu, K. (2015). The Internet of Things: A survey. *Information Systems Frontiers*, 17(2), 243–259. https://doi.org/10.1007/s10796-014-9492-7.
26. "What technologies are used in IOT – Technology behind internet of things." www.avsystem.com/blog/iot-technology/ (accessed May 27, 2023).
27. Simmons, Adam. "Internet of Things (IoT) Technology: Quick and Easy Guide." Dgtl Infra, December 19, 2022, https://dgtlinfra.com/internet-of-things-iot-technology/ (accessed Mar. 2, 2023).

2 6G for Intelligent Internet of Things

Deepakshi Dabas, Pawan Singh Mehra, Diksha Chawla, Jatin Sharma, and Aatif Jamshed

2.1 INTRODUCTION

The wireless communication landscape is undergoing an extraordinary transformation, teetering on the cusp of ushering in a new era of limitless possibilities and unparalleled connectivity. Within this rapidly evolving milieu, the concept of the sixth-generation (6G) wireless communication networks emerges as a visionary leap poised to redefine the landscape of customer services and applications through the transformative lens of the Internet of Things (IoT). The fusion of 6G and IoT is a beacon of enormous promise, offering to lead us toward a future inhabited by fully intelligent and autonomous systems as our linked world continues to weave a complex tapestry of innovation. In this chapter, we embark on an immersive journey into the dynamic crossroads where 6G and IoT intersect, delving deep into the fertile grounds of emerging opportunities that this symbiotic relationship brings to the forefront. Through a comprehensive survey, we will illuminate the foundational technologies that underpin the 6G revolution, exploring their catalytic role in empowering the IoT ecosystem. From the dawn of edge intelligence and the canvas of reconfigurable intelligent surfaces to the boundless realms of space-air-ground-underwater communications, terahertz marvels, and the bedrock of massive ultra-reliable and low-latency communications, we will carefully unfold the intricate tapestry of 6G technologies that are poised to shape the trajectory of future IoT networks and applications. What sets this chapter apart is its departure from surface-level discussions. Here, we plunge into the heart of IoT domains, where the impact of 6G's integration is set to reverberate most profoundly. We venture into the domains of Healthcare Internet of Things, Vehicular Internet of Things and Autonomous driving, Unmanned Aerial Vehicles (UAVs), Satellite Internet of Things, and Industrial Internet of Things (IIoT). By doing so, we unveil the transformative role that 6G is poised to play, envisioning the metamorphosis it will facilitate for each facet of our interconnected world. As we navigate this uncharted terrain, we face intriguing challenges that beckon innovative solutions. Spectrum utilization, dynamic network slicing, and security and

DOI: 10.1201/9781003405535-3

privacy considerations fortifications emerge as pivotal signposts on this expedition. This chapter will crystallize a comprehensive understanding of the symbiotic dance between 6G and IoT, illuminating pathways for groundbreaking research, unprecedented disruptions, and the limitless horizons that await exploration. This chapter examines the theoretical underpinnings of 6G and IoT convergence and delves into real-world implementations across diverse sectors. We will traverse the landscapes of healthcare, transportation, satellite communication, and UAVs, vividly showing how 6G for Intelligent IoT will transform industries and societies. Indeed, the world stands on the brink of a remarkable transformation, where the fusion of cutting-edge technology and transformative connectivity can reshape industries, elevate societies, and redefine the essence of how we live, work, and interact. The convergence of 6G and IoT acts as a lighthouse guiding us through uncharted waters toward a horizon ablaze with innovation and possibility. The papers to come will unravel the intricate details of this convergence, unveiling a path that promises a brighter, more connected future.

The arrangement of the sections is structured as follows: Section 2.2 provides the Literature Review. Section 2.3 delves into the rise of 6G technology. Section 2.4 mentions the Internet of Things. The requisites of the 6G for Intelligent IoT are discussed in Section 2.5. Section 2.6 reviews the Out-and-Out network. Section 2.7 unveils contemporary technologies in conjunction with Intelligent IoT. Section 2.8 discusses the employment of 6G for Intelligent IoT. Section 2.9 outlines potential avenues for future exploration within the context of 6G for Intelligent IoT. Lastly, Section 2.10 encapsulates the concluding remarks.

2.2 LITERATURE REVIEW

The continuous evolution of wireless communication technologies has led to the emergence of the 6th-Generation (6G) networks, poised to redefine the landscape of the Internet of Things (IoT). This literature review critically examines a comprehensive array of studies and research papers centered around the topic of '6G for Intelligent Internet of Things,' elucidating the multifaceted dimensions, challenges, and opportunities that this convergence entails.

Guo, Fengxian, et al. [1] delve into the network infrastructure, highlighting the need for a versatile architecture capable of accommodating the diverse connectivity demands posed by intelligent IoT devices. Tomkos, Ioannis, et al. [2] emphasize the significance of flexible and adaptable networks to cater to the dynamic requirements of IoT applications. Lu, Yang, and Xianrong Zheng [3] underscore the potential of utilizing higher frequency bands, such as terahertz, to accommodate the massive connectivity needs of IoT devices. In contrast, Slalmi, Ahmed, et al. [4] propose innovative communication strategies involving beamforming and advanced Multiple-Input Multiple-Output (MIMO) technologies to address the spectrum scarcity challenge. Porambage, Pawani, et al. [5] highlight the vulnerability of IoT devices to various cyber threats and advocate for robust security mechanisms, including blockchain and quantum cryptography. Nguyen, Dinh C., et al. [6] explore integrating homomorphic encryption techniques to ensure data privacy without compromising utility. Malik, Usman Mahmood, et al. [7] propose innovative techniques for energy harvesting and optimization to enable sustainable and long-lasting operation of IoT devices. Additionally, Qadir, Zakria, et al. [8] delve into edge computing strategies to minimize energy consumption by processing data closer to the source. Zhou, Yiqing, et al. [9] discuss the potential of artificial intelligence (AI)-driven predictive analytics to optimize IoT device behavior, enhancing network efficiency. Nguyen, Tri, et al. [10] explore the role of blockchain in ensuring data integrity and security in 6G for Intelligent IoT ecosystems. Laghari, Asif Ali, et al. [11] investigate the role of 6G in healthcare, enabling real-time patient monitoring and telemedicine. Similarly, Noor-A-Rahim, Md et al. [12] delve into the potential of 6G in transforming transportation systems by allowing intelligent and autonomous vehicles. Hazra, Abhishek, et al. [13] highlight the need for standardized protocols and communication interfaces to facilitate interoperability between heterogeneous IoT devices. Letaief, Khaled B., et al. [14] emphasize the importance of addressing hardware constraints to ensure seamless connectivity and efficient operation.

In light of the diverse studies and research papers mentioned above, it becomes evident that the blend of 6G and IoT presents a host of challenges and opportunities, spanning network architecture, security, energy efficiency, and advanced technologies such as AI and blockchain. Exploring real-world applications further emphasizes the transformative potential of this synergy, guiding future research endeavors in this dynamic field.

The rapid evolution of technology continues to pave the way for groundbreaking advancements in communication networks. As 5G networks gain momentum, the anticipation for the emergence of 6G technology is palpable. Envisioned as the next paradigm shift, 6G holds the potential to redefine the way we connect and interact in the digital age. With a focus on addressing the increasing demands of the Internet of things (IoT) and other data-intensive applications, 6G is set to push the boundaries of speed, connectivity, and intelligence. This section provides a glimpse into the rise of 6G technology and its pivotal role in shaping the future of connectivity and innovation.

2.3 THE RISE OF 6G TECHNOLOGY

2.3.1 The Transition from 5G to 6G Innovation

5G networks are currently being developed, while experts foresee the emergence of 6G networks. These 6G networks are designed to manage sophisticated applications such as augmented reality, virtual reality, and self-driving vehicles, achieving terabit speeds using the terahertz spectrum. The introduction of 6G brings features such as full-duplex communication, artificial intelligence, edge computing, and improved energy efficiency. These elements aim to bolster support for the IoT, sensing capabilities, and high-speed connectivity, positioning 6G as a central platform for upcoming technological innovations.

2.3.2 6TH GENERATION

Driven by widespread mobile device usage and increasing traffic, advancements in wireless communication are key to future services. While 5G improved Quality of Service (QoS), upcoming IoT demands pose challenges. By 2030, 5G may struggle with traffic surge to 5,016 EB/month from 62 EB/month in 2020 due to video apps, M2M, and edge services. New technologies are vital for data-intensive apps. Evolving intelligent systems reveal 5G latency issues, needing <0.1 ms for real-time needs. 6G emerges as a solution, with innovations such as edge intelligence, terahertz, and satellites. It will empower IoT, enable a connected digital realm, and align with the 2030 landscape.

6G promises a new echelon of network capabilities characterized by critical attributes:

- Achieving data speeds of 1 terabit per second (Tb/s) or more to support widespread IoT connection and balance mobility, spectrum availability, and mobile traffic.
- Increasing the density of IoT devices and the capacity of mobile traffic to up to 1 gigabit per square meter (Gb/s/m^2).
- It supports 107 devices per square kilometer (devices/km^2) of exceptionally dense device connections, which is advantageous for dense IoT network installations.
- Providing network latencies as low as 10–100 ps to satisfy the demands of applications like e-health and autonomous driving.

5G networks are in progress, while researchers envision 6G, which is set to handle advanced apps such as Augmented Reality, Virtual Reality, and autonomous vehicles with terabit speeds via the terahertz spectrum. It brings full-duplex Artificial Intelligence, edge computing, and energy efficiency to support IoT, sensing, and high-speed connectivity, making 6G a hub for future tech advancements.

2.4 INTERNET OF THINGS

The Internet of Things (IoT) integrates devices into wireless systems, enabling seamless data sensing, computation, and communication connectivity. Ubiquitous devices such as sensors and smartphones gather IoT data. Ultra-reliable Low-Latency Communications (URLLC) in IoT are set to grow significantly, with Cisco projecting 500 billion IoT devices by 2030, up from 26 billion in 2020. IHS Markit forecasts 125 billion connected IoT devices by 2030. By 2025, the worldwide 5G-IoT market would have increased from USD 694 million in 2020 to USD 6,285.5 million [15]. The Internet of Nano-Things envisions nano-devices for sensing, transmitting, and processing data, requiring novel network architectures. 6G emerges as a catalyst for IoT networks, integrating functions such as sensing, communication, computation, and autonomy. Massive IoT connects numerous devices and sensors, demanding robust infrastructure. Device density can be as high as 1 million devices per square kilometer, leading to extensive data flow, task execution, and data storage and analysis.

2.5 REQUISITES OF 6G FOR INTELLIGENT IOT

The transition to 6G for Intelligent IoT will necessitate more demanding preconditions than its 5G-IoT predecessor to fully realize the goal of an intelligent information society by 2030, characterized by comprehensive intelligence, pervasive device connection, data-driven services, and autonomous systems. The following details underline these increased requirements:

a. *Vast IoT Connectivity:* The rise of smart devices and wireless tech has boosted mobile connectivity. Predictions show mobile data could reach 5,000 exabytes by 2030, 80 times more than 2020. Future connectivity relies on broadband and satellite networks, using low earth orbit (LEO) satellites and UAVs. UAVs ensure continuous IoT connectivity in wide areas. A distributed model with 33 UAVs improved load balancing and cut deployment time by 40%.
b. *Extensive Ultra-Reliable Low-Latency IoT Communications:* Enhancements in URLLC are needed for 6G for Intelligent IoT. Beyond 5G, it's vital for autonomous IoT and aerial systems. Autonomous transportation requires URLLC for real-time video and signaling. Timely data is crucial for mission-critical IoT services and real-time communication in the tactile internet.
c. *Enhanced 6G for Intelligent IoT Communication Protocols:* Emerging IoT applications reshape network architecture. 6G for Intelligent IoT needs integrated standards, edge/cloud computing, and wireless tech. Initiatives such as ETSI Multi-access Edge Computing leverage edge tech. IEEE's 802.11 addresses high throughput for 6G for Intelligent IoT applications.
d. *Extended IoT Network Coverage:* Space, air, ground, and undersea networks are required for the IoT civilization of the future. UAVs and edge intelligence are crucial. UAVs extend 2D networks to 3D, linking terrestrial, satellite, and aerial. High-altitude UAVs aid maritime coverage for ships. Agile UAVs provide distributed broadband services.
e. *Next-Generation Intelligent IoT Devices:* Future 6G for Intelligent IoT needs device-centered networks. Smart devices such as smartphones, vehicles, and robots manage networks. They generate data and control processes. Device-focused solutions include device-to-device and multi-hop cellular communication. Wearables, replacing phones, drive human-centric 6G with wearable IoT networks.

2.6 THE OUT-AND-OUT NETWORK

The space-air-ground-underwater/sea network combines space, air, and underwater networks for seamless connectivity across different domains. It offers high-speed communication for data-intensive IoT and critical tasks using terahertz and visible light spectrum. Edge computing supports small IoT devices by enhancing computation and bridging cloud and IoT. It addresses concerns such as latency and scalability, making it ideal for mobile applications and ultra-low latency needs.

The various tiers of the network are as follows:

a. *Space Communication Tier:* This tier provides wireless coverage using satellites in geostationary, medium, and low Earth orbits (LEO, MEO, and GEO), especially for places without access to terrestrial networks. Space communication technology enables high-capacity networks for satellite-ground communication at various altitudes.
b. *Air Communication Tier:* This tier offers coverage in disaster-affected regions and emergency scenarios that call for ultra-reliable low-latency communication (URLLC), using UAVs and balloons as flying base stations. In 6G for Intelligent IoT contexts, UAVs also serve as aerial users, creating direct air-ground linkages and enabling cooperative sensing and data transfer.
c. *Terrestrial Communication Tier:* This tier links actual base stations, mobile devices, and computer servers, focusing on wireless coverage and connection for human activities on the ground. The terahertz range improves spectral efficiency and transmission rates, especially in heterogeneous networks with a high density of users.
d. *Underwater Communication Tier:* This tier connects IoT devices submerged in water, such as submarines operating in deep waters. Bidirectional communication supports the linking of underwater IoT devices and control systems.

2.7 CONTEMPORARY TECHNOLOGIES IN CONJUNCTION WITH INTELLIGENT IOT

In the dynamic landscape of technology, the fusion of contemporary innovations with the realm of the smart Internet of Things (IoT) has paved the way for unprecedented possibilities. As we stand on the cusp of a new era, the convergence of cutting-edge technologies with intelligent IoT systems promises to revolutionize how we interact with the digital world. This section offers a glimpse into contemporary technologies seamlessly integrated with the fabric of Intelligent IoT, presenting an exciting tapestry of advancements reshaping how we live, work, and connect. Figure 2.1 depicts the mind map of the contemporary technologies mentioned in Section 2.7 in conjunction with IIoT.

2.7.1 Mobile Edge Computing

In the 1990s, content delivery networks (CDNs) improved online content delivery. Edge computing extends this idea by placing computing resources closer to users for faster data processing. Paramvir Bahl's work in 2009 laid the groundwork for edge computing, emphasizing its importance in mobile networks. Edge computing reduces delays, processes data quickly, and enhances responsiveness. With the upcoming 6G networks, mobility-enhanced edge computing (MEEC) will be crucial in handling resource-intensive tasks for AI algorithms. Combining edge computing with AI supports real-time decision-making in applications such as autonomous driving and smart cities. Edge computing also benefits the Internet of Things (IoT), aiding computationally intensive applications such as virtual reality and smart factories. It

FIGURE 2.1 Mind map of contemporary technologies in conjunction with IIoT.

conserves bandwidth and improves privacy by processing data closer to devices. As 6G and IoT advance, edge computing becomes even more essential for efficient data management and processing.

2.7.2 Fog Computing

Fog computing empowers edge devices with nearby computational and storage capabilities through distributed fog nodes. Unlike centralized clouds, fog computing offers decentralized services, low latency, optimized spectrum use, and enhanced reliability. In the 6G era, fog computing will be vital for massive IoT, supporting data exchange and applications. Fog nodes are essential for 6G's connectivity and low-latency goals. Transitioning to fog-based 6G for Intelligent IoT introduces resource allocation, energy efficiency, and security challenges.

2.7.3 Federated Learning: An AI Component

Federated Learning (FL) is reshaping intelligent, privacy-centric IoT systems. FL decentralizes AI training, coordinating devices and a central server without sharing raw data. Workers improve a global model by locally computing updates. FL excels in the following:

- Data Privacy: It avoids exposing raw data and aligns with stringent regulations.
- Network Efficiency: FL reduces latency and conserves resources by avoiding central data transfers.
- Learning Quality: Diverse datasets and resources boost learning accuracy and convergence.

FL's integration spans healthcare, transportation, and UAVs. It empowers collaborative healthcare by preserving patient data privacy and enhances smart transportation services such as autonomous driving and safety prediction. The flourishing adoption of FL in recent IoT scenarios underscores its potential as a vital research avenue.

2.7.4 Quantum Cryptography

Quantum cryptography secures 6G for Intelligent IoT via Quantum Key Distribution (QKD) and entanglement principles. QKD establishes secret keys for secure communication and resisting eavesdropping. Quantum Key Generation encodes data in qubits, which are hard to intercept or measure. Protocols such as QKD and Quantum Secure Direct Communication (QSDC) leverage quantum mechanics for data integrity and authenticity. Quantum cryptography enhances 6G for Intelligent IoT security, addressing cyber threats and vulnerabilities. It ensures confidential and private data exchange, protecting the evolving 6G for the Intelligent IoT landscape.

2.7.5 Incorporation of Wireless Information and Energy Transfer

The incorporation of Wireless Information and Energy Transfer (WIET) is one of the 6G network's groundbreaking advancements. This innovative method uses a range of fields and waves similar to those used in wireless communications. Notably, WIET has enormous potential for improving the battery life of wireless systems, providing support for battery-free devices inside the 6G framework. By efficiently recharging and extending the battery life of wireless networks and related devices, WIET is strategically designed to progress the development of intelligent devices without batteries.

2.7.6 The Amalgamation of Sensing and Communication

A key facilitator of autonomous wireless networks involves the capability to perceive dynamically evolving environmental conditions and enable efficient information interchange among diverse nodes. The 6G networks will empower autonomous or self-directed systems by seamlessly merging sensing and communication through an extensive array of sensing entities, intricate communication resources, multi-level computing resources, and multi-level cache resources. However, achieving this integration presents a formidable challenge.

2.7.7 The Dynamic Network Slicing

Dynamic network slicing holds significant importance for network operators as it ensures the dedicated operation of virtual networks and offers comprehensive support for efficiently delivering services to diverse user categories such as vehicles, industries, and machines. Consequently, dynamic network slicing is pivotal in 5G communication systems, particularly managing multiple users across heterogeneous networks. Applying software-defined networking and network function virtualization approaches is necessary to realize dynamic network slicing. These methods contribute to cloud computing network management by optimizing network performance through centralized control of dynamic routing, monitoring of traffic flow, and effective resource allocation.

2.7.8 Proactive Caching: A Solution for Base Station Traffic Congestion

The primary concern within the domain of 6G networks revolves around the widespread implementation of small-cell networks. This strategic deployment aims to enhance network attributes encompassing coverage, capacity, and mobility management. However, this expansion also raises the potential challenge of significant downlink traffic congestion at base stations (BS). Proactive caching emerges as a solution to address this issue, effectively mitigating these constraints. Proactive caching facilitates reduced access delays and traffic offloading, ultimately contributing to an elevated quality of user experience. Furthermore, comprehensive research endeavors are necessary to establish 6G networks successfully. These efforts should focus on the holistic optimization of multiple facets, encompassing proactive caching, interference management, intelligent coding methodologies, and scheduling techniques that are pivotal for the effectiveness of 6G networks.

2.7.9 Terahertz Communication

Spectral efficiency improves by expanding bandwidth and using advanced MIMO tech. 6G extends into the terahertz (THz) range, which is vital for meeting future demands. ITU-R defines cellular THz as 275 GHz to 3 THz, adding capacity to 6G and combining THz and mm-Wave boosts network capacity by 11 times. THz suits uHSLLC, uMUB, and uHDD, enabling sensing, imaging, and positioning. THz's shorter wavelength allows multiple antennas, OAM multiplexing, and beamforming. Narrow beams manage interference, and atmospheric attenuation in THz can be handled with pencil beam antennas.

2.7.10 Optical Wireless Communication

A range of well-known optical wireless communication (OWC) technologies, including visible light communication (VLC), light fidelity, optical camera communication, and optical band-based free-space optical (FSO) communication, have diverse applications such as vehicle-to-everything (V2X) communication, indoor mobile robot positioning, virtual reality (VR), and underwater OWC. These OWCs are also being considered for integration into 6G networks, focusing on the potential of FSO for network-to-backhaul/fronthaul connectivity. Deploying optical fiber-based backhaul networks faces challenges in complex or remote areas. Optical fiber links for small-cell networks might only sometimes be practical or cost-effective. Additionally, 6G's high user density requires strong backhaul and access network integration. The emergence of FSO fronthaul/backhaul networks is anticipated in upcoming 6G communications. Moreover, FSO systems share characteristics with optical fiber networks, suggesting autonomous and self-directed data transmission through FSO.

2.7.11 MIMO-Cell-Free Communication

Intelligent surfaces are divided into large intelligent surfaces (LISs) and intelligent reflecting surfaces (IRSs), holding promise for 6G. LIS integrates antenna arrays into extensive MIMO systems, offering beamforming and correlation reduction advantages. However, LIS consumes significant power, impacting energy efficiency. In 6G, diverse communication technologies will seamlessly integrate, enabling effortless transitions between networks. This shift from orthogonal to non-orthogonal, cell-free communications enhances network selection and mitigates handover challenges. Multi-connectivity and hybrid techniques further enhance 6G's cell-free capabilities.

2.7.12 Big Data Analytics: A Component of Machine Learning

Big data analytics is a sophisticated and elaborate method used to examine enormous and varied databases, revealing hidden patterns, correlations, and consumer behaviors to guarantee thorough data management. Big data is gathered from various sources, including movies, social networks, photographs, and sensors. Big data analytics is efficiently used in 6G communication networks to handle and analyze massive data volumes. Integrating large amounts of data, deep learning protocols, and big data analytics within the framework of 6G networks would improve the total network, mainly because of the inherent characteristics of automation and self-optimization [16]. The decrease of end-to-end (E2E) latency through integrating big data and machine learning is an example of such development.

2.7.13 Blockchain Technology

Blockchain, rooted in hash tree theory, provides an immutable and secure database structure. In 6G, its decentralized nature offers efficient spectrum management and data security. For example, privacy-preserving blockchain ensures data privacy in 6G for Intelligent IoT. Complexities of IoT in 6G include diverse devices, interoperability, and privacy issues. Blockchain addresses these:

- *Decentralized Management:* Blockchain records network events for cost-efficient auditing.
- *Interoperability Enhancement:* Blockchain enables unified authentication, authorization, and billing across IoT systems.
- *Data Traceability:* Blockchain's timestamps ensure data traceability and digital signatures secure integrity.
- *Distributed Network Management:* Blockchain aligns with 6G's distributed approach, aiding resource management and data sharing.

Applications in IoT for 6G include the following:

- *Distributed Ledger:* Blockchain revolutionizes resource management, e.g., spectrum sharing and improving efficiency.
- *Decentralized Storage:* Blockchain offers secure decentralized storage and data pointers for efficient retrieval.
- *Smart Contracts:* Smart contracts automate access control, identity management, energy trading, and more.

In summary, blockchain's attributes align with 6G's challenges, providing secure and efficient IoT solutions in future mobile communication technology.

2.7.14 Blockchain and Artificial Intelligence: The Prospective Combo

AI and blockchain, both highly disruptive technologies, possess inherent limitations that constrain their application within the realm of 6G. To unlock the full potential of these two methodologies and address future requirements for intelligence, distribution, and security, a synergistic integration, leading to mutual augmentation, is imperative.

2.7.14.1 Blockchain Empowering AI

Current AI techniques rely heavily on a client-server training setup, leading to data vulnerabilities and delays. Distributed AI emerges as a solution, preserving privacy and reducing congestion. Yet, it requires complex interactions among agents and a central server, posing challenges. Blockchain offers a remedy by replacing the server with a blockchain, using smart contracts for model sharing. This approach incentivizes sharing and ensures security, building trust in distributed AI decisions.

2.7.14.2 AI Enhancing Blockchain

AI and blockchain synergize for dual benefits. Firstly, AI enhances blockchain efficiency by optimizing resource-intensive tasks such as mining through intelligent algorithms. It also improves smart contracts, boosting overall blockchain performance. Secondly, AI operates on encrypted blockchain data, ensuring security and enabling data utilization. This integration promises to reshape 6G systems, addressing constraints and leveraging combined strengths for intelligent, secure, and distributed networks.

2.8 EMPLOYMENT OF 6G FOR INTELLIGENT IOT

Leveraging the foundational technologies outlined in the preceding section, 6G is poised to bring novel applications to life within the IoT realm. This section delves deeply into the burgeoning applications of 6G across various pivotal IoT domains, encompassing Health (HIoT) [17], Vehicular IoT (VIoT), autonomous driving, satellite communication, and UAVs.

2.8.1 Healthcare

6G's integration transforms Healthcare IoT (HIoT) with mURLLC and THz communications. Rapid health data transmission empowers wearables and remote medical services. For real-time monitoring, applications demand ultra-low latency (<1 ms) and high reliability (>99.999%). 6G introduces robotics for telesurgery, enabled by UAVs and blockchain. UAVs act as relays for emergency medical supplies, bypassing delays. mURLLC in the THz spectrum supports real-time medical communication, aiding nano-devices and sensors. Telestration in hospitals allows remote guidance during surgery, and connected ambulances stream real-time high-resolution video for accurate diagnosis. AI augments 6G healthcare, analyzing historical records for insights.

2.8.2 Transport

6G's emergence reshapes Vehicular IoT (VIoT), revolutionizing Intelligent Transportation Systems (ITSs). Utilizing mMTCs, 6G VIoT enables autonomous Vehicle-to-everything (V2X) connectivity. A novel vehicle discovery mechanism optimizes V2X by adjusting signatures, minimizing false positives, and allocating radio resources. Predicting data rates in 6G VIoT is complex due to factors such as mobility and channels. Machine Learning (ML) predicts network-assisted throughput through historical network load analysis. Edge intelligence in Roadside Units (RSUs) enhances vehicular insight, estimating traffic and weather from local observations exchanged wirelessly. Deep Learning (DL) plays a pivotal role in 6G VIoT, modelling vehicular channels and optimizing resource allocation via Deep Reinforcement. Learning (DRL). A trust broker entity ensures transparency, and DL optimizes data transmission scheduling. A trust broker entity ensures transparency, and DL optimizes data transmission scheduling. In autonomous driving, 6G enhances safety, energy efficiency, and cooperation via DL-controlled inter-vehicle distances. Edge servers with DL processors aid AV system control. Federated Learning (FL) fosters cooperative vehicular learning and communication while preserving privacy and reducing network overhead.

2.8.3 Satellite Communication

In the 6G era, integrating satellite communications into wireless networks is key for widespread massive IoT coverage, giving rise to satellite IoT (SIoT). SIoT operates on a three-tier satellite network: LEO, MEO, and GEO, providing global services. 6G introduces advanced satellite tech, allowing multiple satellites in varied orbits for global coverage and efficiency via frequency reuse. Terahertz bands facilitate inter-satellite links, enhancing performance over 5G's mm-Wave and optical communications. A hybrid of NOMA and cognitive radio is proposed to tackle spectrum scarcity in 6G for Intelligent IoT. NOMA boosts spectrum efficiency with varied power levels on a carrier, while cognitive radio enables dynamic spectrum sharing. A balanced architecture in 6G SIoT assures flexibility, network performance, and energy management. It offers remote connectivity, supports data services for terrestrial base stations, and ensures energy sustainability via space-based IoT devices using renewables. Satellite integration into 6G advances IoT, especially SIoT. This blend offers extensive coverage, efficiency, and inventive solutions for IoT, laying a robust foundation for an interconnected and capable future.

2.8.4 Unmanned Aerial Vehicles

Drones, or UAVs, are essential in 6G communication networks. UAVs are used in various circumstances to offer fast data rates and wireless communication. Through Base Station (BS) devices, they enable cellular connectivity while showcasing qualities including simple deployment, robust line-of-sight links, and quick mobility. Fixed BS infrastructure is deficient in several qualities. UAVs are helpful instruments in crises or natural disasters where regular terrestrial infrastructure might not be practical. Due to their adaptability, UAVs may also fulfill the requirements of wireless networks' uMUB, uHSLLC, mMTC, and HDD. Their usefulness extends to improving network connectivity, security and surveillance, emergency services, and fire detection [18]. UAVs are now recognized as pillar technologies in the 6G environment. Backscatter communication with a UAV has been investigated. It helps with tasks such as optimizing channel conditions for remote sensors and ambient power supply. UAVs and non-coherent detection systems can provide customized air interfaces for 6G networks. An efficient resource allocation method based on deep reinforcement learning may be used to incorporate UAV intelligence into 6G networks.

2.9 EXPLORATION OF FUTURE AVENUES

This section illuminates various compelling research complexities and outlines potential avenues for future exploration within the context of 6G for Intelligent IoT.

2.9.1 Reliability and Confidentiality of 6G for Intelligent IoT

6G technology incorporation into IoT networks offers increased dependability, extremely low latency, and broad coverage. However, this connection also introduces security flaws, such as unapproved data access, dangers to data integrity, and potential denial of service assaults. Security issues are brought about by the wide variety of IoT devices and access techniques, including higher attack risks during technology transfers. Additionally, the expanding connections between devices and edge computing nodes increase security and privacy risks, with worries about data communications being compromised by eavesdropping, hijacking, spoofing, and DoS assaults. The deployment of AI functions in distributed edge nodes for intelligent 6G for Intelligent IoT networks raises security concerns, potentially enabling malicious manipulation of data training within spectrum access systems. Strategies for risk mitigation are crucial to ensuring robust security and privacy. One approach involves perturbation techniques such as differential privacy, introducing artificial noise to neural network gradients to protect data while maintaining convergence. Privacy-preserving data aggregation solutions within fog computing architectures adhere to ε differential privacy principles, enhancing data protection [19]. Another option for safe, decentralized communication in UAV-IoT networks is blockchain. Blockchain nodes may be operated by UAVs, allowing for decentralized data exchange and communication. While considering UAVs' resource limitations, lightweight mining algorithms can achieve low-latency data consensus. The importance of blockchain is growing as 6G for Intelligent IoT networks progress toward decentralization.

2.9.2 Energy Efficiency in 6G for Intelligent IoT

In the emerging landscape of 6G for Intelligent IoT networks, achieving optimal energy efficiency is a pressing concern. Activities such as data communication and service delivery, including autonomous driving data exchange and UAV-based space communications, demand substantial energy. The energy consumption of wireless cellular base stations, ranging from 2.5 to 4 kW, raises environmental concerns. Extensive 6G for Intelligent IoT network deployment exacerbates energy consumption and carbon emissions. Addressing this, energy-efficient communication protocols via optimization techniques are crucial for eco-friendly 6G for Intelligent IoT networks. An optimization approach considering QoS and energy consumption is integrated into 6G-driven smart automation systems [20]. This involves creating a multimedia data model aligned with QoS measures for energy-efficient multimedia transmission. Integrating energy harvesting uses renewable energy sources for environmentally friendly 6G for Intelligent IoT systems. IoT devices use vibration, thermal, solar, wind, and mechanical energy for computing and communication. Implantable sensors in Healthcare IoT (HIoT) networks transmit sensory data via Bluetooth low-energy modules using solar power. Implantable solar-powered sensors have been used in tests to demonstrate energy harvesting and extend the life of HIoT healthcare systems. The issues of energy efficiency in higher-altitude 6G for Intelligent IoT networks, including satellite networks with flying devices and base stations, should be the subject of future study. Energy harvesting and space communications should be coordinated with the trajectory and environment of the devices.

2.9.3 Hardware Limitations of IoT Devices

Hardware limitations in 6G for Intelligent IoT networks impact communication and computation, particularly in scenarios such as intelligent healthcare. Wearable sensors and devices executing AI functions for edge intelligence and ultra-reliable low-latency communication face constraints. Interaction between IoT sensors and servers adds communication overhead, requiring novel hardware designs to enhance IoT capabilities. A software-based DL accelerator for mobile sensor hardware is introduced. It optimizes hardware utilization using processors (e.g., Graphics Processing Units (GPUs)) for specific DL model inference phases. Algorithms such as run-time layer compression and deep architecture decomposition maintain accuracy while improving execution time and energy consumption. This approach enhances on-device IoT for edge intelligence in wireless networks. Tiny-transfer learning (TinyTL) tackles memory-efficient on-device sensor learning. The lite residual module boosts model capacity with minimal memory overhead. Simulations confirm competitive accuracy (above 90%) and up to a 12.9 reduction in training memory footprint compared to traditional methods. Future research should prioritize hardware-based AI training solutions for nano-IoT devices and wearables in intelligent 6G for Intelligent IoT networks, especially for applications such as enhanced living assistance services.

2.9.4 Standard Pre-requisites for 6G for Intelligent IoT

6G technologies are poised to transform IoT markets and ecosystems, offering advanced wireless capabilities. However, the success of 6G for Intelligent IoT systems relies on rigorous standard specifications. Collaboration among stakeholders such as network operators, service providers, and customers is vital for defining these standards, ensuring seamless integration of 6G functions. Vertical 6G for Intelligent IoT use cases introduce architectural shifts, demanding adaptable mobile networks for diverse requirements such as autonomous driving and e-healthcare. Network standards are pivotal for deploying 6G for Intelligent IoT ecosystems, enabling crucial services between 6G servers and IoT devices. One such is MODBUS, a communication protocol used in the Internet of Things to link sensors, industrial machinery, and computer servers. MODBUS enables industrial control by utilizing mesh networking and RTU, TCP/IP, and UDP protocols. Several IoT applications, including video analytics and multimedia distribution, are supported by initiatives such as ETSI's Multi-access Edge Computing, which combines edge computing with communication. Stakeholders should concentrate on creating new standard specifications for cutting-edge space-air-ground-underwater communications, including IoT satellite communications, shortly. These standards are necessary for implementing commercial Internet of Things applications, such as deep-sea services and space flight.

2.10 PROS AND CONS OF CONTEMPORARY TECHNOLOGIES IN 6G FOR INTELLIGENT IOT

Here is a Table 2.1 summarizing contemporary technologies and their pros and cons in the context of the 6G-based Internet of Things (IoT):

TABLE 2.1
Contemporary Technologies in Conjunction with IIoT

S.No	Technology	Pros	Cons
1.	Mobile Edge Computing	Enhances real-time processing for IoT devices.	Limited computing resources at the edge.
2.	Fog Computing	Low-latency data processing closer to data source.	Scalability challenges for massive IoT deployments.
3.	Federated Learning	Collaborative AI training without centralized data.	Communication overhead and synchronization complexities.
4.	Quantum Cryptography	Ultra-secure key distribution and data exchange.	Practical implementation challenges and hardware requirements.
5.	Wireless Information and Energy Transfer	Simultaneous data and energy transfer.	Energy efficiency and distance limitations.
6.	Sensing and Communication	Integration of data acquisition and transmission.	Limited bandwidth and power constraints.
7.	Dynamic Network Slicing	Customized network segments for diverse IoT needs.	Complex orchestration and management.
8.	Proactive Caching	Reduced latency through cached data availability.	Cache management and synchronization issues.
9.	Terahertz Communication	High data rates and bandwidth for IoT connections.	Limited propagation range and penetration through obstacles.
10.	Optical Wireless Communication	High-speed data transmission using light.	Susceptibility to interference and signal blockage.
11.	MIMO-Cell-Free Communication	Enhanced coverage and capacity with distributed antennas.	Complex interference management and system calibration.
12.	Big Data Analytics	Insights from large IoT datasets for informed decisions.	Data privacy concerns and processing overhead.
13.	Blockchain Technology	Immutable data ledger for secure transactions.	Scalability, energy consumption, and latency.
14.	Blockchain and Artificial Intelligence	Enhanced trust and autonomy in AI systems.	Integration challenges and computational overhead.

These contemporary technologies play diverse and crucial roles in developing 6G for Intelligent IoT networks. While they offer significant contributions in enhancing various aspects of IoT, they also come with limitations that must be addressed to fully harness their potential within the context of 6G for Intelligent IoT applications.

2.11 CONCLUSION

The evolution of wireless communication networks brings new technologies and features to meet communication demands. While 5G shows promise, its limitations have led to the emergence of 6G networks. Extensive research explores the facets of 6G, promising its utility. This overview covers 6G network evolution, enabling technologies, and recent advancements. It urges researchers to delve into enabling technologies defining 6G's essence and IoT applications. Smart 6G networks integrate 5G intelligence with Next-Gen IoT via edge computing, enabling AI algorithms. Challenges include real-time dynamics and edge constraints. The journey leads to the Internet of Everything, reshaping life through intelligent devices at the network edge. 6G's allure captivates academia and industry. The chapter explores 6G's potential in IoT networks, surveying research activities. It delves into Quantum Cryptography, Federated Learning, Edge Computing, Blockchain, and other contemporary technologies with IoT integration, essential for 6G for intelligent IoT harmony. The chapter also explores 6G's role in healthcare, vehicular, autonomous driving, UAVs, satellite, and industrial IoT applications. As the chapter concludes, it encourages further research in 6G for intelligent IoT. It illuminates integration's complexity, driving researchers and stakeholders to intensify their efforts. 6G's transformative power in enhancing IoT networks and user experiences is a central theme, envisioning a future enriched by amplified connectivity.

REFERENCES

1. Guo, Fengxian, et al. "Enabling massive IoT toward 6G: A comprehensive survey." *IEEE Internet of Things Journal* 8.15 (2021): 11891–11915.
2. Tomkos, Ioannis, et al. "Toward the 6G network era: Opportunities and challenges." *IT Professional* 22.1 (2020): 34–38.
3. Lu, Yang, and Xianrong Zheng. "6G: A survey on technologies, scenarios, challenges, and the related issues." *Journal of Industrial Information Integration* 19 (2020): 100158.
4. Slalmi, Ahmed, et al. "Toward 6G: Understanding network requirements and key performance indicators." *Transactions on Emerging Telecommunications Technologies* 32.3 (2021): e4201.; Nguyen, Dinh C., et al. "6G Internet of Things: A comprehensive survey." *IEEE Internet of Things Journal* 9.1 (2021): 359–383.
5. Porambage, Pawani, et al. "The roadmap to 6G security and privacy." *IEEE Open Journal of the Communications Society* 2 (2021): 1094–1122.
6. Nguyen, Dinh C., et al. "Federated learning for the Internet of Things: A comprehensive survey." *IEEE Communications Surveys & Tutorials* 23.3 (2021): 1622–1658.
7. Malik, Usman Mahmood, et al. "Energy-efficient fog computing for 6G-enabled massive IoT: Recent trends and future opportunities." *IEEE Internet of Things Journal* 9.16 (2021): 14572–14594.

8. Qadir, Zakria, et al. "Towards 6G Internet of Things: Recent advances, use cases, and open challenges." *ICT Express* 9.3 (2023): 296–312.
9. Zhou, Yiqing, et al. "Service-aware 6G: An intelligent and open network based on the convergence of communication, computing and caching." *Digital Communications and Networks* 6.3 (2020): 253–260.
10. Nguyen, Tri, et al. "Privacy-aware blockchain innovation for 6G: Challenges and opportunities." *2020 2nd 6G Wireless Summit (6G SUMMIT)* (2020): 1–5, Levi, Finland, March 17–20, 2020.
11. Laghari, Asif Ali, et al. "A review and state of the art of Internet of Things (IoT)." *Archives of Computational Methods in Engineering* 29.2 (2021): 1–19.
12. Noor-A-Rahim, Md, et al. "6G for vehicle-to-everything (V2X) communications: Enabling technologies, challenges, and opportunities." *Proceedings of the IEEE* 110.6 (2022): 712–734.
13. Hazra, Abhishek, et al. "A comprehensive survey on interoperability for IIoT: Taxonomy, standards, and future directions." *ACM Computing Surveys (CSUR)* 55.1 (2021): 1–35.
14. Letaief, Khaled B., et al. "The roadmap to 6G: AI-empowered wireless networks." *IEEE Communications Magazine* 57.8 (2019): 84–90.
15. Chawla, Diksha, and Pawan Singh Mehra. "A roadmap from classical cryptography to post-quantum resistant cryptography for 5G-enabled IoT: Challenges, opportunities and solutions." *Internet of Things* 24 (2023): 100950. http://doi.org/10.1016/J.IOT.2023.100950.
16. Lv, Zhihan, et al. "Big data analytics for 6G-enabled massive internet of things." *IEEE Internet of Things Journal* 8.7 (2021): 5350–5359.
17. Chawla, Diksha, and Pawan Singh Mehra. "QSMAH: A novel quantum-based secure cryptosystem using mutual authentication for healthcare in the internet of things." *Internet of Things* 24 (2023): 100949. http://doi.org/10.1016/J.IOT.2023.100949.
18. Sharma, Jatin, and Pawan Singh Mehra. "A survey of security challenges and existing prevention methods in FANET." in Intelligent Data Analytics, IoT, and Blockchain, Bashir Alam and Mansaf Alam (eds.). New York: Auerbach Publications eBooks, 2023, pp. 252–262. http://doi.org/10.1201/9781003371380-24.
19. Stergiou, Christos L., Konstantinos E. Psannis, and Brij B. Gupta. "IoT-based big data secure management in the fog over a 6G wireless network." *IEEE Internet of Things Journal 8.7* (2020): 5164–5171.
20. Mao, Bomin, Yuichi Kawamoto, and Nei Kato. "AI-based joint optimization of QoS and security for 6G energy harvesting Internet of Things." *IEEE Internet of Things Journal* 7.8 (2020): 7032–7042.

3 A Systematic Study of Intelligent Toll Payment Gateways

Deepika Pantola, Madhuri Gupta, and Umesh Gupta

3.1 INTRODUCTION

Tolls, which serve as a method of generating revenue for the nation's transport network, are becoming exceedingly common. To avoid stopping or slowing down and paying a toll, motorists and state transportation officials would prefer that vehicles not have to do either of those things [1]. In the last 15 years, numerous technologies, commonly called electronic toll collection (ETC), have allowed cars to enter and depart toll systems without interruption. Open Road Tolling (ORT) with e-tolling is the favored approach now because it is more efficient, environmentally friendly, and secure than conventional toll collection [2]. In the past 10 years, growing urbanization and the prevalence of Internet-of-Vehicles (IoV) have led to the development of what are known as Intelligent Toll Gates (ITGs), which provide a reactive and computerized method for making toll payments at ITGs [3]. Toll collections can be made significantly more accessible with the more accessible ETC technology, which enables electronic transactions via device monitoring nodes. The sensing device statistics about smart vehicles (SVs), toll consumption, and high-occupancy vehicle (HOV) lane availability are based on radio transponder components [1]. They interact with one another using low-powered wireless connections (Table 3.1).

The sensor readings are recorded when a CSV is determined to have passed a toll unit. They are linked to the automotive identification number and the lanes' utilization, enabling an intelligent and seamless money transfer between the vehicle owners (VO) and government authorities (GA) using 3rd access points [4]. Even though legitimate toll accumulation through ETC makes it possible to have low latency, high quality of user experience, and lesser queue build ups at intermediate toll booths, it still experiences challenges regarding security, privacy, and location-based attacks on VO-sensitive statistics in ecosystems that are based on IoV. A specific attacker has the potential to initiate device threats such as jamming, Sybil, and DDoS and document erroneous notifications at ITGs. This leads to monetary failures for the VO and the GA [5–7]. Cybersecurity Ventures forecasts that the expenditures of cybersecurity threats will have multiplied by 2021, reaching $6 trillion worldwide. In 2015, these expenses were estimated to be $3 trillion [2].

DOI: 10.1201/9781003405535-4

TABLE 3.1
List of Acronyms Used in This Chapter

Abbreviation	Description
ALP	Anonymous Loop-Back Proxy
BC	Blockchain
SVs	Smart Vehicles
DDoS	Distributed Denial of Service
DSRC	Dedicated Short-Range Communication
ETC	Electronic Toll Collection
ERP	Electronic Road Pricing System
GA	Governing Authorities
GNSS	Global Navigation Satellite System
HOT	High Occupancy Toll
HOV	High Occupancy Vehicle
ITGs	Intelligent Toll Gates
IHMCL	Indian Highways Management Company Ltd.
IoV	Internet of Vehicles
ISRO	Indian Space Research Organization
ITS	Intelligent Transportation System
ALP	Anonymous Loop-Back Proxy
BC	Blockchain
CSVs	Connected Smart Vehicles
DDoS	Distributed Denial of Service
DSRC	Dedicated Short-Range Communication
ETC	Electronic Toll Collection
ERP	Electronic Road Pricing System
GA	Governing Authorities
GNSS	Global Navigation Satellite System
HOT	High Occupancy Toll
HOV	High Occupancy Vehicle
ITGs	Intelligent Toll Gates
IHMCL	Indian Highways Management Company Ltd.
IoV	Internet of Vehicles
ISRO	Indian Space Research Organization
ITS	Intelligent Transportation System
LPI/R	License Plate Identification and Recognition system
LoS	Line of Sight
LTA	Land Transport Authority
NETC	National Electronic Toll Collection
NHAI	National Highways Authority of India
NLoS	Non-Line of Sight
NPCI	National Payments Corporation of India
OBU	On-Board Unit
OCR	Optical Character Recognition
QoE	Quality of Efficiency

TABLE 3.1 (Continued)
List of Acronyms Used in This Chapter

Abbreviation	Description
PAYD	Pay As You Drive
RFID	Radio Frequency Identification device
RFP	Request for Proposals
TRAC	Traffic Reporting and Control
VICS	Vehicle Information and Communication System
VMT	Vehicle Miles Travelled
VO	Vehicle Owner

ITGs, an IoV ecosystem component, connect nearby urban centers to smart highways [2]. ITGs new tactics combine blockchain technology and deep learning algorithms to provide connected, inventive automobiles with reliable and responsive analytics help [8]. This support is available for integrated and innovative vehicles. The eradication of 3rd intermediaries and the safeguarding of pay-outs between governing authorities (GA) and VO is made possible by blockchain [9]. On the contrary side, deep learning makes it easier to make a reliable estimate for a wide variety of complex urban traffic patterns. Nevertheless, variable delays at separate lanes caused by dynamic congestion situations concern VOs because of set toll fee structures based on linked, innovative vehicle types [10]. Road pricing or tolling is implemented for three main purposes such as Revenue / Finance Generation [11, 12], Demand Management [13], Congestion Management [14, 15].

This chapter focuses on the discussion of tolling technology and procedures, exploring their implementation and operation. The chapter also investigates potential connections between tolling systems and other Intelligent Transportation Systems (ITS) installations, considering possible advancements and refinements in the field. Moreover, the research study aims to provide recommendations for vehicle detection and registration systems that not only serve the purpose of tolling but also contribute to additional transportation system management activities, thereby enhancing the overall efficiency and effectiveness of the system.

This study is divided into 6 major sections: introduction, evolution of tolling, existing toll collection systems, architecture of ETC, upcoming technologies and conclusion. The detailed organization of the research paper is shown in Figure 3.1.

3.2 EVOLUTION OF TOLLING

Road pricing is anticipated to grow increasingly widespread over time [16]. According to Table 3.2, the evolution of tolling involves four major techniques which are corridor tolling, cordon tolling, area-wide or vehicle-miles-traveled tolling, and an integrated system management. Each technique increases the system's efficiency but also its

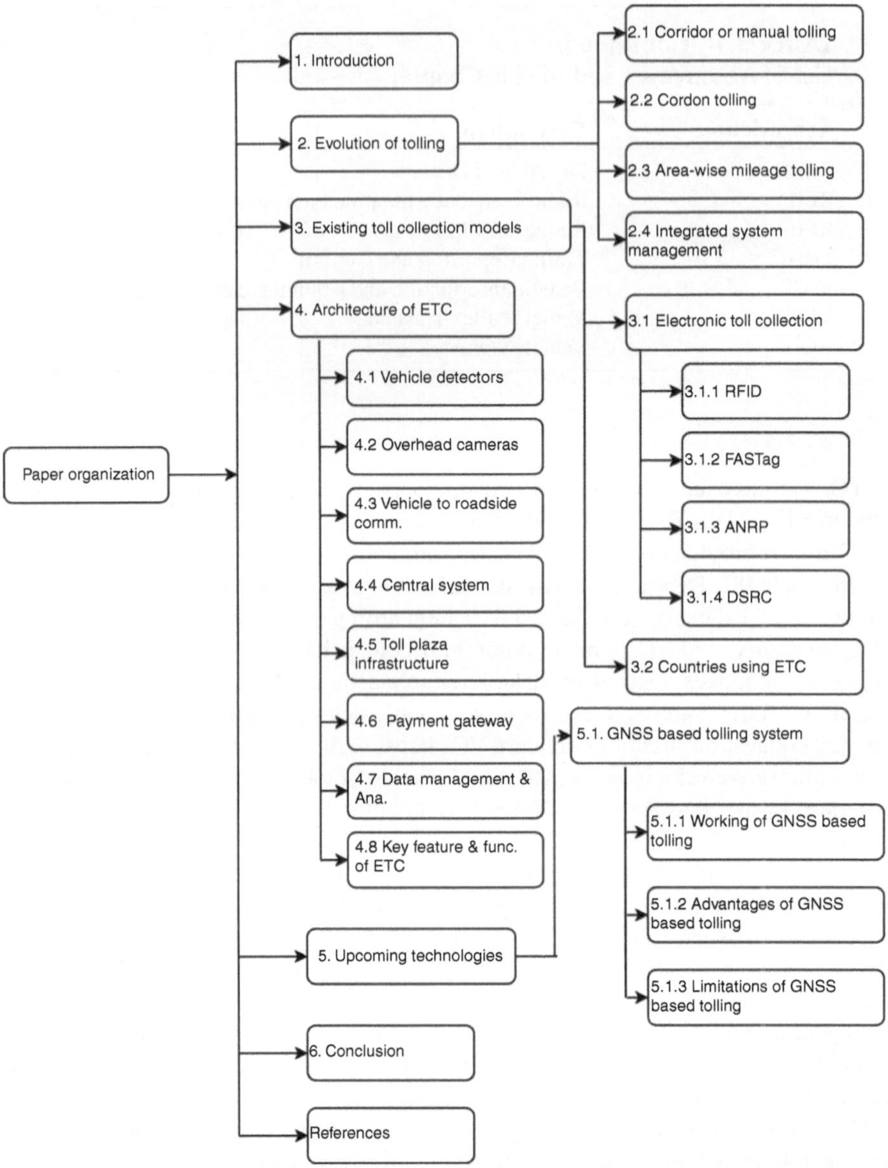

FIGURE 3.1 The organization of the paper.

complexity. In addition, each technique has prerequisite criteria before implementation [17] (Table 3.2).

3.2.1 Corridor or Manual Tolling

Corridor or Manual toll collection is a system whereby the toll collector calculates the toll, which is paid to him. This system is often used for flat tolls where there is no

TABLE 3.2
A Comprehensive Summary of Tolling Techniques and their Goal with Efficiency and Conditions

Tolling Techniques	Technique Goal	Technique Efficiency	Necessary Conditions
Corridor or ManualtTolling System	Repayment for facility	Low	Only those who pay for the privilege should be allowed to use the road.
Cordon Tolling System	Demand management	Medium	The citizenry trust that the possible advantages will be greater than the potential costs.
Area-mileage Tolling System	Revenue generation	High	Uniformity and interoperability
Integrated System management	Demand & congestion management	Very high	Accessibility to information as well as versatility across modes

necessity for calculating toll fees. This is the most prevalent tolling, where a vehicle pays a charge to utilize a specific highway segment or bridge. High Occupancy Toll lanes are designated for vehicles with several occupants but allow solo drivers to pay to use them [18]. The primary purpose of the toll is to recoup the expense of constructing and maintaining the infrastructure. Although most systems are moving toward ORT, others, including those requiring the vehicle to stop and pay cash upon arrival, add an unnecessary layer of complexity. Although the corridor may help ease congestion, it is predicted to be underutilized compared to other non-tolled roads [19]. Otherwise, consumers may not feel obligated to pay, and the software may not generate sufficient income.

The size of the toll roadways and, consequently, the number of exits and entry points determine the progression through the three phases of the human collecting process.

a. Simple manual toll collection
b. Manual Toll Collection with Ticket
c. Manual Toll Collection with Computerized Toll Calculation System

The key benefit of this system is that it requires little equipment and little initial investment on the part of the toll operator. Additionally, users of the shoulder of the road do not need any special equipment. However, there are also some drawbacks because this system depends entirely on manual labor. The toll collection process takes longer when the driver pays a hefty bill at the toll booth, and the toll collector needs to calculate and make a change.

3.2.2 Cordon Tolling

Cordon tolling is also known as the cordon-pricing technique. There is a fee for entering a particular place and the primary purpose of this tolling is to limit vehicle entry. Every entrance point must be able to recognize cars and verify that they have paid, have compensated, or will pay [20]. The toll applies to automobiles that either breach the cordon's border or drive within its confine during high traffic volume. "Cordon" refers to a geographical boundary unequivocally delineated and encircles the tolling area [21]. The tolling scheme could consist of a flat price only applicable during weekly corporate travel times, defined as 7:00 a.m. to 6:00 p.m., on weekdays. This would be the most basic example. More complex systems may charge varying rates for vehicles entering the cordoned-off area at different times of the day, with varying types of cars, and at various points along the route (e.g., higher along the busiest thoroughfares than on less crowded streets). Cordon pricing can increase revenue while easing traffic congestion in the cordon region. It is typically seen to be the most suitable for places with constrained road capacity, high travel demand, and drivers' options to move to other modes (like transit) or adjust their travel plans to avoid peak hours when toll charges are high. Countries like London, Stockholm, and Singapore have adopted variations on the cordon-pricing principle [22].

3.2.3 Area-Wide Mileage Tolling

The complexity of distance-based tolling is significant and necessitates the implementation of uniform standards across the whole region and interoperability across countries. A road user tax is calculated using this method, which charges cars based on their VMT (vehicle miles traveled) [23]. The German truck toll is one example of this system; under this system, all trucks operating inside Germany must pay tolls depending on the distance they have gone within the country [24]. This tactic is comparable in some ways to the gas tax levied in the United States, in that, in principle, the amount of tax owed by each vehicle is determined by the number of miles traveled. The primary goal is to control the amount of driving, while the secondary goal is to create income for the transportation system. The quantity of driving will be regulated to a lower degree.

3.2.4 Integrated System Management

In this far-reaching idea, the demand for transportation would be handled by the dissemination of information; passengers would have their pick of transportation modes and routes, as well as a variety of options for paying for their trips [25]. The price would provide an incentive for selecting the mode of transportation that is the most efficient, and the market should determine how much capacity is made available. It will be necessary to use highly complicated systems, such as communications between roadside infrastructure, vehicles, and travelers; nonetheless, it is anticipated that system users will be highly productive. The flexibility of the market and access to information are necessary prerequisites [26].

3.3 EXISTING TOLL COLLECTION MODELS

This section discusses the existing toll collection systems worldwide. Section 3.3.1 describes the current toll collection systems in detail, and Section 3.3.2 presents information on various countries using different ETC systems.

3.3.1 Electronic Toll Collection

ETC is a technique which is currently being applied all over the globe. Many nations, including Canada, Poland, Japan, Italy, and Singapore, are a few that have chosen to use the ETC system [11]. Certain other governments have also embraced the approach [27, 28]. In this sub-section, we have discussed four ETC systems: radio frequency identification (RFID), FASTag, automatic number plate recognition (ANPR), and dedicated short-range communication (DSRC).

3.3.1.1 RFID

Electronic Toll Collecting is a system that facilitates the digital processing of toll payments. This enables toll collection and tracking of traffic relatively steadily [29]. Stickers with RFID chips already embedded in them are affixed to the vehicles as part of the system, which enables the automatic deducting of toll fees at checkpoints. The first ever compatible RFID technology based on Electronic Tolling System in India was unveiled at the Charoti Toll Plaza in Dahanu, in the Thane region of Maharashtra [29]. Vehicles would be able to breeze past a total of six toll plazas, thanks to an inter-operable RFID-based system being developed by three separate companies in India that specialize in the construction of roads: L&T, IRB Infrastructure, and National Highways Authority of India (NHAI) are the three companies that make up this consortium. They include the NHAI plaza at Narmada Bridge, the L&T IDPL Plaza in Karjan, Vadodara, and the IRB toll plazas at Charoti, Bhagwada, Boriach, and Choryasi. Both plazas are in Vadodara. In addition to this, the trial program for it has also been started on National Highway 5 (NH-5) between Chandigarh and Parwanoo. The functionality of the RFID is illustrated in Figure 3.2.

3.3.1.2 FASTag

The National Highway Authority of India maintains and administers the FASTag automated payment accumulating program. The FASTag is a passive RFID tag that can make toll payments straight from the linked prepaid or savings/current account of the consumer [30]. It is fixed on the car's front windshield and permits the user to pass through toll plazas without stopping and paying the toll at any station. The user's linked account automatically charges the appropriate amount for the tolls incurred. FASTags are also vehicle-specific, meaning they cannot be moved to another vehicle once attached to a particular vehicle. Any of the National Electronic Toll Collection (NETC) Member Banks can sell FASTag to interested customers. In case, a FASTag is linked to the primary account, the record must be refilled or capped complete based on the user's consumption of the FASTag. If the FASTag is not tied to the record, the summary does not require any action. If the user does not keep an acceptable balance on their FASTag, the FASTag will be banned at the toll plaza. If a user goes through

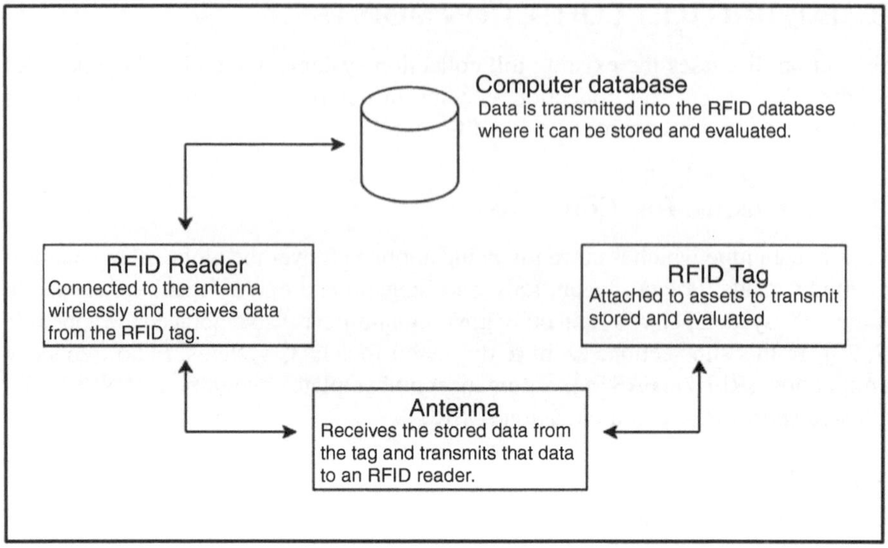

FIGURE 3.2 How RFID works.

a toll plaza without recharging his NETC pass, he cannot use the NETC services and must pay the toll fare in cash [31]. In this case, the user cannot use the NETC services. Toll booths along the section of the Golden Quadrilateral that runs between Chennai and Bengaluru were the first to begin receiving FASTag payments in July 2015 [32, 33]. By the time April 2015 came around, the FASTag system had been implemented at 247 toll booths on national highways across India. This accounted for 70% of the nationwide toll booths. As of November 23, 2016, 347 of the 366 charge entrances on expressways around the nation accept FASTag transactions.

3.3.1.3 Automatic Number Plate Recognition (ANPR)

ANPR, also known as Automatic Number Plate Recognition, is an image processing technology used to capture vehicle images and extract license plate information from them. This technology converts the extracted license plate data into machine-readable formats, such as text strings, making it easier to process and store in a database. ANPR is widely employed for various purposes, including vehicle identification, monitoring, and enforcement. By automating the recognition and analysis of license plate data, ANPR enables efficient management and tracking of vehicles. ANPR technology originated in 1976 when it was invented by the UK police. The first practical implementation of ANPR as a pilot system occurred in 1979, specifically for monitoring traffic on highways. Initially, ANPR was primarily used in policing and security applications. However, over time, it has progressed and changed along with other Information and Communications Technologies (ICTs). As a result, there has been a significant surge in new methods and groundbreaking applications of ANPR across various research fields.

3.3.1.4 Dedicated Short-Range Communication (DSRC)

DSRC stands for Dedicated Short-Range Communications. It is a wireless communication technology designed for short-range communication between vehicles and roadside infrastructure. DSRC operates in the 5.9 GHz frequency band and is specifically developed for ITS applications. DSRC enables vehicles and infrastructure to exchange data and information in real-time, facilitating communication for various purposes such as vehicle safety, traffic management, and cooperative driving. It supports high-speed, low-latency communication, making it suitable for applications that require quick response times. Some key features and applications of DSRC include:

- **Vehicle-to-Vehicle (V2V) Communication:** By utilizing DSRC technology, vehicles can communicate their current speed, acceleration, and braking information to nearby vehicles. This enables cooperative driving and safety applications such as collision avoidance and emergency vehicle warning systems.
- **Vehicle-to-Infrastructure (V2I) Communication:** DSRC enables vehicles to communicate with roadside infrastructure such as traffic signals, toll booths, and parking facilities. This facilitates applications like traffic management, toll collection, and parking guidance systems.
- **Vehicle-to-Pedestrian (V2P) Communication:** DSRC can also support communication between vehicles and pedestrians, enhancing safety by providing warnings and alerts to pedestrians in proximity to vehicles.
- **Signal Phase and Timing (SPaT) Broadcast:** DSRC can transmit real-time signal phase and timing information, allowing vehicles to receive data about traffic signal status. This can be used for applications like optimizing traffic flow and providing driver assistance.

Overall, DSRC technology plays a significant role in enabling connected vehicle systems and improving transportation safety and efficiency. It has been widely used and tested in various pilot projects and research initiatives worldwide. Table 3.3 presents a comprehensive comparison of the different ETC techniques, focusing on their methodology, efficacy, efficiency, and major challenges.

3.3.2 Countries Using ETC

The ETC technology has already been used in several countries' transportation systems. The ETC technologies and related parameters are included in Table 3.4 along with the nations that correlate to them.

3.4 ARCHITECTURE OF ETC

Electronic toll collecting relies on several elements to carry out a transaction successfully. The detection of vehicles and the identification of accounts are, by far, the two most critical components. Recognition of cars can be achieved by in-road and above

TABLE 3.3
Comparative Analysis of Existing ETC Technologies Worldwide

ETC Technique	Methodology	Efficacy	Efficiency	Major Challenges
RFID (Radio Frequency Identification)	RFID uses radio waves to communicate between a transponder (tag) attached to the vehicle and a reader at the toll station. This technique is used worldwide.	RFID-based systems are widely adopted and proven to be effective in toll collection. They offer high accuracy in detecting and identifying vehicles.	RFID-based systems enable seamless and quick toll transactions, as the tag is detected automatically without requiring the vehicle to stop.	The main challenge with RFID-based systems is the need for infrastructure setup, including tag installation and reader deployment. Additionally, issues like tag interoperability and signal interference can impact performance.
FASTag	FASTag, on the other hand, is a specific implementation of RFID technology designed for electronic toll collection in India.	FASTag has significantly improved the toll collection process by eliminating the need for manual cash transactions, reducing human errors, and enabling smoother flow of vehicles through toll plazas. FASTag offers users the ability to track their toll transactions, check their account balance, and manage their accounts conveniently through online portals or mobile apps.	FASTag enables non-stop tolling, allowing vehicles to pass through toll plazas at highway speeds without the need to stop and pay in cash. The streamlined toll collection process with FASTag helps save time for drivers and reduces fuel consumption by minimizing idling and congestion at toll plazas.	One of the major challenges is ensuring universal adoption of FASTag among vehicle owners. While it has been made mandatory for certain categories of vehicles, ensuring compliance across all vehicles on the road remains a challenge. Technical glitches or system failures in the FASTag infrastructure, such as reader malfunctions or communication issues can pose challenges and lead to disruptions in toll collection.

ANPR (Automatic Number Plate Recognition)	ANPR uses optical character recognition technology to capture and interpret vehicle license plates at toll stations.	ANPR systems can effectively identify vehicles and associate them with registered accounts, enabling accurate toll collection.	ANPR-based systems provide non-stop tolling, reducing congestion and enabling seamless passage through toll plazas.	Challenges include handling variations in license plate formats, maintaining accurate and up-to-date vehicle databases, and managing system robustness under different lighting and weather conditions.
DSRC (Dedicated Short Range Communication)	DSRC uses short-range wireless communication between the vehicle's onboard unit and the toll plaza's roadside equipment. This is widely adopted across Europe.	DSRC-based systems are known for their reliability and accuracy in toll collection, providing secure and fast transactions.	DSRC-based systems allow for non-stop tolling, improving traffic flow and reducing congestion by enabling vehicles to pass through toll plazas at highway speeds.	Challenges include the need for infrastructure deployment, interoperability among different DSRC systems, and potential security vulnerabilities.

TABLE 3.4
A Country-Wise Summary of ETC-based Technology, Factors, and Related Technologies

Country	Technology	Factors and Related Technologies
America	Technology on the cutting edge, such as Edge Service Public Toll and AuditToll DBE.	RFID, ETC, DSRC, ANPR
China	With 5G in mind, China is developing C-V2X networks.	AV-CV technology, ETC device
Germany	GPS-based mobile phone toll compendium in Germany	GSM, OBU, Automatic Payment Method
Hong Kong	Automatic toll gathering was pioneered in Hong Kong by DSRC Technologies in 1993.	OBU, DSRC (Dedicated short-range communication), ITS
India	The National Payments Corporation of India is responsible for creating the National Electronic Toll Collection project. This was done to fulfill the demands of the Indian market for computerized toll booths.	FASTag, RFID, BookMyToll System
Japan	The vehicle Information and Communication System is a Japanese satellite navigation system.	DSRC, ITS, ETC, Ministry of Land, Infrastructure and Transport of Japan
Russia	In 2015, Russia introduced the Platon toll system based on ETC for collecting tolls using federal roads by vehicles weighing over 12 tons.	Onboard Units (OBUs), SkyToll
Singapore	Singapore will implement Satellite-based Electronic Road Pricing System (ERP) in the first half of 2021.	GNSS, RFID, Cash card, Smartcard, Land Transport Authority (LTA)
Stockholm	Congestion pricing system implemented in Stockholm, Sweden.	Stockholm Congestion Tax, Traffic congestion

sensors, cameras, communication between vehicles and roadside infrastructure, or any combination of these methods. The categorization of cars is an element of vehicle identification in jurisdictions that apply various fees to different classes of vehicles. The typical architecture of ETC is shown in Figure 3.3. The user account is identified by linking the vehicle ID to the user account, often performed using a relational database.

3.4.1 Vehicle Detectors (In-Road Sensors)

Subsurface, roadside, or overhead installation options are all possible for sensor systems. The presence of an automobile can be detected via inductive sensors that

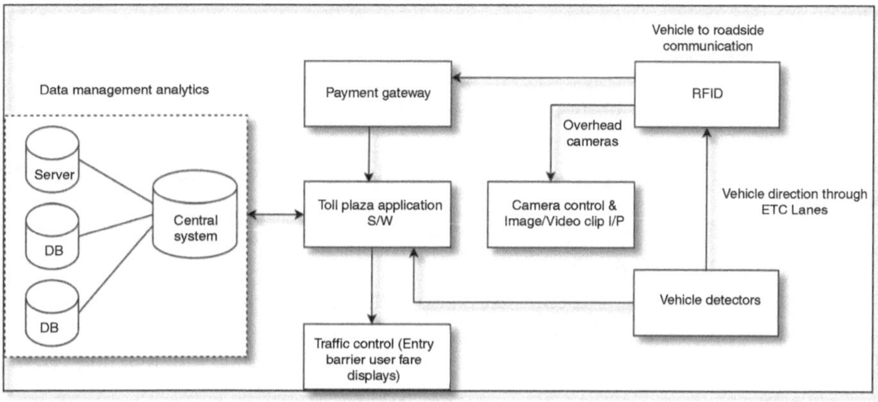

FIGURE 3.3 The architecture of ETC.

have been incorporated into the pavement. When a vehicle drives over treadles, the treadles note the number of axles on the vehicle. Offset treadle installations may detect dual-tire cars and axle counts. Vehicle separation light curtain can capture the vehicle's contours, allowing for easier differentiation between trailers and trucks. Sensors can also identify the distance between cars to offer data regarding the number of vehicles through a specific area.

The Traffic Reporting and Control (TRAC) system is an example of an in-road sensor system. This system combines cutting-edge inductive loop detection technology with sophisticated signal processing. Both manned and unmanned toll lanes can take advantage of the pre-classification capabilities of the TRAC system. It is also possible to use it for post-classification purposes to impose tolls. It just takes 0.1 seconds to complete the entire categorization procedure. The TRAC system is contained within a conventional inductive loop detector, which is characterized by the following:

- Categorizes automobiles into twenty-three different categories.
- Provides precise measurements of the speed of a single loop.
- Determines the vehicle's length and the total number of axles.
- Gives an overview of traffic flow at the point, toll segment, and regional toll levels.
- Provides online access to all data.

3.4.2 Overhead Cameras

"Video tolling" refers to a vehicle identification method that uses cameras mounted on overhead structures. License plate identification and recognition systems (LPI/R) are utilized in video tolling. Cameras mounted on gantries overhead snap an image of a vehicle's license plate as it travels through the checkpoint. The license plate image is scanned by Optical Character Recognition (OCR) software and then the number is looked up in a database to locate the vehicle's owner linked with the plate. When utilized as a method of enforcement for cars that do not have a toll account, video tolling is typically implemented in conjunction with other systems.

Despite its widespread application, LPI/R suffers from the following drawbacks:

1. A low resolution of the image is frequently caused by the plate needing to be in focus.
2. Images are characterized by blurriness, particularly motion blur, most likely caused by increased vehicle speeds.
3. Inadequate lighting and contrast can be attributed to overexposure, reflection, shadows, or the color or style of the plate's background.
4. A distinctive typeface, as seen on out-of-state license plates and vanity licenses.
5. A variety of license plate designs, like those used on federal vehicles.
6. Circumvention techniques (such as reflective plates).

In addition to being used for tolling, LPI/R manufacturers have highlighted several other applications for their technology. A truck's license plate can be verified via a directory of permitted plates to decide whether it can skip a weigh post or access a secure area, which is helpful for practical road vehicles and protected regulation. This helps determine whether a vehicle can enter a specified location. License plates can be examined vs. a "hot cars" list at global checkpoints to find stolen vehicles and plates and those assigned to defendants, outlaws, or trafficking concerns. By displaying a plate number and the vehicle's speed on a variable messaging sign, LPI/R can issue speeding citations to drivers who break the speed limit or serve as a friendly reminder to drivers who fail the speed limit. Either it can help detect and fine polluters by recording the alphanumeric sequence of a plate while automatically evaluating the effluents from the exhaust, or it can simplify the process of testing vehicles' emissions.

3.4.3 Vehicle-to-Roadside Communication

The use of transponders to identify vehicles has grown rampant. An electronic tag is a device or sticker that contains an RFID chip. This tag is attached to the inside of the windshield of a vehicle, close to where the rear-view mirror is located. It can respond to radio signals as the tag travels underneath a gantry with a transmitter mounted. Although laser and infrared signals have also been examined, it has been determined that the radio frequency spectrum offers the highest level of precision. On the other hand, some windshields are coated with metal, preventing the electronic tag from being read. It is estimated that just approximately 0.5% of all windshields are of this type, which indicates that the percentage of cars with such windshields could be higher.

3.4.4 Central System

The central system is the core backend infrastructure that manages and operates the entire toll collection system. It typically consists of servers, databases, and software applications that handle transaction processing, account management, reporting, and customer service.

3.4.5 Toll Plaza Infrastructure

This includes physical infrastructure such as toll booths, barriers, and lanes equipped with sensors and communication systems. It facilitates the entry and exit of vehicles and ensures the smooth flow of traffic during toll collection.

3.4.6 Payment Gateway

The payment gateway is the interface between the toll system and the various payment methods available to users. It enables transactions through cash, credit/debit cards, electronic wallets, or other electronic payment systems.

3.4.7 Data Management and Analytics

This component involves the storage, management, and analysis of toll-related data, including transaction records, vehicle information, and traffic patterns. Data management systems and analytics tools provide valuable insights for operational optimization and decision-making.

3.4.8 Key Features and Functionalities of ETC

The key features and functionalities of ETC are as follows:

a. **Transaction Type:** Analysis of transactions is done remotely and near virtual environments because toll plazas communicate their activities at frequencies of 10 minutes.
b. **Interoperability:** NETC ecosystem supports multiple issuers and multiple acquirers', i.e., underneath the NETC initiative, a tag that any account holder granted is appreciated at all toll plazas received securely and confidentially by any banking institution.
c. Users can link their FASTag to their present savings, regular checking account, or a preloaded arrangement, depending on the kind of account the sponsor financial institutions are issuing. This provides clients more freedom over basic credit instruments. It is unnecessary to have a good rapport with the provider bank to open a preloaded login with them.
d. **Tag Issuance:** Can be issued by member banks authorized for NETC Program.
e. **Offering without Cash:** The FASTag system allows online payments at the toll plaza while the car is moving.
f. **Save Time and Fuel:** Customers can travel without stopping at the toll plaza using the FASTag, thus reducing congestion at plazas and saving fuel and travel time.
g. **Reload Your FASTag Account Online:** Clients can refill their FASTag accounts online by using UPI, credit cards, debit cards, NEFT, RTGS, or

online banking through the portals offered by the financial institutions that granted their FASTags.

3.5 UPCOMING TECHNOLOGIES

Tolls or the notion that the "user pays" are efficient methods to increase sustainable mobility. Our nation's road infrastructure is an essential component in the effort to grow our economy. This is true both from the perspective of concessions and users. Governments must provide a dependable road network that safely and effectively transports passengers and products.

3.5.1 GNSS-Based Tolling System

GNSS stands for Global Navigation Satellite System. India has a functional NETC system known as FASTag. This system was provided by the Indian Highways Management Company Ltd. (IHMCL) and is now in use. RFID technology, more than 720 electronic toll plazas located nationwide, and On-Board-Units (OBU) installed in cars are the system's foundation. FASTag was first released in 2016, and its usage became mandatory in February 2021. By March 2021, the number of users had reached 30 million. At nationwide electronic toll plazas, drivers without a FASTag are subject to paying double the average toll amount. According to Gadkari, most automobiles in India use FASTag to pay their tolls.

The NHAI has moved up the timeline for introducing a GNSS-based tolling system by sending proposals to find a technical consultant who can prepare a roadmap outlining the tolling system that may be implemented. In addition, the evaluation will investigate the current standing of the ETC system and how the GNSS-based tolling system might be used to supplement it. It is necessary to provide a detailed description of the technological solution and the conditions the infrastructure must meet to implement such a solution.

3.5.1.1 Working of GNSS-based Tolling System

Numerous different uses are possible for GNSS. The amount of dependability that is required can be used to classify these applications into one of the following three categories [11]:

- **Deemed security:** Applications are deemed security when an inaccuracy has the possibility of causing bodily damage or fatalities. Applying GNSS-based solutions in air transport is a notable example of the application of this subject.
- **Responsibility:** Mistakes in responsibility systems can have severe economic and legal repercussions. One example is the use of GNSS in tolling systems, where failure would not cause any actual injury but would result in financial damages.
- **No liability:** This module includes those apps that do not fit into the other two groups. For instance, the use of GNSS for navigational reasons in the modern day is included in this category.

As was previously said, utilizing GNSS for road pricing is an application that places a significant emphasis on liability. Because of this, such things must be developed with caution, and prior to being made available to the public, the system's trustworthiness needs to be extensively evaluated. As a result, substantial study has been conducted on GNSS-based toll collection.

The tolling system based on GNSS technology will comprise a simulated toll plaza that will record the arrival and leave of drivers' cars on the tolled stretch of the roadway. This virtual toll booth, which will receive data from the vehicles equipped with an internal GNSS device, will eventually replace the standard toll-collecting booth that is now outfitted with high-resolution cameras. After that, the data of the user's bank account and car, including the make and model of the vehicle as well as its registration number, will be entered into the device. This information is communicated between the automobile and the online toll booth whenever a car does this. Therefore, an automatic tolling procedure will be initiated whenever a vehicle goes by the virtual toll booth. The Indian Space Research Organization (ISRO), which consists of seven satellites, has also constructed its own regional GNSS and has given it the name Navigation Indian Constellation (NavIC). The satellite-based ETC would likely use the regional Indian GNSS solution and other well-known GNSS systems like Galileo (the GNSS used in Europe).

3.5.1.2 Advantages of GNSS-based Tolling System

Compared to the electronic toll-collecting gems in India, tolling based on GNSS might offer numerous advantages. Processes can be enlarged and scaled up with low investment, and the OBUs assembled on automobiles for GNSS can be used for increased value-added services to the end-users. GNSS-based processes make a greater variety of charging strategies technically feasible.

The adaptability offered by GNSS technology for tolling networks allow for developing a tolling mechanism that is both fair and intake. Due to the ever-shifting nature of customer demand, several pricing structures may be necessary for various contexts. GNSS-based toll collection solutions are versatile and can be integrated with various payment methods. According to the information presented in [33], there are generally two billing schemes: discrete and continuous. The application of discrete pricing models is predicated on occurrences that may or may not occur. Tolls on bridges are a typical illustration of a discrete tolling scheme.

Regarding continuous road pricing schemes, the toll is calculated using a cumulative parameter, like cumulative distance. However, genuinely unlimited plans appear rare in practice because the two types of schemes can be combined. Consequently, every system will contain a distinct part, at minimum once leaving a charging area [33]. The charging process is initiated when a car travels through a charging point. Typically, these charging outlets will be arranged as virtual gantries. Multiple virtual gantries can form charging zones by positioning them at all conceivable entry and exit points and elsewhere within the zone. A further illustration of a discrete scheme is provided by the fact that this form of charging area relies on virtual gantries, which, when passed, cause a charge to be triggered [10].

By increasing the electronic cartography needed to calculate tolls, GNSS-based toll collection solutions can be scaled swiftly and with minimal or no extra costs. This is achievable since there is no need for roadside units. Because of this, the routes subject to tolls can be theoretically redefined instantly [11, 31]. After establishing a GNSS-based toll-collecting solution, it is straightforward to incorporate various other activities that leverage the location details supplied by the OBU. Such VAS may include immediate emergency services with geolocation, legitimate traffic updates, and PAYD insurance premiums, and it has been deployed by several companies, notably Hollard in South Africa. Because everyone can develop these value-added services, they present many new business prospects [20].

Due to the effective implementation of GNSS-based toll-collecting systems, many of these benefits have been realized in Europe. Some examples of these systems are the Stockholm congested tax, the Austrian truck toll, the German TollCollect, the toll system in Slovakia, and the French Eco-tax. Every GNSS-based tolling system in Europe has a compliance (payment) rate of more than 98% or 99%.

3.5.1.3 Limitations of GNSS-based Tolling System

However, such technologies also come with their own set of difficulties, including those regarding accuracy, confidentiality, and consistency. In the case of India, the most prominent issue that still must be answered is whether or if the recently adopted electronic toll-collecting system known as FASTag would be changed shortly after it becomes obligatory. Gadkari also stated that while all newly manufactured commercial cars would come equipped with vehicle tracking systems, the state will need to devise a method to retrofit GNSS technology into older vehicles so that it may have a broader impact. Gadkari claimed this would be necessary.

The concept of charging reliability refers to the need for a toll operator to provide evidence that a computed charge was determined accurately [32]. The idea of a system's reliability is typically broken down into several different aspects, the majority of which, depending on the source, is comprised of the following: accuracy, availability, continuity, and integrity.

A system's continuity can be evaluated based on the likelihood that it will be able to carry out all its intended functions for the entirety of the period for which it was designed. In these scenarios, poor precision results in vehicles being positioned on the incorrect roadways, leading to inaccurate charging of those vehicles. It is straightforward to make a false detection of parallel service roads and highways [6]. As a result, the cars and receivers are typically encircled by tall structures in urban settings. Because of this, the visibility of satellites is restricted, and direct Line-of-Sight (LoS) signals are either extremely rare or non-existent, which reduces the accuracy of position calculations.

The term "availability" refers to the amount of time conveyed as a percentage of the total during which a GNSS-based system is accessible. The availability of positioning services is equivalent to the availability of satellite services. They employ a monetary charge to assess a system because [32] it concentrates on a performance metric to compare various tolling systems. Because the availability of charging guarantees the functionality of a system, this measure seeks to safeguard the toll operator from the risk of being unable to collect the tolls [33].

The level of confidentiality afforded to users is essential in determining whether a GNSS-based tolling system will be used. Because the position data of the vehicle is continuously transferred to the data center, exceptionally when thin OBUs are used, it is technically conceivable to develop movement profiles without the users' knowledge. Ref. [4] discusses a GNSS-based system architecture that is friendly to users' privacy and uses thick OBUs. Utilizing a proxy, which does not associate data with specific automobiles, is still another method that can be used to protect users' privacy. The Anonymous Loop-Back Proxy (ALP), which is favored by [30], is one of the potential solutions that can be achieved by employing proxies. The OBU is responsible for sending trip data that is opted to the ALP. The ALP is then responsible for calculating the proper fee and sending the correct back to the OBU. The OBU can then either communicate the lead to a network infrastructure or debit the fee from a smart card. Since the proxy does not save any information, the details relating to the trip is stored in the automobile directly and may be erased by the consumer after payment is received. Another benefit of the ALP is that the data center has no exposure to the proxy and, as a corollary, does not have the ability to view the real trip details. While examining various dimensions of fairness, it is essential to distinguish between the fairness of a toll and the fairness of a system. It is essential to apply the same rate of toll to all motorists and other people who use the road to set up a system that will be regarded as fair [22]. In locations with high vehicle traffic volumes and less populated regions, users may be required to pay a fee to access public roads.

Nevertheless, to have an equitable charging system, there needs to be a differentiation between the various categories of vehicles in terms of their impact on the environment. The enforcement system of a pricing system is the primary factor determining how fairly the system is regarded by its users. The enforcement system ensures that all cars must pay the toll.

3.6 CONCLUSION

This study has outlined various ETC technologies and their worldwide effects on transportation sectors. The identified pertinent ETC technologies used in many nations are included at the beginning of this research. Through an in-depth analysis and examination of existing literature and research, we have gained valuable insights into the key components, technologies, and challenges of developing intelligent toll payment gateways. The chapter highlights the importance of efficient and secure toll collection systems in modern transportation networks. We have identified various intelligent technologies employed in toll payment gateways, including RFID, ANPR, and DSRC. These technologies enable automated and contactless payment processing, enhancing the overall efficiency and user experience at toll plazas. We also discussed upcoming technologies such as GNSS-based tolling systems, its advantages, and limitations. Every day, there will be a rise in the number of automobiles. Therefore, for every country, it is high time to start planning for socioeconomic growth over the next 10 years and to take the necessary actions to implement an intelligent ETC system.

REFERENCES

1. S. Matti, "Delivering transportation infrastructure through public-private partnerships: Planning concerns," *Journal of the American Planning Association,* vol. 76, no. 1, 2009, 43–58.
2. B. Andrea and C. Gertz, "Tolling heavy goods vehicles: Overview of European practice and lessons from German experience," *Transportation Research Record*, vol. 2066, no. 1, 2008, 106–113.
3. Z. Haibo, W. Xu, J. Chen and W. Wang, "Evolutionary V2X technologies toward the Internet of vehicles: Challenges and opportunities," *Proceedings of the IEEE,* vol. 108, no. 2, 2020, 308–323.
4. L. Ching-Huang, P. Hsiao, Y. Yang, S. Lin and S. Lung, "Effects of the manual and electronic toll collection systems on the particulate pollutant levels on highways in Taiwan," *Atmospheric Pollution Research*, vol. 12, no. 3, 2021, 25–32.
5. L. Ming-Yeng, Y. Chen, D. Lin, B. Hwang, H. Hsu, Y. Cheng, Y. Liu and P. Tsai, "Effect of implementing electronic toll collection in reducing highway particulate matter pollution," *Environmental Science & Technology*, vol. 54, no. 15, 2020, 9210–9216.
6. Q. Wei, H. Chen and B. An, "Dynamic electronic toll collection via multi-agent deep reinforcement learning with edge-based graph convolutional networks," In *IJCAI'19: Proceedings of the 28th International Joint Conference on Artificial Intelligence*, Macao, China, August 2019, pp. 4568–4574.
7. W. Jian, V. Goh, T. Yap and H. Ng, "Vehicle classification using convolutional neural network for electronic toll collection," In *Computational Science and Technology*: *6th ICCST 2019*, Kota Kinabalu, Malaysia, vol. 29, August 30, 2019, pp. 169–177.
8. J. Sharma and P. S. Mehra, "Secure communication in IOT-based UAV networks: A systematic survey," *Internet of Things*, vol. 23, 2023, 100883. http://doi.org/10.1016/J.IOT.2023.100883.
9. S. Sharma and B. Kaushik, "A survey on internet of vehicles: Applications, security issues & solutions," *Vehicular Communications,* vol. 20, 2019, 100182.
10. D. Ahmed, A. Bahadir, Y. Hillali, A. Rivenq and M. Faruque, "Eve, you shall not get access! A cyber-physical blockchain architecture for electronic toll collection security," In *2020 IEEE 23rd International Conference on Intelligent Transportation Systems (ITSC)*, pp. 1–7. IEEE, Rhodes, Greece, September 20–23, 2020.
11. M. Chakraborty, S. Stapleton, M. Ghamami and T. Gates, "Safety effectiveness of All-Electronic Toll Collection systems," *Advances in Transportation Studies*, 2020, 127–142.
12. C. Ernesto, L. Mannini, B. Montemarani, M. Nigro and M. Petrelli, "Congestion pricing policies: Design and assessment for the city of Rome, Italy," *Transport Policy*, vol. 80, 2019, 127–135.
13. H. Daniel, R. Singh and D. J. Graham, "Social distancing in public transport: Mobilising new technologies for demand management under the Covid-19 crisis," *Transportation,* vol. 49, no. 2, 2022, 735–764.
14. R. Seshadri, P. André and B. A. Moshe, "Congestion tolling—Dollars versus tokens: Within-day dynamics," *Transportation Research Part C: Emerging Technologies*, vol. 143, 2022, 103836.
15. T. Walter, "Congestion control in Singapore," *International Transport Forum Discussion Paper No. 2020/10*, OECD Publishing, Paris, 2020.
16. "Evolution of tolling," Available at: https://assets.kpmg/content/dam/kpmg/pdf/2015/06/kpmg-toll-benchmarking-study-2015-v2.pdf. (Accessed on Jan 12, 2023).

17. S. Dutta, S. Kar and T. Sadhu, "Evolution of public private partnership in India: A case study in roads & highway structures," International Journal for Research in Applied Science and Engineering Technology, vol. 9, no. 3, pp. 1022–1027, Mar. 2021, doi: 10.22214/ijraset.2021.33345.
18. "Corridor tolling." Available at: www.virginiadot.org/vtrc/main/online_reports/pdf/22-R5.pdf. (Accessed on Jan 12, 2023).
19. D. Gabor, Uses of dynamic tolling net revenues for corridor investments. . Transportation Research Board 98th Annual Meeting, Transportation Research Board, Washington DC, United States, 2019, 12p. https://trid.trb.org/view/1572991
20. T. Jyh-Faq1 and S. Li, "Cordon tolling for mixed traffic flow," *Transportmetrica A: Transport Science,* vol. 15, no. 2, 2019, 1662–1687.
21. V. Shaghayegh, H. Poorzahedy and R. Lindsey, "Predictive cordon pricing to reduce air pollution," *Transportation Research Part D: Transport and Environment,* vol. 88, 2020, *102564.*
22. M. Johari and H. Haghshenas, "Modeling the cordon pricing policy for a multi-modal transportation system," *Case Studies on Transport Policy,* vol. 7, no. 3, 2019, 531–539.
23. "Area-wide wide milage trolling." Available at: www.rand.org/content/dam/rand/pubs/reprints/2011/RAND_RP1423.pdf. (Accessed on Jan 28, 2023).
24. B. Singichetti, J. L. Conklin, K. H. Lich, S. S. Nasim and R. B. Naumann, "Congestion pricing policies and safety implications: A scoping review," *Journal of Urban Health,* vol. 98, no. 6, 2021, 754–771.
25. I. H. Sung, Y. G. Park and J. W. Jang, "A study on the factors Causing the Intention to Use a Smart Tolling System," *International Journal of Software Innovation (IJSI)*, vol. 10, no. 1, 2022, 1–17.
26. I. H, Seong, Y. G. Park and J. W. Jang, "Case study for smart tuning effectiveness verification by comparing high pass and smart tolling," *In 2021 21st ACIS International Winter Conference on Software Engineering, Artificial Intelligence, Networking and Parallel/Distributed Computing (SNPD-Winter),* Ho Chi Minh City, Vietnam, 28–30 January 2021, pp. 267–269. IEEE.
27. L. Xiang, Y. Zhao, D. Li and W. Guan, "Error Serial Episodes Discovery from Mobile Payment Log in Distributed ETC," *In Algorithms and Architectures for Parallel Processing: 21st International Conference, ICA3PP 2021, Virtual Event, December 3–5, 2021, Proceedings, Part I,* pp. 210–221. Cham: Springer International Publishing, 2022.
28. R. Malalatiana, A. Cabani, H. Chafouk and G. Fremont, "Geolocation process to perform the electronic toll collection using the ITS-G5 technology," *IEEE Transactions on Vehicular Technology,* vol. 68, no. 9, 2019, 8570–8582.
29. Z. J. Wong, V. T. Goh, T. Timothy and Hu Ng, "Vehicle classification using convolutional neural network for electronic toll collection," In *Computational Science and Technology: 6thICCST2019,* KotaKinabalu, Malaysia, 29–30August2019, pp. 169–177.
30. A. Sontakke, A. Diwakar and G. Kaur, "Intelligent automatic traffic challan on highways and payment through FASTag card," *Indian Journal of Science and Technology,* vol. 12, no. 44, 1–6, 2019. doi: 10.17485/ijst/2019/v12i44/146468
31. S. Oza, "A descriptive Study on FASTag: Electronic Toll, Standing Tall," *Journal of Information Technology & Management,* vol. 3, no. 1, 2020, 26–29.
32. J. Bharavi, K. Bhagat, H. Desai, M. Patel and J. K. Parmar, "A comparative study of Toll Collection Systems in India." *International Journal of Engineering Research and Development,* vol. 13, no. 11, 2017, 68–71.
33. S. Chandrappa, M. S. Guruprasad, HN N. Kumar, R. Kumar and D. K. S. Kumar, "An IOT-based Automotive and Intelligent Toll Gate Using RFID," *SN Computer Science,* vol. 4, no. 2, 2023, 154.

4 IoT-Based Healthcare System for Early Prediction of Alzheimer's Disease

Zuha Khan and Reshu Agarwal

4.1 INTRODUCTION

Alzheimer's disease (AD) is the leading cause of dementia in the elderly. There is now interest in using machine learning to identify metabolic diseases such as AD and diabetes that affect large populations worldwide. Their condition is increasing at an alarming rate every year. In AD, the brain is affected by neurodegenerative changes. As the elderly increase, more people, their families, and their health will suffer from diseases that affect memory and work. These impacts will be social, financial, and economic. AD is a neurodegenerative disease characterized by cognitive impairment and memory impairment. As the global population ages, the prevalence of AD is becoming a public health concern. Early detection and prediction of AD plays an important role in intervention and effective treatment. In recent years, data mining techniques and IoT have attracted attention as an important tool for exploring big data to identify AD patterns and determinants.

Moreover, The Internet of Things (IoT) is a network of interconnected devices that helps to collect, communicate, and analyse data over the cloud. Now a days, a lot of industries which includes manufacturing, agriculture, transportation, and smart cities are implementing IoT devices. However, healthcare is one of the most exciting and significant fields where IoT may have an influence. Through remote monitoring and virtual visits, IoT in healthcare may provide patient care outside of a clinic or hospital, optimize the pharmaceutical production process, and enable healthcare practitioners to access patient data easily and rapidly. By lowering mistakes, boosting diagnosis, and enabling personalized treatment, IoT may also increase the quality and safety of healthcare services.

Further, data mining is identifying and extracting hidden patterns, variables, and trends from large data sets using statistical, mathematical, and artificial intelligence tools. Data mining is part of the broader field of knowledge discovery in data (KDD), which is the process of looking at data to identify any hidden patterns and trends of information that can be used to categorise the data. Data mining is used to convert

 DOI: 10.1201/9781003405535-5

raw data or information into data that is used for business purposes. Organizations frequently collect and combine data from various enterprise data warehouses. These organizations use complex statistical and mathematical algorithms to evaluate data sets, calculate the predictions of a future event, and make informed business decisions. Data mining can help businesses analyse large amounts of data to uncover patterns and trends. Organizations use data mining to extract information from large databases or sets to solve various business questions or healthcare problems. It can help with various purposes, such as understanding customer preferences, detecting fraud, and observing present and future trends. These findings can be used as a data input for further analysis. For example, data mining can identify customer preference based on their behaviour, which can be used to improve marketing strategies or product design. Data analysis and data mining are related terms but different. Data analysis is interpreting data to find trends and patterns that can help answer specific questions or test hypotheses. Data analysis is a process that includes investigating and interpreting data that can help answer specific questions and help gather important and useful information to make necessary decisions. Data mining is a part of data analysis that aims to extract hidden patterns and gain valuable insights from large datasets. Data analysis can be done on any data, regardless of the size or complexity, structured or unstructured whereas data mining can only be done in large and structured data.

This chapter aims to explore the potential of data mining in the early prediction of AD by analysing a variety of data, including genetic data, clinical data, and neuroimaging data. Further, IoT devices are suggested for monitoring patients' health conditions. Findings from this study have the potential to help improve diagnosis, risk assessment, and individualized treatment strategies for individuals at risk for AD.

4.2 LITERATURE SURVEY

The use of data mining tools for the early prediction of AD has advanced because of several case studies and research findings. This research has investigated the possibilities for developing precise prediction models using various data sources, including clinical data, genetic markers, neuroimaging, and cognitive tests. Researchers monitored an older cohort for 10 years while detailed clinical data were gathered, including demographics, medical histories, and cognitive evaluations. Data mining techniques were used to find AD predictive trends. The study discovered that certain demographic characteristics, including age and education level, together with tests of cognitive performance significantly and accurately predicted the onset of AD [1]. Further, deep learning methods were used to analyse neuroimaging data to forecast the course of AD. A convolutional neural network (CNN) was used in the study to extract features from brain magnetic resonance imaging (MRI) data. Using the CNN-based model, it was possible to distinguish between healthy controls and people who would eventually develop AD with astonishing precision. The results demonstrated the promise of deep learning and neuroimaging for early AD prediction [2]. Many researchers used machine learning methods to forecast AD risk using a huge dataset of genetic markers. They discovered gene variations linked to greater susceptibility to the disease by comparing the genetic profiles of people with and without AD. The predictive model showed the potential of genetic information in early AD prediction and had a high accuracy rate.

Further researchers investigated the utility of ensemble approaches in AD prediction utilizing cognitive tests. The study used ensemble approaches to increase prediction accuracy and combined numerous cognitive assessments, including memory, attention, and executive function measures. The results showed that employing ensemble methods to combine various cognitive measures improved prediction performance over performing individual tests alone [3]. Data fusion approaches were used in a case study to combine clinical, genetic, and neuroimaging data for AD prediction. The study created a thorough predictive model by combining data from several modalities, such as genetic markers, MRI scans, and clinical factors. The accuracy of the multi-modal analysis was higher than that of models that only used one data source, demonstrating the value of data fusion in the prediction of AD [4]. Further researchers examined the longitudinal examination of cognitive deterioration in those who were at risk for AD. Over a number of years, the study gathered data to monitor changes in cognitive function. The research team found particular cognitive paths connected to the development of AD by using data mining approaches, such as time series analysis and pattern recognition algorithms. The results highlighted the possibility of longitudinal data analysis for predicting early AD [5]. Together, these case studies and study findings show how data mining approaches have significantly advanced the field of early AD prediction. Researchers have found predictive patterns and markers linked to AD risk by combining multiple data sources and cutting-edge algorithms. We are getting closer to accurate and prompt detection of AD for early intervention and better patient outcomes, thanks to the research that have paved the path for additional investigation and improvement of predictive models.

4.3 EFFECTS OF ALZHEIMER'S DISEASE

AD is a neurodegenerative disease characterized by cognitive impairment, memory impairment, and behavioural changes. It is the most common cause of dementia, including in a large proportion of patients worldwide. Over time, people with AD experience reduced memory, thinking, and ability to perform daily tasks. AD affects memory, cognition, and daily function. Amnesia is a common symptom that begins with mild forgetfulness and progresses to an inability to remember recent events or recognize familiar people.

As the disease progresses, people may have problems thinking, judging, and making decisions. Difficult tasks like managing money or following instructions become difficult. AD can also affect language skills, making communication difficult. Changes in mood, attitude, and behaviour are common, along with restlessness, confusion, and withdrawal from activities. These cognitive and behavioural deficits affect daily functioning, increasing reliance on caregivers for critical tasks. AD is very common worldwide and affects an increasing number of people. The ageing of the global population is a major factor in the progression of AD. As life expectancy increases, the number of people at risk of developing AD increases. AD raises public health concerns as it affects individuals, families, and communities as a whole. This disease has a huge impact on healthcare, caregivers, and business.

The costs associated with AD, including medical care, long-term care, and loss of productivity, are significant. In addition, AD has social and emotional consequences that affect the quality of life of individuals and their families. Addressing the problems caused by AD requires a concerted effort in research, early detection, support services, and the development of effective interventions to reduce the burden on people and people.

4.4 SYMPTOMS OF PEOPLE WITH AD

People with AD usually have many symptoms that gradually worsen over time. Mental retardation is a common condition, particularly affecting the handling of recent events and new information. This may appear as forgetting the name, giving the wrong item, or repeating the question. Cognitive decline is not limited to memory but also includes language difficulties such as finding the right words or following a conversation. Problem-solving and decision-making skills are reduced, which leads to problems planning and organizing activities. Behavioural changes such as confusion, irritability, and mood swings are also common in people with AD. Some major symptoms of AD are as follows:

- Loss of memory, especially recent events and new information.
- Difficulty in language and communication, such as finding the right words or following a conversation.
- Critical thinking and problem-solving skills.
- Difficulty planning, organizing, and executing tasks.
- Confusion and disorientation, especially related to time, place, and familiar surroundings.
- Mood swings, sensitivity, and personality changes.
- Loss of initiative and motivation.
- Difficulty performing daily activities of daily living, such as dressing, bathing, and eating.
- Withdrawal from social interaction and entertainment.
- Problems with spatial awareness and perception.

4.5 DIFFERENT STAGES OF AD DEVELOPMENT AND COGNITIVE DECLINE

The progression of AD can be divided into three stages, each characterized by specific cognitive decline and functional impairment:

4.5.1 Initial Stage

a. Declining educational attainment often goes unnoticed by individuals and their families.
b. Occasional memory loss, such as forgetting names or misplacing things.
c. Have problems summarizing or finding the right words.

d. Difficulty with problem-solving and decision-making.
e. Mild impairment in daily activities and social interaction.

4.5.2 Middle Stage

a. Moderate cognitive decline is more detrimental to daily life.
b. Memory loss, such as forgetting personal information or recent events, becomes more apparent.
c. Difficulty in language and communication, struggling to follow conversations or convey ideas.
d. Disorientation, confusion, and disorientation in familiar places.
e. Changes in behaviour such as anger, agitation, or anxiety.
f. Difficulty organizing and performing tasks that require assistance in activities of daily living.
g. Increasing dependence on others for care and support.

4.5.3 Final Stage

a. Cognitive decline occurs with significant impairment of memory and thinking ability.
b. Forgetting the name or personal history of loved ones, memory loss.
c. Difficulty understanding or using language, limited ability to speak.
d. Complete confusion of time, place, and environment.
e. Loss of physical coordination and movement.
f. Dependence on caregivers for all daily life activities such as eating and toileting.
g. Personality changes, withdrawal from social relationships, and loss of awareness of self and others.

4.6 THE IMPORTANCE OF EARLY DETECTION FOR EFFECTIVE MANAGEMENT AND INTERVENTION

Early detection of AD plays an important role in effective management and intervention. Early detection of AD allows timely access to appropriate medical care, therapy, and support services. Early diagnosis allows individuals and their families to plan, make informed decisions, and optimize treatment strategies. This allows participation in clinical trials and research, contributing to advances in the treatment and understanding of AD. In addition, early intervention can help people maintain cognitive function longer, manage symptoms more effectively, and slow disease progression. By emphasizing early detection, healthcare providers can improve outcomes and improve the overall quality of life for AD patients and their caregivers. Early detection of AD offers several benefits to individuals, families, and healthcare systems:

a. **Improving Clinical Management:** Early diagnosis improves symptom management, appropriate treatment, and therapies. It allows healthcare professionals to design personalized treatment plans, optimize medication regimens, and effectively monitor disease progression.

b. **Improve Quality of Life:** Early detection enables people and their families to plan care, make informed decisions, and access support services. It allows them to participate in legal and financial planning, provide security measures, and access resources such as support groups.
c. **Potential for Disease-Modifying Therapy:** Early detection of AD offers the potential for disease-modifying therapy. With early intervention, researchers can focus on developing therapies that target the underlying pathological processes of AD, slowing or halting disease progression.
d. **Advances in Research:** Early detection facilitates recruitment of participants for clinical trials and research studies. This contributes to a better understanding of the disease, risk factors, and possible interventions, ultimately driving progress in AD research.

4.7 APPROACHES AND METHODS FOR EARLY DETECTION AND PREDICTION

Numerous procedures and techniques are hired for the early detection and prediction of AD:

a. **Biomarkers:** Biomarkers, which include neuroimaging (e.g., MRI, puppy scans) and cerebrospinal fluid evaluation, can provide precious insights into brain adjustments associated with AD. Those biomarkers aid in identifying early symptoms of the ailment and predicting its progression.
b. **Cognitive Exams:** Cognitive tests, such as neuropsychological tests, can evaluate reminiscence, language, interest, and other cognitive functions. These exams assist come across subtle cognitive decline and become aware of people vulnerable to developing AD.
c. **Genetic and Molecular Markers:** Genetic factors which includes gene mutations (e.g., Apolipoprotein E-ε4 allele) are related to an increased hazard of growing AD. Molecular markers, consisting of beta-amyloid and tau proteins, may be detected via imaging or fluid-based exams, imparting insights into disorder pathology.
d. **Machine Studying and Records Mining:** Superior computational strategies, including gadget getting to know and facts mining, are utilized to analyse big datasets and discover patterns that correlate with advert. These techniques permit the improvement of predictive fashions that may verify a character's risk of developing advert based on various factors.
e. **Cognitive and Behavioural Tests:** Monitoring changes in cognitive characteristics and behaviour over the years aids in the early detection of AD. Tools such as informant interviews and self-record questionnaires can offer valuable information about modifications in reminiscence, behaviour, and everyday functioning.

4.8 BENEFITS OF IOT IN HEALTHCARE

IoT in healthcare may help patients, healthcare professionals, and society at large in several ways. Among the key advantages are:

- **Improved Patient Outcomes:** This might assist in spotting any irregularities or changes in their state of health and notify the medical team or carers as necessary. IoT can also speed up prompt intervention and treatment, particularly in cases of emergency or chronic illnesses. As an illustration, ZOLL Medical Corporation provides a range of tools and wearables to help patients with their cardiac demands when they are on the road. Clinical arrhythmias are continually recorded and tracked by the company's cardiac monitor. By automatically transmitting signals of a cardiac event to their healthcare practitioner, the gadget helps individuals maintain contact with their doctors.
- **Reduced Healthcare Costs:** By minimizing needless hospital visits, maximizing resource utilization, and improving operational efficiency, IoT can help lower the costs associated with providing healthcare. Preventative and predictive healthcare can also be made possible by IoT, which can reduce the risk of problems and illnesses. For instance, AliveCor manufactures cardiac monitoring equipment. An EKG card that may fit within a wallet is a component of the company's KardiaMobile programme. Patients may give their doctors a summary report provided by the FDA-approved gadget, which also archives patient data. The patient's EKG equipment can identify six of the most prevalent arrhythmia types.
- **Increased Patient Satisfaction:** Patients will have access to their medical records whenever and wherever they are, will find it simple and safe to connect with their doctors, and will be better equipped to manage their health, thanks to the help of IoT [6]. IoT may enhance the quality of life for those with mobility issues or other constraints by providing remote assistance and support. For instance, the startup Aclima utilizes IoT to track the environment's health by keeping an eye on things like noise, temperature, humidity, and air quality. The business sells wearable sensors that gather information about a person's environment and transmit it to a cloud platform for analysis. The consumers might then receive personalized knowledge and guidance on enhancing their well-being.
- **Innovative Healthcare Solutions:** The IoT have made it feasible to develop smart objects with unique capabilities, including wearables, implants, and sensors. Blockchain, big data, cloud computing, and artificial intelligence are just a few of the technologies that can be more easily integrated with IoT to develop more sophisticated and smarter healthcare systems. One business that creates digital medications is Proteus Digital Health, which combines ingestible sensors, wearable patches, mobile applications, and data analytics. The company's solutions may track a patient's behavioural habits, physiological reaction, and adherence to medicine. The business asserts that the value

of healthcare, clinical results, and patient involvement may all be enhanced through digital pharmaceuticals.

4.9 TECHNIQUES FOR ALZHEIMER'S DISEASE PREDICTION

Data mining professionals use various methods or techniques to extract, transform, analyse, and understand data to derive useful information or conclusions and support decision-making. These strategies play a vital role in predicting Alzheimer's ailment (AD) through uncovering styles and relationships in huge datasets. Those techniques make use of gadget studying algorithms and statistical techniques to research various information assets, consisting of clinical, genetic, imaging, and cognitive assessments. Also, these methods are based on statistical, machine learning, and database management techniques, enabling data mining professionals to handle different data types and problems; some standard data mining techniques include classification, association, prediction, clustering, regression, and sequential patterns. Each technique has advantages and limitations depending on the data and the goal of the analysis.

a. **Classification:** It is a technique used in data mining to categorise data points based on their attributes. This helps to obtain accurate predictions and analysis in large data sets. To group items, businesses use this, which relies on machine learning. This involves allocating each item to a predefined group. In this method, mathematical techniques such as linear programming, decision trees, and neural networks are implemented effectively. Classification is used as an input to the results of other techniques, such as clustering, to identify clusters based on common attributes in different classifications. For example, a supermarket can use classification to categorize the items that customers purchase into specific categories, such as meat and dairy products. This allows the supermarket to understand better customer buying preferences, sales trends, inventory management, etc. The advantage of these algorithms is that they can cope with complex datasets with several variables, considering the identification of applicable predictors. They could cope with lacking information and outliers, making them appropriate for real global medical datasets. Those algorithms can offer interpretable fashions, supporting researchers identify key threat factors.
b. **Clustering:** This data mining technique identifies data points in a data set with many similarities with the help of clustering analysis. This method will assist businesses in identifying various similarities and differences in the data set. Unlike classification, which assigns data to predefined categories, clustering does not require prior knowledge of the data structure. A business analyst may acquire information about the general demographics and behavioural patterns of customers. Clustering is the process of categorizing a dataset into groups based on their similarities to each other yet different from other clusters; these groups are known as clusters. Clustering explores and discovers patterns in

large and complex datasets. For example, a supermarket uses this technique to segment customers based on their preferences, demographics, and purchase behaviours. This can help the supermarket gain useful insights and make necessary plans.

c. **Association Rule:** This technique is about finding connections among different data points within a data set. Using this technique, it can uncover hidden patterns or trends in the data that may not be easily noticeable otherwise. An association rule is a conditional statement that uses if-then statements to reveal the possible outcomes of interactions between different data points in large data sets. For example, retailers use this technique to understand what products customers tend to buy together and give their recommendations accordingly. This method is known as the relation technique because it focuses on the connection between items within a single transaction. The algorithm used for this technique may differ depending on the dataset and the intended result.

d. **Outlier Detection:** Outlier detection is a technique that aims to find data points that deviate significantly from the general patterns or trends in a dataset. Outer detection is also called as outlier mining or outlier analysis. Outliers can occur owing to several reasons, such as malicious activities, data entry errors, natural variability, or measurement errors. Outliers can affect the performance and accuracy of data mining algorithms, such as clustering, classification, or regression. Therefore, it is important to identify and handle outliers before applying data mining techniques. There are various applications of this technique, such as the detection of debit and credit card fraud, the identification of outliers in wireless sensor network data, as well as the identification of network interruptions. For example, in businesses, in the case of fraud detection, outliers can indicate suspicious transactions or activities that deviate from normal customer behaviour. This technique involves observing data items that don't match an expected pattern or behaviour within a data set.

e. **Sequential pattern:** It is another technique used in data mining, which specializes in discovering sequential patterns or frequent subsequence in sequential data. Sequential data refers to data with a temporal or sequential order, such as customer transaction sequences and customer searched lists. Sequential pattern mining aims to identify frequent and systematic patterns within the sequences. These patterns provide valuable insights into the interdependencies and time-related relationships present in the data. To find patterns, sequential pattern mining scans through a dataset to locate frequent item sets. These are groups of items that appear together more frequently in sequences. This data mining process allows users to recognize or discover different patterns, either small or big, in the transaction data over a certain period.

f. **Prediction:** Prediction uses data mining techniques that can help us understand the past and anticipate the future. It involves building and applying models that can learn from historical data and make accurate forecasts for new or unknown data instances. The goal of prediction is to find the hidden patterns and relationships in the data that can explain the variation in the target variable, which is the variable we want to predict. For example, suppose the target variable is numeric and continuous, such as a product's price or a city's temperature. We can use regression models to estimate its value or range based on the input variables. Suppose the target variable is categorical and discrete, such as the type of customer or the game's outcome. In that case, we can use classification models to assign it to one of the predefined classes based on the input variables. Prediction models can also use different algorithms and techniques to learn from the data and make predictions.

g. **Deep Learning:** Deep learning strategies, mainly deep neural networks, have shown promise in AD prediction. These models can robotically extract hierarchical representations from complex records, along with mind imaging or genomic information. Deep studying fashions, together with CNNs and recurrent neural networks (RNNs), have tested excessive accuracy in identifying styles related to AD. Deep learning models can capture tricky relationships in massive-scale datasets, consisting of high-dimensional mind photos or genomic facts. They have the potential to find novel biomarkers or capabilities that might not be identified by using conventional statistical techniques.

h. **Feature Selection and Dimensionality Reduction:** Feature selection technique's goal is to become aware of the maximum informative functions or variables for AD prediction. These techniques help lessen the dimensionality of the dataset and remove beside the point or redundant variables, enhancing the efficiency and interpretability of the prediction models; unusual function selection strategies consist of correlation evaluation, recursive feature removal, and essential aspect evaluation. Feature selection techniques improve version performance by focusing on the most applicable variables, reducing overfitting, and improving generalization. These techniques beautify the interpretability of the prediction fashions by way of identifying the key capabilities related to AD chance; by lowering the dimensionality of the dataset, feature choice strategies can expedite the computational process.

i. **Ensemble Methods:** Enhancing strategies integrate a couple of prediction fashions to enhance accuracy and robustness. Bagging and boosting are popular ensemble strategies utilized in advert prediction. Bagging combines predictions from a couple of models skilled on exclusive subsets of the statistics, whilst boosting specializes in interactively refining the version by giving greater weight to misclassified samples. These techniques can mitigate the unfairness and variance related to man or woman models, mainly to progress prediction performance. These methods are sturdy in opposition to noise and outliers within the dataset. Ensemble techniques can offer uncertainty estimates, allowing researchers to evaluate the reliability of the predictions.

j. **Data Integration and Fusion:** Data integration and fusion strategies combine data from more than one sources, inclusive of scientific, genetic, and imaging records to enhance the accuracy and comprehensiveness of AD prediction fashions. By way of leveraging heterogeneous records, those strategies can seize distinct factors of the disease and provide a more holistic information of AD risk. These techniques allow for a comprehensive evaluation of AD-related facts from numerous resources. By using combining complementary statistics, these techniques improve prediction accuracy and decrease false positives or negatives. They provide a complete view of AD threat, allowing personalized interventions and remedy techniques. In conclusion, facts mining strategies provide effective tools for predicting AD and assisting in early detection.
k. **IoT Technology:** The Internet of Things (IoT) provides how intelligent objects can be interconnected in computing environments everywhere [7]. With the development of the IoT network framework due to expanding requests, the Internet of Things is the most recent promising arising innovation that associates everything on the planet through the Internet. An enormous number of IoT gadgets and organizations have been utilized in ongoing years for various areas of purposes specially in healthcare [8].

4.10 DATA MINING ARCHITECTURE

a. **Data Source:** Data mining requires much historical data to discover patterns and insights. Organizations usually store their data in databases or warehouses, which can contain multiple data sources. Text files and spreadsheets can also be used as data sources, as they may contain relevant information. The data from these sources is then collected and processed for data mining analysis. Some of the sources of data that can be used for data mining are listed below:

    - **Database:** refers to collecting organized data or information that can be easily searched and manipulated.
    - **Data Warehouse:** A centralized repository of data from multiple sources that supports analytical queries and reporting.
    - **Internet:** The Internet is a vast network of interconnected computers offering global information access.
    - **Text Files:** A simple data format that various applications can read and write.

b. **Knowledge Base:** Knowledge Base is a crucial component of the data mining process that helps to guide the search for meaningful patterns. It may also provide data mining engine inputs based on domain knowledge and user feedback. A data mining engine relies on a knowledge base to guide its search for relevant patterns and to evaluate their interestingness. The knowledge base contains domain knowledge derived from user experiences or other sources. The goal of the knowledge base is to enhance the accuracy and reliability of the results. The knowledge base interacts with the pattern evaluation module to obtain and update information. The pattern evaluation module uses the knowledge base to measure the discovered patterns' significance and direct the search. The knowledge base contains data from user experiences, user beliefs, and other sources that are relevant to the data mining process.

c. **Graphical User Interface (GUI):** GUI is a module component of the data mining architecture that facilitates communication between the user and the graphical components in a system. Our system offers a user-friendly interface that simplifies the data mining process, allowing for efficient interaction with the system. The GUI module also supports the user in defining and executing data mining queries or tasks and analyse and visualize the results. The GUI module collaborates with the data mining system to retrieve and display the relevant information based on the user's input.
d. **Pattern Evaluation Modules:** This type of architecture is vital in assessing pattern quality using a threshold value. It works with the data mining engine to direct the search toward meaningful patterns. This module uses various stake measures that help the data mining modules to find relevant patterns in the data. It may apply a stake threshold to eliminate uninteresting patterns. For effective data mining, it is highly recommended to incorporate the evaluation of pattern stakes into the mining process as much as possible to limit the search to only significant patterns.
e. **Data Mining Engine:** The data mining engine is an important component of any data mining system. It has multiple modules that perform different data mining tasks, such as characterization, association, prediction, clustering, and classification. These modules help extract knowledge and hidden patterns and gain useful insights from the data stored in the data warehouse. The data mining engine is the base of the architecture. It provides tools that can analyse and learn from any data available on the server or warehouse.
f. **Data Warehouse Server:** A data warehouse server is essential to data mining architecture. It is a central repository that stores and organizes large amounts of data from various sources. The data in the Data Warehouse Server is structured and formatted to facilitate efficient queries, analysis, and reporting. The role of a data warehouse server in data mining is to provide a unified and integrated view of the data, making it easier for data mining algorithms to access and process the information. It is a storage and management system that merges data from different operational systems, databases, and external sources. The data warehouse server usually involves several steps in its architecture. First, data extraction is done to collect data from different sources. Then, data transformation and cleaning processes are applied to ensure consistency, resolve conflicts, and handle missing values. After the data is preprocessed, it is loaded into the data warehouse server. Figure 4.1 shows the data mining architecture.

FIGURE 4.1 Data mining architecture.

4.11 FUTURE DIRECTIONS IN EARLY DETECTION AND PREDICTION OF AD

a. **Integration of Digital Technologies:** Incorporating digital technology, which includes wearable devices and telephone apps, can offer real-time records for monitoring cognitive function and conduct. Those technologies offer the capacity for continuous monitoring and early detection of adjustments related to AD.
b. **Big Statistics and AI:** Leveraging massive records and artificial intelligence (AI) can facilitate the evaluation of massive-scale datasets and discover novel styles and risk factors for AD. AI algorithms can assist in identifying excessive-threat individuals and improving predictive accuracy.
c. **Multi-domain Tactics:** Combining more than one domain of data, together with genetics, neuroimaging, cognitive exams, and lifestyle elements, can

enhance the precision and reliability of early detection models. Integrating diverse data resources allows for a complete understanding of AD threat.

d. **Intervention and Prevention Trials:** Carrying out intervention and prevention trials targeting people at an excessive chance of developing AD can offer precious insights into the effectiveness of early interventions. Identifying modifiable hazard elements and trying out interventions in preclinical degrees are critical for a hit prevention strategy.

4.12 DATA SOURCES FOR ALZHEIMER'S DISEASE PREDICTION

Accurate prediction of Alzheimer's sickness (advert) is predicated at the usage of numerous information resources that provide treasured facts about the disorder, its risk elements, and its progression. By leveraging multiple data kinds, researchers can increase strong prediction models that capture the complex nature of AD. This segment discusses the numerous statistics sources commonly used for AD prediction, highlighting their specific contributions to early detection and risk evaluation.

1. **Genetic Data:** Genetic data plays an important role in understanding the genetic elements underlying AD susceptibility. Genome-wide association research (GWAS) and complete-genome sequencing offer insights into specific gene versions and genetic markers associated with expanded AD danger. Genetic records resources assist in identifying potential biomarkers, genetic threat profiles, and gene-environment interactions relevant to AD prediction.
2. **Clinical Data:** Clinical data includes demographics, medical history, cognitive evaluations, and clinic diagnoses taken from patient medical records. This information offers crucial insights into the patient's general health status, cognitive function, and probable comorbidities. Clinical data builds baseline measures for AD prediction models, tracks disease development, and identifies risk factors.
3. **Neuroimaging Data:** Neuroimaging techniques like MRI, positron emission tomography (PET), and functional magnetic resonance imaging (fMRI) capture the brain's structure, function, and metabolism in detail. These techniques provide important insights into neuroanatomical modifications, the existence of amyloid plaques and tau tangles, and the functional connectivity changes related to AD. When used to identify neuroimaging biomarkers for early AD detection, neuroimaging data offers objective assessments of brain health.
4. **Cognitive Assessment:** Cognitive assessment evaluates people's cognitive abilities, such as memory, attention, language, and executive skills. The Mini-Mental State Examination (MMSE) and the Montreal Cognitive Assessment (MoCA), both neuropsychological examinations, evaluate cognitive ability and look for early indications of cognitive decline. Longitudinal cognitive tests enable the discovery of patterns linked to the evolution of AD by tracking cognitive changes over time.
5. **Biomarker Data:** Biomarkers are quantifiable markers of biological functions or disease conditions. Biomarkers used to predict AD include those found

in cerebrospinal fluid (CSF), blood, and neuroimaging-derived indicators (such as amyloid-beta and tau protein levels). These biomarkers assist in risk assessment and disease progression monitoring and offer insightful information on the underlying pathological changes linked to AD.

6. **Environmental and Lifestyle Data:** AD risk is influenced by variables like education, occupation, physical activity, food, and social interaction. It is possible to evaluate these elements' effects on AD development and include them in predictive models by gathering information about them. For a better knowledge of how environmental and lifestyle factors affect AD prediction, large-scale epidemiological research and lifestyle surveys provide useful information.

4.13 MACHINE LEARNING MODELS FOR ALZHEIMER'S DISEASE PREDICTION

By examining intricate patterns and interactions in large-scale datasets, machine learning approaches have demonstrated considerable promise in predicting AD. These models use statistical methods and algorithms to find predictive traits and create precise prediction models. In this section, we review the popular machine learning models for predicting AD and highlight their benefits for early detection.

a. **Logistic Regression:** A popular machine learning model for predicting AD is logistic regression. Based on a collection of input features, it is a binary classification algorithm that calculates the likelihood that a person has AD. To predict the likelihood of developing AD, logistic regression models can combine various data sources, including clinical information, genetic markers, and cognitive tests. They can be interpreted, giving details about how each factor affected the forecast.

b. **Random Forest:** A strong prediction model is produced using the ensemble learning technique known as random forest, which mixes various decision trees. Each decision tree is trained using a subset of the data, and the trees' predictions are combined to get the final forecast. Large and varied datasets, such as those from medicine, genetics, and neuroimaging, can be handled by random forest models. They can discover the most pertinent predictors by capturing complicated interactions between characteristics and providing changeable relevance rankings.

c. **Support Vector Machines (SVM):** Support Vector Machines are strong classification models for problems like AD prediction. SVM aims to find the best hyperplane to divide data points into distinct classes. SVM models may identify people with AD and healthy controls using a variety of data sets, including neuroimaging, genetic, and cognitive tests. Using kernel functions, they can effectively handle high-dimensional data and non-linear relationships.

d. **Neural Networks:** Due to its capacity to automatically extract features and learn complicated patterns, neural networks, and more specifically deep learning models, have become increasingly popular in AD prediction. RNNs can handle sequential data, such as longitudinal cognitive tests, whereas CNNs are frequently employed for neuroimaging data analysis. Deep learning models may combine many data kinds and produce accurate predictions, but they need much computer power and training data to be effective.

e. **Gradient Boosting Models:** Gradient boosting models are ensemble learning methods that iteratively integrate weak predictors to produce a strong predictive model. Examples are XGBoost and LightGBM. These models have shown great performance in AD prediction using several data sources. Gradient boosting models can capture non-linear correlations between characteristics and manage missing data, contributing to precise predictions.

f. **Long-Short-Term Memory (LSTM):** This RNN variant excels in processing sequential data. LSTM models can analyse longitudinal cognitive evaluations and spot minor changes that could be signs of developing AD. Temporal dependencies can be captured by LSTM models, which can offer important insights into how diseases develop.

4.14 RESULTS AND DISCUSSION

The plan has three crucial steps. To start, the pandas are loaded with the dataset for Alzheimer's illness. This study used longitudinal data; hence, a long-term study is required to fully comprehend the results. The first stage is analysing the data at a certain point in time or at the baseline. The data were then thoroughly reviewed, including a comparison between the primary study section and the pertinent data gathered during each visit.

Our primary data in this investigation are longitudinal MRI results. The study included 150 patients' MRI data, who ranged in age from 60 to 96. Each patient has at least one MRI scan. Everybody is correct. Seventy-two participants in the study were labelled as "not demented." Sixty-four individuals were initially categorized as having dementia, and they stayed in this category for the duration of the study. The dataset description for the MRI data is displayed in Table 4.1.

Early disease is now more fully understood when machine learning techniques are used on datasets related to AD. The key data on AD are inconsistent and repeated, which reduces the algorithm's accuracy. It is required to prepare good data for analysis by removing undesired features, irrelevant data, and redundant data once more before analysing machine learning algorithms. Data must be divided into training and testing sets before a machine learning model can be built. The training data are used to create a model in the following data preparation stage, which is subsequently used to predict AD using the test data.

The test data assess the missing data after the model has been trained on the training data. The dataset is divided into three subsets for cross-validation. Make model predictions using a set of data (the test data) and then use other datasets (the training and validation data) to assess the model's performance. Pre-processed data is divided 80:20 randomly into 80% for training and 20% for testing. Working techniques for early prediction of AD are shown in Figure 4.2.

TABLE 4.1
Data Set description

S. No	Attributes	Description
1	ID	Identification
2	M/F	Gender (M if Male, F if Female)
3	Hand	Handedness
4	Age	Age in years
5	EDUC	Years of education
6	SES	Socio-Economic Status
7	MMSE	Mini-Mental State Examination
8	CDR	Clinical Dementia Rating
9	eTIV	Estimated Total Intracranial Volume
10	nWBV	Normalize Whole Brain Volume
11	ASF	Atlas Scaling Factor
12	Delay	Delay

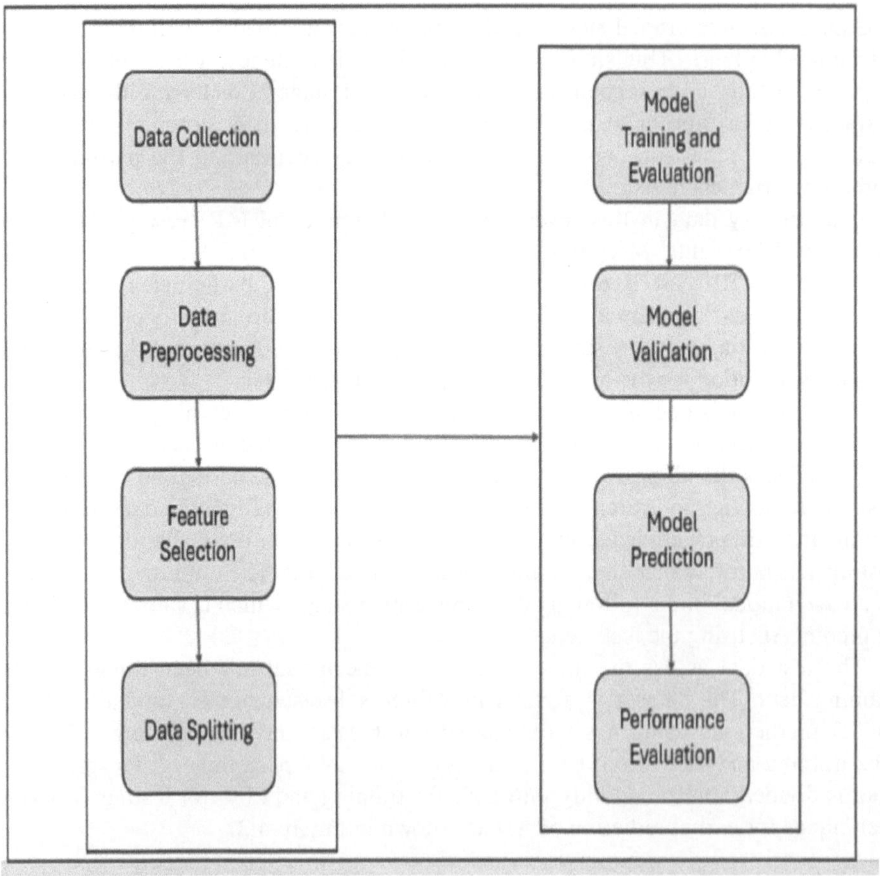

FIGURE 4.2 Proposed workflow.

TABLE 4.2
Min, Max, and Median Values of Each Attribute

Attributes	Min	Max	Median
EDUC	7	22	14.2
SES	2	6	2.3
MMSE	16	30	26.2
CDR	0	1	0.3
eTIV	1,120	1,990	1,450
nWBV	0.55	0.81	0.7
ASF	0.87	1.43	1.3

Data processing methods are used only at this stage to clean and prepare data. Handle missing values, delete properties, modify properties, and more as part of this. In the Standard Error of the Slope (SES) line, nine non-significant lines were found. There are two approaches to deal with this issue. Taking out the missing lines is the simplest fix. Replacing them with the appropriate values is another technique to fill in the blanks. Since there are only 140 measurements, evaluating the model should improve performance. The mean was used to load nine lines from the SES property that had missing values.

In data analysis, we review how each MRI test relates to dementia. Graphs illustrate the relationship between the data; before collecting or analysing data, we conducted a data analysis technique to forecast associations. The nature of the data may then be deduced, and the analytical techniques employed can be determined using this information. The minimum, maximum, and average values for each attribute are displayed in Table 4.2.

Further, feature selection is crucial in machine learning. Clinical data on AD were specifically selected for this study, and hundreds of samples are available. There are three ways to choose features: the filter approach, the wrapping method, and the embedding method. One technique frequently utilised in the pre-treatment stage is water filtration. Another technique that concentrates on subsets of attributes is wrapper methods. The last step is the embedded method, which unifies filter and wrapper methods. The correlation coefficient, data increment, and chi-square are this study's best and most frequently used options.

4.15 IMPLICATIONS AND FUTURE DIRECTIONS

For both individuals and society, the early diagnosis and prediction of AD utilizing data mining techniques has important consequences. Early detection, which identifies people at risk of developing AD before symptoms appear, permits prompt interventions, individualized care, and better results. The ramifications of early AD prediction are covered in this section, along with future directions for both research and application.

a. **Personalized Interventions:** Early AD prediction enables the deployment of individualized therapies catered to the unique requirements of each individual.

Healthcare practitioners can offer targeted therapies, lifestyle changes, and cognitive training programs to high-risk individuals to slow disease progression and improve quality of life.

b. **Research Advancements:** Early AD prediction advances AD research by offering important insights into the course of the disease and underlying mechanisms. Predictive models can be used to find new biomarkers, genetic variations, and cognitive markers linked to AD to better understand the condition and identify prospective therapeutic targets.
c. **Resource Allocation:** Early detection encourages the best use of resources by focusing medical attention on high-risk individuals. Healthcare systems can more effectively and efficiently deploy resources, such as diagnostic tests, imaging technology, and specialist consultations, to those most benefit from them by identifying those more likely to develop AD.
d. **Integration of Multi-Modal Data:** To create complete prediction models, future research should combine data from several sources, including clinical, genetic, neuroimaging, and lifestyle. Combining many modalities can improve prediction accuracy and dependability and give a more comprehensive knowledge of AD risk.
e. **Longitudinal Studies:** Studies that follow people who are at risk for AD over an extended period are essential for improving prediction models. Researchers can analyse the temporal dynamics of AD biomarkers, cognitive decline, and risk factors by gathering data over long periods, which increases the models' precision and prognostication ability.
f. **Advancements in Machine Learning:** The accuracy of AD predictions may be improved by future developments in machine learning techniques, such as deep learning, reinforcement learning, and transfer learning. These methods may identify intricate correlations and patterns in massive datasets, enabling more accurate and trustworthy predictions.
g. **Prevention Strategies:** The creation and assessment of preventive measures based on early AD prediction should also be among the future directions. To develop effective preventative strategies and ultimately lessen the impact of AD on people and society, it is important to identify modifiable risk factors and evaluate therapies in the preclinical phases.

Using data mining approaches to predict AD early has the potential to improve patient outcomes and advance AD research in conclusion significantly. Early detection affects healthcare planning, scientific developments, public health initiatives, and individual patient treatment. We can improve predictive models and make early prediction strategies easier to implement, ultimately improving outcomes for people at risk of AD. To do this, we must integrate multi-modal data, advance machine learning techniques, address ethical issues, and concentrate on clinical translation.

4.16 CONCLUSION

The use of data mining techniques for the early diagnosis and prognosis of AD has been examined in this chapter. The importance of early identification and prediction in enhancing intervention and treatment options was revealed using various IoT devices. Creating trustworthy AD prediction models is possible by utilizing various data sources and applying data mining approaches such as feature extraction, selection, and machine learning models. The results of this study add to the growing body of information in the area and provide new opportunities for AD prediction research and development. Early detection ultimately could improve people's lives and the functioning of public healthcare systems by allowing for prompt treatments and individualized care for those who are at risk of acquiring AD.

REFERENCES

1. Bardach, S.H., Parsons, K., Gibson, A., Jicha, G.A. (2020). From Victimhood to Warriors": Super-researchers' Insights into Alzheimer's Disease Clinical Trial Participation Motivations. *Gerontologist*, 60(4), 693–703, https://doi.org/10.1093/geront/gnz096.
2. Li, Z., Jiang, X., Wang, Y., Kim, Y. (2021). Applied machine learning in Alzheimer's disease research: Omics, imaging, and clinical data. *Emerging Topic in Life Sciences*, 5(6), 765–777, https://doi.org/10.1042/ETLS20210249.
3. Bilderbeck, A.C., Penninx, B.W.J.H, Arango, C., Wee, N., Kahn, R., Rossum, I.W., Hayen A., Kas, M.J., Post, A., Dawson, G. R. (2019). Overview of the clinical implementation of a study exploring social withdrawal in patients with schizophrenia and Alzheimer's disease. *Neuroscience & Biobehavioral Reviews*, 97, 87–93, https://doi.org/10.1016/j.neubiorev.2018.06.019.
4. Li, X., Ji, M., Zhang, H., Liu, Z., Chai, Y., Cheng, Q., Yang, Y., Cordato, D., Gao, J. (2023). Non-drug Therapies for Alzheimer's Disease: A Review. *Neurology Therapy*, 12(1), 39–72, https://doi.org/10.1007/s40120-022-00416-x.
5. Oskouei, R.J., MousaviLou, Z., Bakhtiari, Z., Jalbani, K. B. (2020). IoT-Based Healthcare Support System for Alzheimer's Patients. *Wireless Communications and Mobile Computing*, 2020, 1–15, https://doi.org/10.1155/2020/8822598.
6. Mitra, S., Sarker, Z., Patel, P., Mehra, P.S. (2023). Multi-factor Assymetric Key Encryption (MAKE) for Mutual Authentication in IoMT," *ICSCCC 2023 – 3rd International Conference on Secure Cyber Computing and Communications*, Jalandhar, India, 26–28 May 2023, pp. 379–384, https://doi.org/10.1109/ICSCCC58608.2023.10176485.
7. Sharma, J., Mehra, P.S. (2023). Secure communication in IOT-based UAV networks: A systematic survey. *Internet of Things*, 23, 100883, https://doi.org/10.1016/J.IOT.2023.100883.
8. Chawla, D., Mehra, P.S. (2023). QSMAH: A novel quantum-based secure cryptosystem using mutual authentication for healthcare in the Internet of Things. *Internet of Things*, 24, 100949, https://doi.org/10.1016/J.IOT.2023.100949.

5 Ethical and Social Implications of IoT-Integrated Autonomous Vehicles

Shreya Sahgal and Dolly Sharma

5.1 AUTOMATED VEHICLE

An autonomous vehicle (AV) is a driverless vehicle that operates on its own and performs all necessary functions using a combination of sensors, cameras, and artificial intelligence (AI) that traverses between different locations without any human intervention. For a vehicle to be qualified as fully automated, the vehicle should be able to reach the predetermined destination without human help. It uses a fully automated driving system to respond to and manage external conditions well in a way that a driver can control. Automated vehicles are revolutionizing the transportation industry. The portrait is presented in Figure 5.1. They possess features such as:

- Self-parking
- Auto steering

Slowing down accurately and relatively to the car moving in front of them.

Companies, including Audi, Ford, BMW, Tesla, Volkswagen, Volvo, and Jaguar Land Rover, are developing and testing AV [1]. California highways have been traversed over 140,000 kilometres via the Toyota Prii and AudiTT.

In addition, as a part of its autonomous driving initiatives, Jaguar Land Rover, based in the UK, has been testing automobiles on the road for 2 and 3 years now, indicating the enormous potential and future AVs growth. It also continues to develop and build new intelligent vehicle designs for connected vehicles and the ability to monitor the occupant state, which is an important parameter in designing, optimizing, and adapting the system for maximum safety and comfort of the passengers [2].

5.1.1 History of Automated Vehicles

Autonomous cars have been the centre of attention for a long time now. Due to the combined effort of car manufacturers and research centres, the concept of

DOI: 10.1201/9781003405535-6

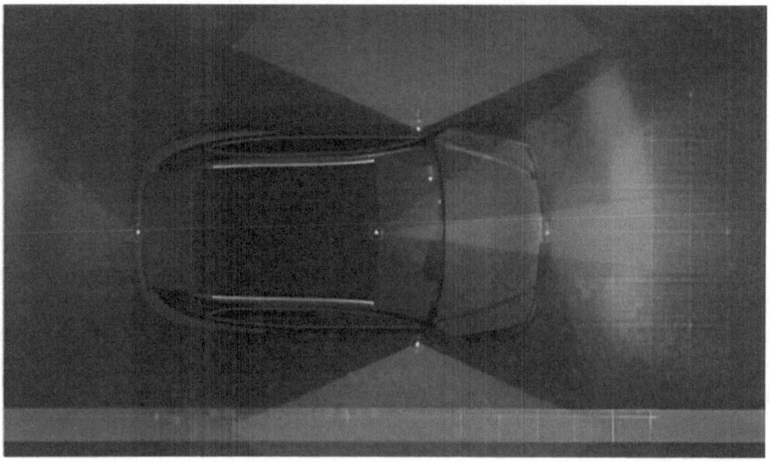

FIGURE 5.1 Visual representation of AV.

transportation is meant to change dramatically in the future. It becomes important for all of us to know from where we get this remarkable discovery that once seemed inconceivable. The following timeline will help to comprehend the development of automated vehicles more easily.

In 1500, Leonardo Da Vinci created a cart that moved back and forth. Springs were placed on the bottom of the cart, which provided power and steering so that the vehicle could automatically follow the predetermined path.

Centuries later, in 1925, Mr. Houdiana initiated the first step. He exhibited a vehicle controlled by radio that could travel freely near Manhattan's streets without the need for any drivers by linking it externally, connected externally to it. It included features like:

- It could be turned off at convenience.
- Gears can be shifted.
- You can make use of a radio to find your way outdoors.

The project went unsuccessful because it crashed twice into other vehicles. However, it was the time when the concept of automated cars triggered people's minds.

General Motors acquired the initial self-driving car model at the World Fair after 14 years of intensive research and study. According to the car's design, the magnetized metal spikes submerged in the road were regulated by a radio-controlled electromagnetic field guiding the vehicle. Only a very basic model existed. But in 1958, it officially became a reality. It had a sensor that, to some extent, used electrical flow for recognizing objects in the vicinity and the path. The sensor might send instructions to the steering wheel to turn left or right based on the passage occurring through a wire hidden in the road. The next few years were filled with the hype to discover space exploration. Researchers thought of creating vehicles that could roam on the moon's surface on their own [3].

In 1961, James Adams invented the "Stanford Cart" with a camera that could follow specified lines on the ground. The use of cameras is a vital event in the journey of automated vehicles.

The Japanese added a camera that could take photos of the road in 1977, analysing the data to allow the vehicle to proceed. This created a dynamic change when the first auto-driven passenger vehicle, capable of travelling at a speed of approximately 20 mph, was developed.

The invention of neural networking and a camera at Carnegie Mellon University at the turn of the century led to an improved steering control system and the creation of the car NavLab 5.

About 3,000 miles were independently driven by automobile from Pittsburgh to San Diego. DARPA supported more AV research in the early 2000s. Ford, Mercedes, BMW, and Uber all became interested in how self-driving technology works.

5.1.2 Features of AVs

Automated Vehicles have the following advantages:

- **Automated Emergency Braking System (AEBS):** The system is designed to detect pedestrians, cyclists, and other vehicles on the road and reduce the risk of collisions in busy urban areas.
- **Night Vision Capability:** AVs have night vision cameras that detect obstacles in low light conditions, keeping the passengers safe.
- **Automated Parking:** It means that cars can navigate tight spaces and can do parallel parking easily. Once the car is in position, it automatically adjusts its speed and steering to ensure it parks safely and accurately.
- **Street Sign Recognition:** It is a software program that enables it to identify road signals and process sensor data.
- **Adaptive Cruise Control:** ACC helps the car to maintain a proper distance from the vehicle which will be moving in front of it. The sensor detects when the car slows down and then automatically adjusts its speed according to it. It prevents rear-end collisions and sudden stops or starts.

5.1.3 Current State of Autonomous Car Industry

The Society of Automotive Engineers classified the vehicles that will enable us to understand the level of AV technology fully. The following classifications have to be understood to comprehend the situation as it stands: Level 0 to Level 5 six phases illustrate the abilities of entirely and partially autonomous cars. These are:

- **Level 0: No Automation:** In this case, the driver is solely responsible for controlling the car.
- **Level 1: Driver Assistance:** It is the most basic grade of assistance. The driver receives aid with acceleration and steering. An excellent example of this is adaptive cruise control.

- **Level 2: Partial Driving Automation:** In certain situations, the Advanced Driving Assistance System (ADAS) in these vehicles can take over the controls for steering, acceleration, and braking.

From this point onwards, human drivers are not required.

- **Level 3: Conditional Driving Automation:** It automatically renders decisions and modifies driving situations automatically using AI and other driving aid systems. No one inside the car has been keeping a watch on anything.
- **Level 4: High Driving Automation:** In this instance, operating the vehicle does not involve interacting with a human. Consequently, steering mechanisms and paddles may not be present.

  In the event of a system breakdown, the program will also terminate automatically. This is mostly utilized for taxis and public transportation, and they will be tightly guided when going from point A to point B.
- **Level 5: Full Driving Automation:** The highest level of automation is at this juncture. The vehicle will drive independently when you've selected a destination, regardless of traffic or weather [1].

5.1.4 AVs as Public Transport

It is well said by Dave Waters that self-driving vehicles automatically choose the most efficient route. AI will improve logistics. Furthermore, AVs can transform public transportation into something more attractive, efficient, and equitable for people across the city. The potential benefits include improved safety, energy efficiency, comfort for the passenger, and traffic flow. Naturally, if public transportation is made more attractive, people can stop using private cars, hence reducing carbon emissions; thus, we can achieve a greener transit system [1].

Considering the case of Rouen, which is one of the largest Transdev networks in France, the authority has integrated a fleet of four AVs into the network, and these vehicles have been supervised by the same control centres that earlier supervised the transportation of 200 buses and tramways running in Rouen. Their focus goals also include the reduction of the carbon footprint of mobility and the electrification of bus fleets and railways. AVs can be developed anywhere with the same tools and the same people, and the only requirement is the collaborative nature of people and patience in combining the existing practices and new technologies, as shown in Figure 5.2.

5.1.5 Ethical and Social Implications of Autonomous Vehicles

- **Technical Reliability:** AI-driven vehicles have the advantage of being more precise as well as consistent, thereby decreasing the likelihood of an accident. After all, a computer program cannot doze off, getting sidetracked, or even exceeding the posted speed limit. From a moral point of view, autonomous

FIGURE 5.2 AV as public transport.

automobiles are more stable than traditional ones since driving is solely up to a thoroughly evaluated and optimized driving algorithm [4].

- **Reduce Driver's Stress and Improve Productivity:** AVs very simply provide comfort to passengers by letting them rest or letting them complete their work while travelling. It provides independent mobility to people who cannot drive or should not drive, such as pregnant women, senior citizens, and so on. It benefits travellers by improving their access to education, health, or employment opportunities. Dedicating sole lanes to AVs will reduce human-operated traffic.
- **Reduction in Carbon Emissions:** AVs will communicate through complex networks and automatically detect traffic and crowded areas. They will also use the navigation system to reach the location in minimum time. Hence, less time on roads would result in less carbon emissions. Furthermore, electric cars are known to emit significantly less carbon dioxide. It will very clearly improve the balance of nature and cause fewer respiratory problems to humans and a better habitat for wildlife. According to MIT News, if every vehicle on the road is autonomous, its control system is designed so that fuel consumption will decrease by 18% and carbon emissions by 25%. It will boost the travel speed up to 20%. A simple portrait is presented in Figure 5.3.
- **Increased Lane Capacity:** State Smart Transport Initiative claims the AVs can boost motorway travel speeds by over 20% and increase highway capacity by 100%. By utilizing cloud-connected telematics systems, Internet of things (IoT) solutions can contribute to safer roads. The stakeholders may monitor their fleets in real time and receive alerts regarding potential safety hazards,

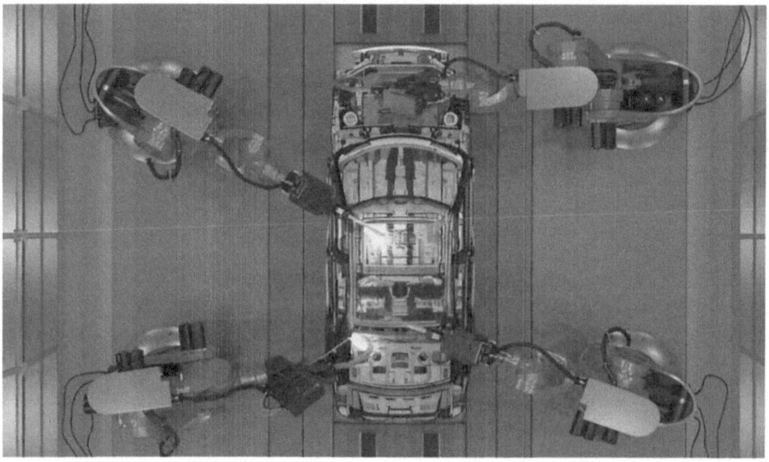

FIGURE 5.3 Construction of AVs.

from fleet management to vehicle diagnostics. An additional boost in capacity can be accomplished by platooning AVs in a "hypothetical continuous train," that may transport 10,000 or 8,000 vehicles per hour at 80 mph, says Eric Sundquist, managing director of SST [5].

- **Contactless Deliveries:** Due to the sudden outbreak of the COVID-19 pandemic in parts of the globe, the urge for contactless deliveries has significantly increased. Delivery companies have started contactless deliveries for consumers using low-speed automated delivery vehicles or ADVs. According to Guidehouse Insights reports, the number of automated deliveries could compound an increasing growth rate of up to 170%. This means that roughly 7 million ADV deliveries were made in 2021, and the estimation is that it will increase to 51 billion by 2030.
- **Consumer Savings:** In line with the KPMG report, connected and automated vehicles – The UK Economic Opportunity shows that driverless vehicle integration in society could save commuters up to 40% in their total travel time. The integration of driverless cars in society will reduce insurance costs, parking fees, and consumption of gas, and most importantly, will save time [5].

With the positive aspect of Automated Vehicles, we have some drawbacks of the emerging technology, which include:

- **Accident Programming:** It is estimated by the Ministry of Road Transport and Highways that every year approx. 1.5 lakh people die in India, which clearly means that 422 deaths occur yearly and 18 deaths every hour, of which 92% are car crashes caused due to human error. A person reacts randomly to the situation instead of giving an analytical and calculated response. Well, that is human nature that cannot be controlled. However, if five AVs are introduced on the roads, almost half of fatal crashes and injury crashes will be reduced. But here, the major question arises: if a human's instinctive response can be dangerous, how can we trust a machine that works on a simple algorithm and is not even capable of taking instinctive decisions? How should a car respond in an accident, especially when there is a no-win scenario?

How does the algorithm function when an AV participates in a crash where there is an elevated probability that someone will get injured? The other cars or pedestrians should be rescued first.

- **Driver Licences and Liability:** Since entirely AVs are moving closer to being a reality, a significant concern is whether or not driver's licences will continue to be required. Could a child ride to school in an autonomous car lack a licenced driver? If so, who is to blame for the mishaps?

The biggest misconception is that AI can think and respond to changing conditions, unfortunately not the case. It works on a mere program. Does it mean the manufacturer is responsible for the accident?

These are the questions that society needs from the government bodies [6].

- **The Issue of Hacking:** On a deeper level, society should rethink that everything in the hands of AI would not be a safer option. Putting a computer in the driver's seat of an automobile that could turn out to be lethal may increase the risks than the comfort it provides to society. Unfortunately, hackers have figured out a way to decode the algorithm. One such case was found in the USA in 2015. Miller and Valasek identified a means for harming the Linux operating system of the multimedia computer.

The vehicles eventually ended up on a crowded interstate highway when the hackers took possession of the entire vehicle, from the windscreen lights to the accelerator becoming disabled [7]. Although the hackers in this scenario were only 10 miles away, they could have accessed vehicles throughout the country via cellphone data, serving as a hacking entry point. The incident was carried out in a non-AV, but the risk would have been significantly greater if it had been more linked and no one had been able to take over the wheel. AVs lack the protection that conventional vehicles get by having to protect themselves by limiting connections to the outside world, but AVs do not have that luxury.

- **Ownership and Operating Costs:** The necessary equipment and services are dissimilar for all AVs and shared AVs, and the required equipment and services are too dissimilar. Since a car's mistake could be fatal, they need trustworthy, redundant parts put in place by competent technicians, raising maintenance costs. The current situation involves expensive optional AV devices, including remote starting, active lanes assist, and safety surveillance cameras. As a result, the price of maintenance, software, and mapping will increase due to autonomous driving capacity. Commercial self-driving vehicles will presently cost approximately $0.80 and $1.20 per automobile mile [7].

The first comparison determines the total cost comparison, as shown in Figure 5.4, whereas the second tells the variable costs comparison mentioned in Figure 5.5. The average cost is the capital the traveller considers when purchasing a vehicle, and variable costs are the amount of money passengers spend on a particular trip.

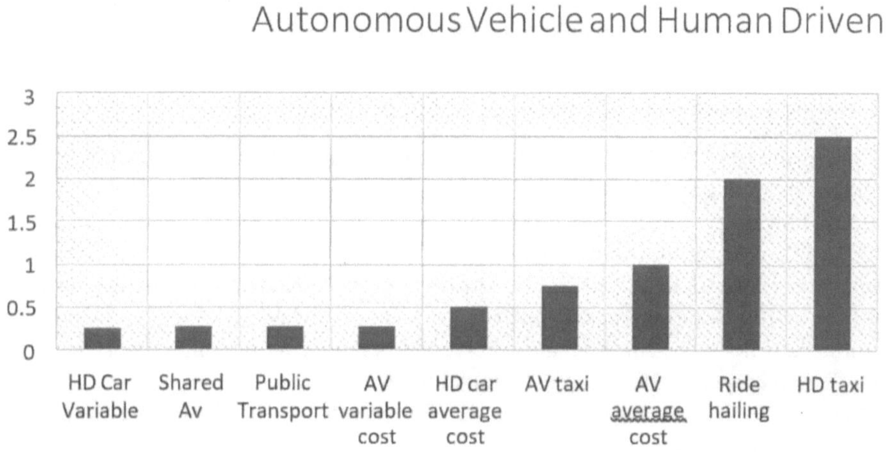

FIGURE 5.4 Total cost comparison for AV and normal cars with respect to the cost of automobile per mile.

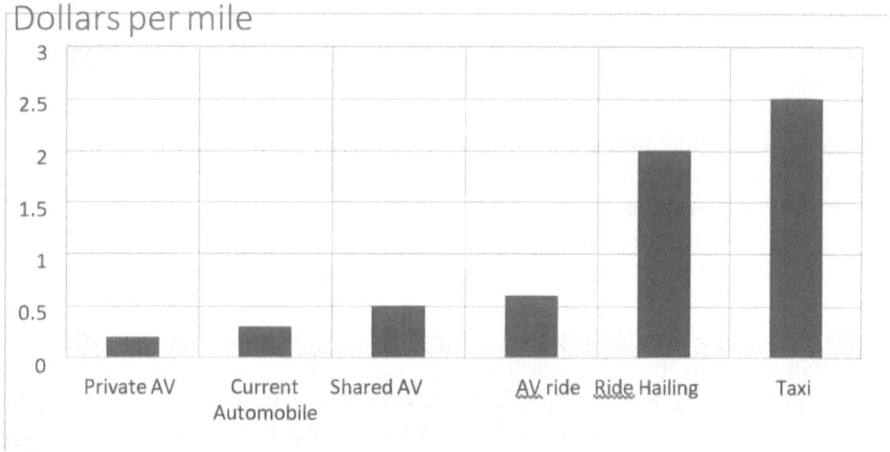

FIGURE 5.5 Variable cost comparison for different cars with respect to dollar per.

- **Horizontal Equity to Road Space:** Road space is a viable and limited source. Horizontal equity means space should be given to vehicles with high occupancy rates at lower costs, such as vanpools and buses so that passengers won't be congested due to single-occupant automobiles. If a proper calculation for road pricing is not formulated, owners of AVs will find it simple to park their cars anywhere or avoid paying for off-street parking, which will add to traffic congestion. To maximize equity, public roadways should be controlled to prioritize space-efficient modes, including both conventional and automated car roadways controlled to prioritize space-efficient modes, encompassing regular and automated cars, for the greatest equity [8].
- **Parking Planning:** Special parking areas with charging stations for electric vehicles and car maintenance and cleaning services will be required for AVs.

5.1.6 Integration of IoT with Automated Vehicles

IoT paves the road for AVs by enabling enhanced transportation efficiency, a greater driving experience, and advanced vehicle management capabilities. This post will examine how automotive IoT solutions will bring automobiles up to date.

By 2025, 470 million linked vehicles will be driving on highways all across the world, according to a Deloitte study. Utilizing cloud-connected telematics systems, automotive IoT solutions can contribute to safer roadways. The stakeholders may monitor their fleets in real time and receive alerts regarding potential safety hazards, from fleet management to vehicle diagnostics [9].

Automobile IoT technologies also enable vehicles to communicate with their surroundings [10]. This includes automated smart parking systems that use cameras and sensors to detect vacant spaces and lane-centering driving systems.

The Internet of Things is crucial to the market forecast for connected vehicles. Market research company Gartner projects that in 2020, IoT goods and services suppliers will increase their revenue by more than $300 billion [11]. However, according to IDC, a different research company, the global market for IoT solutions will reach $7.1 trillion in 2020.

The future automotive era is being shaped by a variety of IoT-based solutions, including:

- Preventing future problems
- ADAS stands for advanced driver assistance systems.
- Wi-Fi capabilities using 3G/4G/5G technology
- Car-to-car communication
- Vehicle information system

For many commuters, leaving everything up to their cars is the ultimate objective. When people get in the car in the morning, they want to drive without worrying about weather, traffic, or traffic lights until they reach their destination. IoT is helping automakers and buyers accomplish this aim by offering creative and cutting-edge solutions.

5.2 FUTURE SCOPE

The way clients experience mobility is going to shift, thanks to AD. Driving will become simpler, less hazardous, and more satisfying as the outcome. Driving time can now be used to devour a book, talk on a cell phone, or even do something productive. Hence, work productivity will increase. Since workers can work from autonomous cars, most work will be done even if they live far from their offices. Hence, it will attract more people from the suburbs and rural areas. AD will improve mobility for elderly drivers and youngsters [12].

It can be clearly observed that AD will generate value for the auto industry. In particular situations, most cars will achieve SAE Level 4 or driverless control. Consequently, ADAS and AD could generate between $300 billion and $400 billion in the passenger car market by 2035, based on a McKinsey consumer survey.

Based on different degrees of innovation availability, user uptake, and regularity support, McKinsey additionally generated data on sales for autonomous passenger cars.

The above scenario projects how, in 2030, 4% of new passenger cars sold are installed with L3+ AD function, and this figure will increase to 17% in 2035 [13].

Furthermore, consumers will benefit from employing AD systems in a variety of ways, including increased safety, and most are very open to the idea. A quarter of respondents to a mobility poll conducted by McKinsey 2021 of more than 25,000 people stated they would choose advanced AD when they bought their next car. The research clearly indicates the pricing inclinations. Twenty percent of poll participants indicated they were willing to purchase ADAS features via subscription, while the remaining 30% claimed they would pay for each time they used a function. The statistics have been mentioned in Figure 5.6 [14].

Most people are fascinated by autonomous driving and have adopted a realistic and cautious attitude towards self-driving cars. Figure 5.7 shows consumers' readiness to

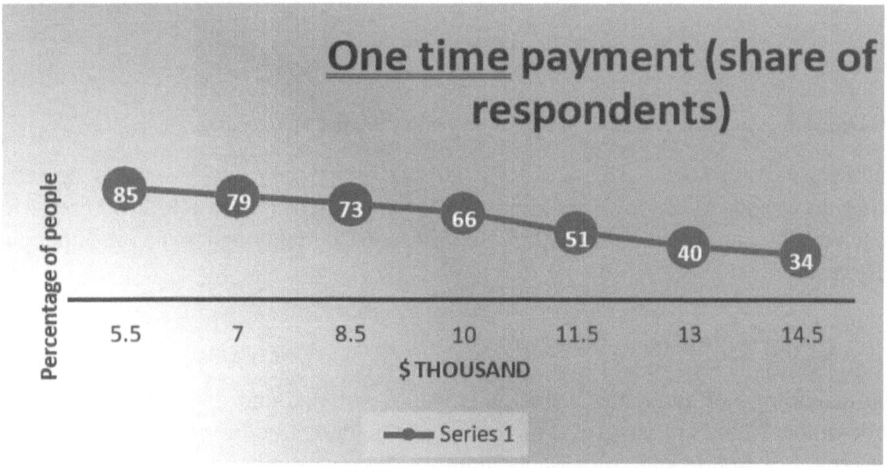

FIGURE 5.6 % share of people to use automobile vehicles on one-time payment.

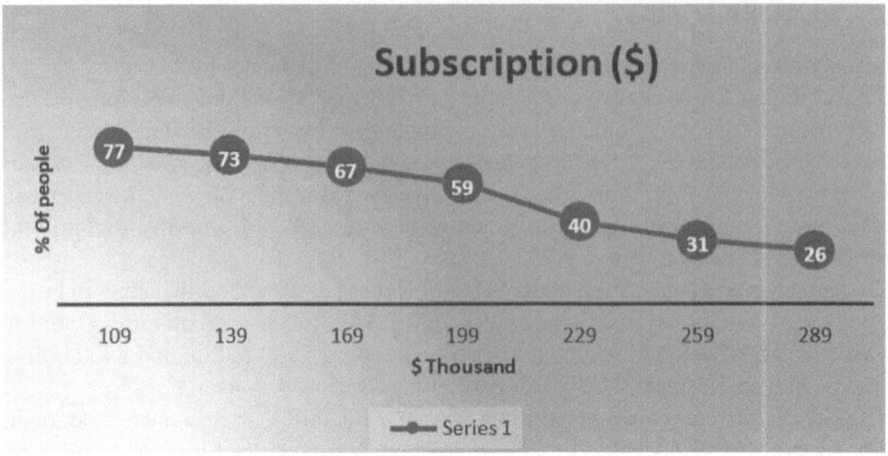

FIGURE 5.7 % share of people to use automobile vehicle on subscription.

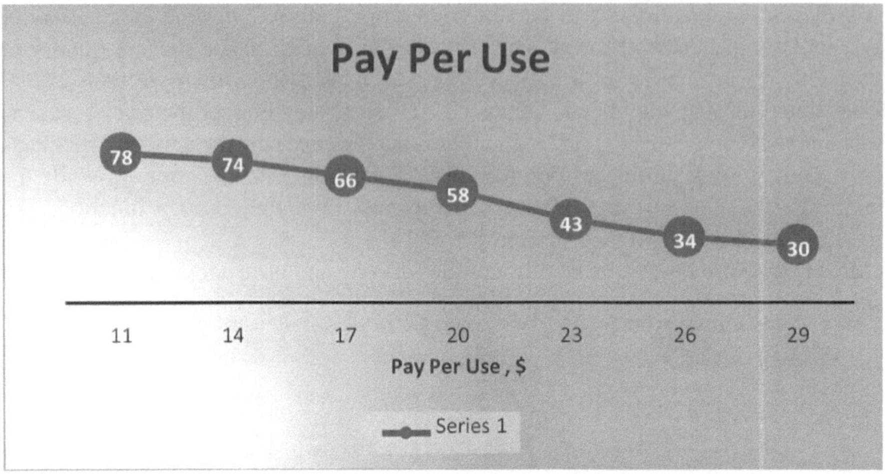

FIGURE 5.8 % share of people to use automobile vehicle as pay per use.

switch to private AV is down by almost 10%, whereas in Figure 5.8, 26% prefer to switch to fully automatic cars in 2021. Reports in 2020 mentioned that 35% of people were ready for it.

5.3 CONCLUSION

The technology of automated vehicles is well-developed and open for commercial deployment. Many automobile companies have made incredible progress in navigation, street mapping, and collision avoidance. The government can easily accelerate the use of automated vehicles by how they regulate them. Authorities are sure that they will address relevant issues and ensure the rules are stringent enough. If we talk about the social and ethical implications, software designers will definitely make choices about how a system will deal with them. Researchers will navigate all the economic and social ramifications and will surely bring solutions to it. These developments have opened up the possibility of AV technology, and the statistics provided will be crucial in helping society understand the emergence and implications of AV, which aims to increase accessibility for travel while reducing accidents, energy use, pollution, and congestion. IoT is increasingly becoming a part of the DNA of modern Tesla vehicles. IoT is revolutionizing the worldwide automobile market, from automated transportation systems to connected autos. It is anticipated to continue performing in the same manner in the future.

REFERENCES

1. "Artificial Intelligence Definitions from TechTarget." www.techtarget.com/searchenterpriseai/definitions/Enterprise-applications-of-AI, Accessed 1 July 2023.
2. D. Mohammed and B. Horváth, "Travel demand increment due to the use of autonomous vehicles," *Sustainability*, vol. 15 (8937) 11, 2023.
3. "Autonomous Vehicles in Rural Communities: Is It Feasible? – IEEE Technology and Society." https://technologyandsociety.org/autonomous-vehicles-in-rural-communities-is-it-feasible/, Accessed 12 January 2023
4. "The Ethical Considerations of Self-Driving Cars | Montreal AI Ethics Institute(montrealethics.ai)Search, Bing, Montreal this." https://montrealethics.ai/the-ethical-considerations-of-self-driving-cars, Accessed 15 April, 2023.
5. "10 Advantages of Autonomous Vehicles." www.itsdigest.com/10-advantages-autonomous-vehicles, Accessed 31 July 2023.
6. "Hacking an Autonomous Vehicle – Common Attacks and Vulnerabilities.". https://roboticsbiz.com/hacking-an-autonomous-vehicle-common-attacks-and-vulnerabilities
7. "Autonomous Vehicle Implementation Predictions: Implications for Transport Planning." www.bing.com/search?q=Autonomous+Vehicle+Implementation+Predictions%3A+Implications+for+Transport+Planning+(vtpi.org)&cvid=df92a8fe0c3d4c4cb0b95d9dd9c3f9f5&aqs=edge.0.69i59.747j0j9&FORM=ANAB01&PC=DCTS.
8. J. Dean, Alexander J. Wray, L. Braun, Jeffrey M. Casello, L. McCallum, and S. Gower, "Holding the keys to health? A scoping study of the population health impacts of automated vehicles," *BMC Public Health*, vol. 19, 1258, Jan. 2019.
9. J. Sharma and P. S. Mehra, "A survey of security challenges and existing prevention methods in FANET," *In Intelligent Data Analysis, IoT, and Blockchain*, Bashir Alam and Mansaf Alam (eds.). New York: Auerbach Publications eBooks, 2023, pp. 252–262. doi: 10.1201/9781003371380-24.
10. J. Sharma and P. S. Mehra, "Secure communication in IOT-based UAV networks: A systematic survey," *Internet of Things*, vol. 23, 100883, Oct. 2023, doi: 10.1016/J.IOT.2023.100883.
11. X. Krasniqi and E. Hajrizi, "Use of IoT technology to drive the automotive industry from connected to full autonomous vehicles," *PapersOnLine*, vol. 49, no. 29, pp. 269–274, Jan. 2016, doi: 10.1016/j.ifacol.2016.11.078.
12. "The Social Implications of Driverless Cars."www.bbvaopenmind.com/en/technology/artificial-intelligence/the-social-implications-of-driverless-cars/, Accessed 15 June 2023.

13. M. Mhamunkar, "Understanding facial recognition and detection with use of EmguCV and haarcascade files," *IJSRD – International Journal for Scientific Research & Development*, vol. 9, no. 3, 2021." www.ijsrd.com/articles/IJSRDV9I30242.pdf, Accessed 15 Jan. 2023.
14. "The economic and social impacts of fully autonomous vehicles – Compact." www.compact.nl/en/articles/the-economic-and-social-impacts-of-fully-autonomous-vehicles/, Accessed 12 May 2022.

Part II

Energy Conservation for Intelligent IoT Networks

6 Green Internet of Things

Pankaj Kumar Dubey, Bindeshwar Singh, Deependra Singh, and Aryan Kumar Dubey

6.1 INTRODUCTION

Numerous research groups have been looking into how to construct and transform society into an intelligent and smart one with applications for things like homes, businesses, cities, buildings, cars, garages, and transports. IoT is currently in its groundbreaking stage and has 27 billion "intelligent" gadgets and machines that are connected. According to estimations, by 2025, over 75 billion IoT equipment or items will be associated globally, representing a nearly five-fold connection expansion in ten years [1]. The concept of integrating detectors and cognition into natural objects was first tackled when college scholars wanted to modify a Coca-Cola vending machine so that they could remotely access tracking of its supply. The apparatus was heavy, and expansion was sluggish. "IoTs" was first introduced in 1999 by research expert Kevin Ashton. While working for Procter & Gamble, Ashton promoted using radiofrequency identification (RFID) tags to track products across a supply network. He supposedly used the then-popular phrase "internet" in his briefing to capture the managers' attention. The adage continued to be used. Over the following decade, as more smart gadgets hit the market, the social advantages of IoT technology (TECH) began to surge. The initial iPhone was launched in 2007, the first smart refrigerator was presented by LG in 2000, and by 2008, there were many more connected devices than humans on the planet. In 2009, Google started working on self-driving vehicles, and in 2011, the firm unveiled its Nest smart thermostat, which allowed for distant management of the thermal infrastructure. However, the Green Internet of Things (G-IoT), an innovative concept that proposes to deliver sustainable (SUST) solutions at the edge of networking by permitting the widespread implementation of automated apps and services, is used to the power challenges. The focus of this evaluation work is on G-IoT and the SUST of IoT products, with an emphasis on energy-saving methodologies.

6.1.1 CONTRIBUTION OF CHAPTER

The following list highlights the paper's significant contributions:

- In this chapter, the basics of green IoT are explained.

DOI: 10.1201/9781003405535-8

- A sophisticated review of several green IoT equipment.
- Tabulated updates on the G-IoT advancements (literature survey).
- G-IoT business situations at present.
- Case study of smartphone emissions reduction.
- The future possibilities of the G-IoT.

6.1.2 ORGANIZATION OF THE CHAPTER

This chapter is divided into five sections. There is an introduction in Section 6.1, a comparison of the various G-IoT system types in Section 6.2, a literature survey of the most recent research trend and the state of the IoT market in Section 6.3, a discussion of opportunities and challenges in Section 6.4, and a conclusion and outlook in Section 6.5.

6.1.3 Internet of Things

The Internet of Things (IoT) is a sophisticated automated and analytical program that utilizes big data, artificial intelligence (AI), connectivity, detecting, and sensor systems to give comprehensive solutions for an item or service. Any business or organization can use these technologies to achieve improved openness, accountability, and efficiency. By attaching to an IoT router or another side gadget, which either sends data to the cloud for analysis or analyses it nearby, IoT applications exchange the sensor information they hold. These technologies occasionally interact (Case Management Model and Notation (CMMN)) with similar products and take action depending on the information they get from each other. IoT TECH, in principle, consists of four key components [2]:

- Internet: to facilitate interaction between anything at any moment, from everywhere. It includes Internet Protocol (IP) for intelligent devices, smart online services, and cloud technology.
- Hardware: hardware for integrated connectivity, including transceivers, identifiers, controllers, and detectors.
- Middleware: for computation, incident management (MNGT), and data MNGT.
- Presentation: to comprehend the various systems' and apps' visualization and interpreting features.

The logical design of IoT equipment is represented in Figure 6.1, a block diagram of the various IoT interfaces [3]. In the IoT, it is utilized for system and server communication. Assume that the client is using his tablets to stream movies with the help of a coordinated effort from the CPU, memory, I/O interfaces, and communication channels.

Advantage: Portable gadgets can assist with the more effective MNGT of homes and cities. It offers safety and stability for the person. By automated processes, we can shorten the period people spent on them. Information can be gathered and updated daily, although we are far from our starting point.

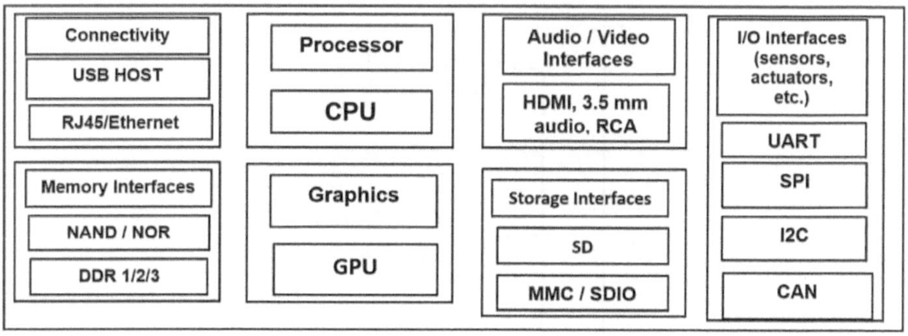

FIGURE 6.1 Block diagram of an IoT device.

Disadvantage: With such an extensive network of gadgets, incorporating encryption and security measures might be challenging. Some firms might employ substandard TECH since they're inexpensive instead of investing time, energy, and money to implement it across all devices.

Due to these limitations, IoT places a strain on the electricity network, thus to address the issue, Green IoT, which is covered below, replaces traditional IoT [3].

6.1.4 Green Internet of Things

In terms of creating a green IoT and intelligent society, energy usage quickly becomes state-of-the-art. Power savings should represent the IoT to minimize (MIN) the greenhouse gases (GHG) and carbon dioxide (CO_2) pollution of sensing, devices, apps, and activities to create a SUST TECH world.

The procedure of developing, generating, disposing of, using, and linking subsystems (like scanners, displays, CMMN hardware, and memory sticks) more effectively and regularly while minimizing the influence on the environment and society is referred to as the G-IoT. IoT adoption of energy-efficient practices to lessen the GHG of current apps and services is known as "green IoT." Numerous green innovations, such as green RFID tags, green sensor networks, and green cloud services networks, should be incorporated into the green IoT [2]. The primary technologies needed to implement a green IoT system are shown in Figure 6.2.

Below is an overview of the many TECH in green IoT and its uses that will influence the era of SUST IoT utilization.

Green RFID

Utilizing the benefits of RFID tags by making them smaller to utilize less non-biodegradable material will help significantly reduce waste since the tags are rarely recyclable. The developers can also create energy-efficient tag estimate techniques.

Green WSN

The prompted activation of sensor nodes only while in use can reduce energy usage, as mentioned above under G-IoT and its usage. Another method in the G-IoT and its

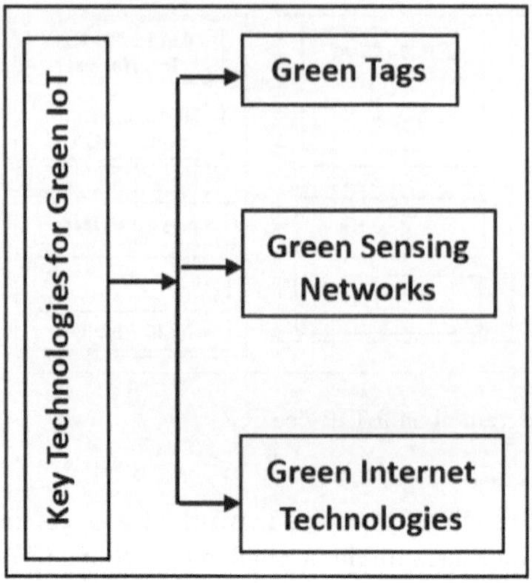

FIGURE 6.2 The essential green IoT solutions.

uses in real time (RT) is to reduce the amount of data that is communicated across the connection link by the aggregate, adaptive sampling, and compressed. Another option is transporting the data using a mechanism that uses little energy.

Green Cloud Computing

Employing energy-efficient virtual machine approaches, integrating energy-efficient hardware and software, adhering to energy-efficient resource provisioning procedures are crucial steps in reducing energy consumption in computing environments.

Green M2M

Intelligently managing transmission (TRANS) power, strategies, creating strong CMMN standards, etc. are some more TECH of G-IoT and its uses.

Green Data Centre

G-IoT and its implementations make use of effective sleep planning, reducing the length of wired and wireless data lines, inventing energy-aware navigation techniques, and other methods to maintain data centres.According to the essential elements, green ICT technologies, and green IoT technologies depicted in Figure 6.3, various green IoT technologies are categorized [4].

An additional crucial innovation for enabling G-IoT is a green wireless system. There are many sensor nodes in a wireless sensor network (WSN), but they have little power and memory. To achieve green WSN, a variety of strategies should be considered:

- After completing the necessary action, the sensor enters an idle.
- Utilize renewable energy (RE) for recharging and usage. Additionally, shocks and kinetic energy can utilize energy-conscious optimization algorithms.

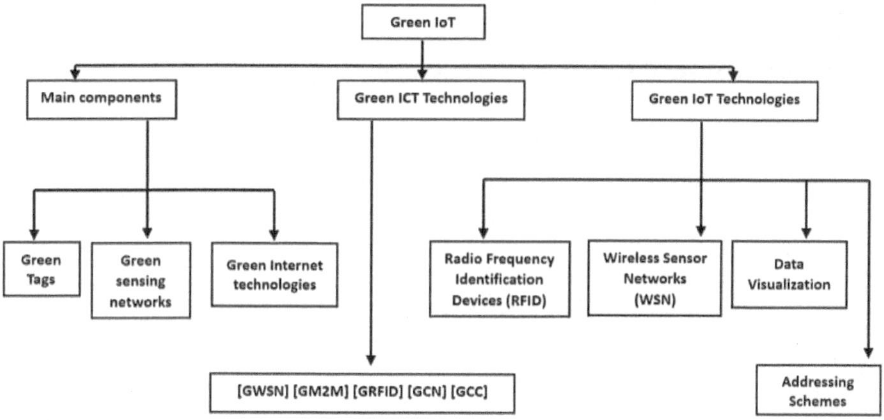

FIGURE 6.3 Categorization of green IoT.

- Minimize the dimensionality of the file and the storage required by using information and context-aware techniques.
- Utilize energy-efficient transit strategies to reduce the power used by mobile devices.

6.2 COMPARISON BETWEEN VARIOUS TYPES OF GREEN IOT TECHNOLOGY

Bluetooth, Bluetooth RE, LoRa, Wi-Fi, Wi-Max, Wireless HART, Zigbee, Z-Wave, radio frequency identification devices (RFID), 2G/3G, 4G/5G, and near-field CMMN are only a few examples of IoT devices (NFC). A comparison of different G-IoT TECH should be done in this area as represented in Table 6.1 [5, 6].

6.3 LITERATURE SURVEY OF RECENT TRENDS IN G-IOT

In this section, we discuss of recent trends through a literature review of 37 publications, and current global market scenarios of IoT are discussed in Sections 6.3.1 and 6.3.2.

6.3.1 New Developments in Green IoT Technology

In this study, we suggest creating applications that depend on the features of the road rock that protect from bugs getting revealed via G-IoT safety using advanced fuzzing logic (AFL) and synthetic corpora (SC) [7]. To maximize net utility for a multi-hop G-IoT Network (GIN), we suggested a two-time-scale resource assignment algorithm (TSRA) that takes into account the stochastic nature in the supply of resource MNGT (RM), energy, and spectrum over time [8]. To create a SUST smart world (SW), numerous data and CMMN TECH (ICTs) and G-IoT-related concerns have been covered in this study [9]. In this research, using the Neumann expansion (NE)

TABLE 6.1
Comparison of Several Green IoT Technologies

Types of G-IoT Parameters	Bluetooth	Bluetooth LE	LoRa	Wi-Fi	Wi-Max	Wireless HART
Frequency	2.45 GHz	2.4 GHz	865 MHz	2.5–5 GHz	5 GHz	2.4 GHz
Data rate	700 Kbps	2–3 Mbps	50 Kbps	12–100 Mbps	15–100 Mbps	250 Kbps
Range	<20 m	5–10 m	1–4.5 km	4–20 m	50 km	800 m
Power usage	Low	Very low	Low	Medium	High	Medium
Cost	Low	Low	Medium	Low	High	Medium
Applications	File transfer.	Data and health.	Logistics.	Internet.	Commercial automated.	Networking for industry detection.
Battery life	Low	Low	1–3 years	Hours	Years	Years
Topology	Mesh, star, tree.	Star	Star	Star	Mesh, star	Mesh, star
Network	PAN	WPAN	WAN	LAN	MAN	LAN
Nodes	8	16	1,000	32	100	64
Standards	IEEE 803.15.1	IEEE 802.15	IEEE SA P1451.5.5	IEEE 802.11	IEEE 802.16	IEEE 802.16
Spreading	FHSS	FHSS	CSS	DSSS, OFDM.	OFDM	DSSS, CCK.
Data protection	16 Bit CRC	16 Bit CRC	128 Bit CRC	32 Bit CRC	128 Bit CRC	64 Bit CRC
Encryption	E0 stream cipher	E0 stream cipher	AES Block cipher	AES Block cipher	3DES, AES Block cipher	3DES

Zigbee	Z-Wave	NFC	RFID	2G/3G	4G/5G
870–920 MHz, and 2.4 GHz.	865.2 MHz	13.56 MHz	300–1,000 MHz	800–2,100 MHz	10–40 GHz
250 Kbps	100 Kbps	400 Kbps	400 Kbps	1.8–7.2 Mbps	1 Gbps
9–295 m	30 m	<10 cm	<3 m	1 km	500 m
Low	Very low	Very low	Very low	High	1,060 W
Medium	Low	Low	Low	High	High
Matrix of sensors.	Domestic lightning and automation.	Pay and share.	Item tracking.	Cell phones.	IoT and cell phones.
Month to years.	10 years	Very low	3–5 years	1–4 days.	2–3 days.
Mesh, Tree, Star.	Mesh	P2P	P2P	Mesh	3GPP
LAN	LAN	PAN	PAN	WAN	WAN
65,000	4,000	2	8	1,00,000	5,00,000
IEEE 802.15.4	IEEE 802.11	ISO/IEC 18092	IEEE 802.15	IEEE 802.11	IEEE 802.11ad
DSSS	CSS, OFDM.	FHSS	FHSS	DSSS	DSSS, CCK
16 Bit CRC	128 Bit CRC	8 Bit CRC	16 Bit CRC	32 Bit CRC	32 Bit CRC
AES Block cipher	AES-128	AES	E0 stream cipher	A5/1	State of art

and nonlinear Perron-Frobenius theorem (NLPFT), we investigate the MIN issue of the complete power in the IoT, where users concurrently interact over numerous independent channels [10].

Researchers provide an introduction to the time-reversal (TR) wireless architecture for the IoT in this research using time-reversal division multiple access (TRDMA) [11]. This study examines the IoT from a green approach [12]. This study suggests a deep reinforcement learning (DRL)-based intrusion detection algorithm (IDA) that follows traffic flow patterns by identifying statistical information from historical network activity for traffic prediction [13]. This work offers a unique latent discriminative low-rank projection (LDLRP) approach with an augmented Lagrange multiplier algorithm (ALMIA), for visual feature MIN, to decrease the data dimensions while maintaining the racist and discriminatory inherent information from the data [14].

In this chapter, we use an exchange-matching-based algorithm (EMA) with access point modelling to study the clustering and power MNGT issue for coordinated multipoint (CoMP) TRANS in green non-orthogonal multiple access (NOMA) networks (APM) [15]. The fundamentals of in-network storage in the IoT are covered in this chapter, along with implementation recommendations for brokers that allow caching (CEB) [16]. The unique cognitive ambient backscatter device (BD) CMMN (CABC) architecture that we suggested in this research may integrate the benefits of conscious radio TECH and ABC TECH. Based on this, we examined the outage probability (OP) and intercept probability (IP) of the CABC model under consideration and created the corresponding analytical formulations for both [17].

This study proposes the distributed energy-efficient computation architecture known as deep-green (DG) for the industrial IoT [18]. This chapter evaluates the joint optimization (OPT) of a diverse array of Unmanned Aerial Vehicle (UAV) framework design variables, like the UAV's pace, altitude, beamwidth, and budget utilization, to enable uplink CMMN between floor IoT devices and a UAV utilizing concise, ultra-reliable, and low-latency (URLLC) packet data (RM) [19]. To increase spectrum usage, network lifespan, and effective G-IoT connectivity while ensuring soft restraints, a dispersed cross-layer layout relying on a none-persistent Carrier Sense Multiple Authorization (CSMA) system has been proposed in this study for cognitive radio (CR)-based G-IoT sensor channels [20]. The study discusses the issue by putting forth design standards for an indoor, permanent, ambient light energy harvesting Bluetooth low-energy (BLE) beacon [21]. This chapter proposes the KPCA (kernel principal component analysis) in the Seasonal Autoregressive Integrated Moving Average (SARIMA) framework as an energy-efficient Cloud-centric system (CCS) to assess the dryness of the present state and forecast the forthcoming timeline [22]. In this chapter, we propose setting up fog networks (fogs) by strategically arranging the entertainment gadgets in the network edge, which are subsequently integrated with the cloud to supply storage and processing resources. This is accomplished using integer linear programming (ILP) frameworks to resolve the fog timing issue within an integrated Cloud-Fog (I-Cloud-Fog) architecture [23].

This chapter empirically investigates the NP-Hardness of the multi-channel multi-radio times' interval co-planning problem (MMCS) and proposes the greedy approach G-Schedule, which includes planning for time slots from base stations to detectors and from ground stations to ground stations [24]. The private keys in the suggested scheme were produced using Finite Difference Hankel Determinant (FDHC). A nondominated sorting genetic approach III (NSGA-III) was also used in order the five-dimensional hyperchaotic map's ideal beginning properties (FDHC) [25]. In this chapter, we suggest a Sink Rotating Joint UAV Data Gathering (SR-UAV-DG) system in answer to the G-call IoT for energy conservation [26]. The workload allocation issue in the Environmentally Intelligent (EI) for towns that are environmentally friendly towns are studied this research. To address this, a green workload placement technique (GWP) employing the multi-action DRL is suggested [27].

Consequently, researchers have proposed cyclic delayed variety orthogonal frequency division multiplexed index modulator (CDD-OFDM-IM) and CDD cyclic-prefixed unit-carrier (CPSC) IM to boost the spectral effectiveness (SEs) of CDD networks with OFDM and CPSC TRANS [28]. To better utilize spectrum resources and tackle the concerns of disturbance synchronization among device-to-device (D2D) CMMN and mobile services for the G-IoT, this study proposed a human interaction-assisted asset platform (RSS) [29]. This work presents a unique dispersed clustering approach (DCA) for the suggested uneven cluster-radius layout after first optimizing the cluster architecture [30].

To improve the effectiveness of the vehicle-to-network (V2N) CMMN utilizing the sequential convex approximate (SCA) approach, the UAV employed the joint information system with the assistance of the satellite and base stations (BS) [31]. For non-quartered and adaptable patterns in upscaling eye issues of IoMT-UAV systems, a new invention of a nonlinear blind edge-guided spatial filter relied on linear MIN mean square error-estimation theories (ALMMSE) estimator was studied in this chapter [32].

To reduce the median lifespan of details and Maximum (MAX) energy effectiveness for various independent and rechargeable sensor networks in the G-IoT, this chapter primarily focused on the mobile elements scheduler and developed multi-objective OPT algorithms like age-optimal mobile elements scheduling (AMES) and optimized mobile aspects scheduling (AMES) [33]. The best factor to take into account while creating a SUST ecosystem for IoT is still green computing [34].

The viability of a tightly unified unmanned gadget with Wi-Fi capabilities is examined in this chapter [35]. The difficulties of deploying strategies to obtain power-efficient IoT TECH are discussed in this research. It outlines a heterogeneous circular packing-based (Ht-CPBDS) installation technique and a homogeneous grid-based installation method (Hm-GBDS) [36]. This study intends to present a paradigm of safeguards for smart urban (SC) utilizing the G-IoT and cloud-integrated data administration (CIDM) [37].

To build a strategic plan while lowering the responsibility in terms of power control, this study offers an intelligent and safe edge-enabled computation (ISEC) architecture for SUST towns utilizing G-IoT [38]. This study examines several green power collecting techniques for SUST and SC programs for edge-based intelligent urban computation (EBIUC) [39].

The density-based spatially cluster of programs with noise (DBSCAN) clustered method for the IoT is presented in this study coupled with power analysis of a data mining approach (DMA) built on a cloud service [40]. Utilizing the Internet of Things (IoT) for Green Environment (G-ENVT), this research demonstrates an Energy-saving Big Data-based Secured (EBDS) approach that aims to decrease energy usage and improve confidentiality and protection for big data [41].

To demonstrate how the quickest route and the fewest number of links on a route (SPLL) and the longer hops (LH) effectively cooperate in sensor-enabled wireless networks, this study provides a power-based route selection and message rescheduling methodology, such as M/M/1 queuing assessment [42]. The clustering approach (CA), which reduces power consumption by grouping mobile phones into groups depending on RSS values, is suggested together with apps Internet Standard (IP), Medium Access Protocol (MAC), and Location Identifiers (LOC) [43]. Analysis of Green IoT technologies are represented in Table 6.2 to Table 6.6.

6.3.2 Current Market Scenarios of Green IoT

- The value of the sector IoT market is increasing from \$399.41 billion in 2022 to \$486.7 billion in 2023, growing at a 21.9% compound annual growth rate (CAGR).
- At a CAGR of 21.4%, the size of the worldwide IoT market is anticipated to increase to \$1,057.55 billion in 2027.
- Major corporations are investing more and more in the creation of smart cars, including Mercedes-Benz, Volkswagen, Volvo, Toyota, and Google Inc.
- UBS estimates that there will be around 26 million driverless taxis operating worldwide by 2030, accounting for 12% of all vehicles sold.
- Many of the largest software vendors in the market competition in the IoT field, including Microsoft with its Azure IoT offerings and Amazon Web Services with its AWS IoT services and solutions.
- India's Internet of Things (IoT) market size was estimated to be about five billion dollars in 2020 and grow to over 9.3 billion dollars in 2025.
- In 2023, the IoT market is anticipated to generate US\$1,102.00 million in revenue. With a projected market volume of \$1,081.00 m in 2023, the largest market segment is Smart Home Technologies.
- Hyperlink Info System, LTI, HCL Tech, TCS, Kelton Tech, TVS Next Private Limited, VVDN Technologies, Mphasis, NTT DATA Services, and H-Data Systems are among the best IoT businesses in India.

As depicted in Figure 6.4, various IoT sectors like health care, electricity, security, agriculture, manufacturing, urban infrastructures, resource extraction, and retail market share have been shown [44]. The healthcare and manufacturing sectors contribute 40% and 30% to the total global IoT market share.

TABLE 6.2
Analysis of Green Internet of Things Technology

Ref. No.	Title	Objective	Component	Drawback	Operational Factor	Methodology	Upcoming Plans	ROC
7	SC to evaluate fuzzing for G-IoT programs.	AFL fast detects bugs faster than AFL.	G-IoT.	Weakness of cycle explosion.	By using 15 binaries, detection of five bugs with AFL fast.	SC, and AFL.	AFL Fast+.	1
8	For G-IoT systems, there are two time-scale RM.	A long time to buy the power, and a short time to acquire the statistics.	G-IoT.	Information queue length shortens but weight increases again.	In the case of $V = 200$ and 300, the initial quantity of sensed data is typically nil.	TSRA.	Investigate the RM of RF-powered GIN.	17
9	G-IoT for SW.	LPWANs.	G-IoT.	Theoretical analysis.	40 km range.	Cloud computing, and ICTs.	A virtualized sensor as a service is envisioned.	41
10	Optimization algorithms for multi-access green CMMN in IoT.	Achieve efficient power and rate allocations.	Green communication with IoT.	More degree of accuracy increase.	NE is 1.4 times more accurate than NLPFT.	NE, and NLPFT.	Artificial intelligence (AI) techniques were used.	8
11	TR wireless paradigm for G-IoT: An Overview.	Improve client safety and confidentiality in IoT.	G-IoT.	Less uploading speed.	Resolution costs reduce from US $329 to US $295.	TRDMA.	6G technology.	12

(continued)

TABLE 6.2 (Continued)
Analysis of Green Internet of Things Technology

Ref. No.	Title	Objective	Component	Drawback	Operational Factor	Methodology	Upcoming Plans	ROC
12	Enabling TECH for G-IoT.	Cloud computing is best in present scenarios.	G-IoT.	Theoretical analysis.	ICT will contribute, on average, between 6% and 8% annually.	Cloud computing and ICTs.	Attention is being drawn from businesses in the field of green IoT.	36
13	A deep determinism strategy gradient-based method, IDA in the G-IoT.	Accurately predicts network activity and finds infiltration.	G-IoT.	Less accurate.	In original dataset I, the two methods' respective real positive rates are 99.98% and 92.48%, while in data set II, they are 100% and 82.76%.	IDA, and DRL.	Improving the accuracy of intrusion detection.	3
14	LDLRP for visual dimension reduction in G-IoT.	High-dimensional information should have fewer dimensions.	G-IoT.	Expensive.	Classification accuracy is 91.30±1.39%.	Subspace learning, ALMIA, and LDLRP.	Creating a unified framework aids in locating the best solution worldwide.	3

TABLE 6.3
Analysis of Green Internet of Things Technology

Ref. No.	Title	Objective	Component	Drawback	Operational Factor	Methodology	Upcoming Plans	ROC
15	CoMP cluster and energy management for G-IoT networks with NOMA support.	After each exchange matching operation, the overall transmit power lowers.	G-IoT.	Less accurate.	Maximum transmit power is 33 dBm, and noise power is –174 dBm/Hz * 180 kHz.	EMA, and APM.	AI techniques were used.	3
16	Caching within a system for the G-IoT	The entire amount of energy used by the sensor nodes with limited energy is known as MIN.	G-IoT.	Large size.	Sensor nodes will reduce energy usage for data transfer by 24%.	CEB principle.	Cleansing, grouping, and forecasting of data.	3
17	Physical layer security of CABC CMMN for G-IoT.	Increase of main-to-eavesdropper ratio.	G-IoT.	Less accurate.	As MER grows, U's safety rises but BD's protection declines.	CABC.	AI.	24

(continued)

TABLE 6.3 (Continued)
Analysis of Green Internet of Things Technology

Ref. No.	Title	Objective	Component	Drawback	Operational Factor	Methodology	Upcoming Plans	ROC
18	Deep-Green: A dispersed energy efficiency computing paradigm for G-IoT.	Reduces the computing load and CMMN overhead.	G-IoT.	Theoretical analysis.	Power varies from 200 to 126 W.	Cloud computing.	Enhance the Deep-Green model's viability.	8
19	Green URLLC in mobility UAV-assisted IoT networking enabled by quasi-OPT of uplink energy: A perturbation-based technique.	IoT usage involves little data transfer.	UAV, and G-IoT.	Limited data transfer capacity.	The relative difference between our suggested method and Shannon's formula for the parameters given is 47.51%.	URLLC, and Shannon's formula.	Equality issue for the devices that require more data transmission.	5
20	Cross-layer spectrum bands in CR G-IoT networks that use less energy.	Reduced power consumption.	G-IoT.	Less accurate.	The backoff policy's dispersion has a negligible impact on energy use.	CR TECH, and CSMA scheme.	AI.	6

21	Architecture, modelling, and field trials for LUXBEACON, a battery-free beacon for the G-IoT.	Allowing continuity of service even within low-light situations.	Battery-less G-IoT.	Power MNGT is missing.	40 lux intensity of light, and once completely recharged, can last more than 17 hours.	BLE Beacon TECH.	Power MNGT of a SUST BLE beacon.	6

TABLE 6.4
Analysis of Green Internet of Things Technology

Ref. No.	Title	Objective	Component	Drawback	Operational Factor	Methodology	Upcoming Plans	ROC
22	Green foundation for IoT-based cloud-based intelligent drought forecasting.	Energy savings and drought prediction.	Green framework with IoT.	Climate dependent.	The battery level can vary by 0.99% when using a very accurate system with a 0.083% error rate.	KPCA, and SARIMA model.	AI-based.	5
23	Green fog strategy for the best routing of IoT activities.	Design with complete connectivity and organising fogs.	Green fog planning with IoT.	Less accurate.	IoT tasks between 25% and 50% of the time don't need RT or mobile capabilities.	ILP models.	Effects of several fog designs, and others, on the efficiency of the system.	4
24	MMCS in G-IoT for a warning on the safety of underground mining.	Reduced transmission latency and packet loss rate effectively cut down on energy use.	G-IoT.	Complex.	Data acquisition times are 100, 300, 600, and 900 ms.	MMCS, and G algorithm.	Recent OPT.	2
25	G-IoT relies on a power-efficient and safe e-health care architecture.	Secure the CMMN in G-IoT networks.	E-health, and G-IoT.	Less accurate.	Energy through single-photon emission is 0.0203 joules.	θ-NSGA-III, and FDHC.	AI-based.	23

26	A crucial DG architecture for G-IoT data gathering and control mechanisms.	The information reception ratio is assured to reach 92%, and the flying time of UAVs is cut by 16.83%.	UAV, and G-IoT.	Theoretical analysis.	Decreases the network's maximum energy usage by 99.21%.	SR-UAV-DG.	Place the SR-UAV-DG architecture in real-world use cases.	2
27	Learning-based GWP for energy internet in smart cities.	Energy conservation and job speed.	Green workload with IoT.	Long-running period.	Power expenditures can be reduced by up to 65.46% and 68.92%, respectively.	GWP, and DRL.	Reduction in the computational period.	1
28	CDD with index modulation for G-IoT.	Using OFDM and CPSC broadcasts, CDD systems can increase their SEs.	G-IoT.	Complex and slow response.	The empirical coding gain was accurately calculated as ten log10(R) = 20 log10(p/p1) = 1.0231 dB	CDD-OFDM-IM and CDD-CPSC-IM.	GA.	2

TABLE 6.5
Analysis of Green Internet of Things Technology

Ref. No.	Title	Objective	Component	Drawback	Operational Factor	Methodology	Upcoming Plans	ROC
29	To move toward the G-IToMN uses asset interchange through socialization.	Enhancements to the G-transmission IoT's success rate.	G-IoT.	Theoretical.	P c s = 1,000 dBm, = 0.58, and = 4.1 are the transmission power settings.	Social Layer Assistance scheme.	NSGA II.	2
30	MIN converge-cast time and energy consumption in G-IoT.	Use of large-scale WSN installation to efficiently track IoT energy effectiveness.	G-IoT.	Less accurate.	MIN the amount of node state transitions by 15%, lowering the electricity costs.	DCA.	AI techniques.	18
31	UAV-aided wireless CMMN architecture for an interconnected space-air-ground G-IoT with energy constraints.	To achieve greener CMMN while maintaining superior efficiency, UAVs are essential.	UAV, and G-IoT.	Theoretical approach.	A greater sum rate is discovered by comparing the various battery capabilities, i.e., set Qmax = 7.25105 J, 7.35105 J, or 7.50105 J.	SCA.	Experimental setting implementation.	9

32	A blind-scaled edge-guided reconstructive filter called BL-ALM is used by G-IoMT-UAV networks to observe the surroundings intelligently.	Low-cost sensing in green Internet of multimedia thing.	G-IoMT-UAV.	Theoretical analysis.	According to bilinear ALM, the Max signal-to-noise ratio and structural similarity are 24.23 dB and 0.9607.	ALMMSE.	Experimental setup.	5
33	Age-appropriate mobile component planning for data gathering and recharging in the G-IoT.	Better energy effectiveness & a greater sensor node successful ratio.	G-IoT.	Complex.	The CMMN radius exceeds 60, and the AMES successful visitation ratio is 1.	AMES, and GA.	AI method.	2
34	Green computing for IoT.	Reduce energy consumption.	Green computing IoT.	Costly.	Growth of IoT being eco-friendly.	Cloud and edge computing.	GA.	3
35	Antennas and photovoltaic (PV) panels: Toward a G-IoT.	Improving the performance.	PV, and G-IoT.	Less accurate.	The side lobe elevation is acceptable, and the main lobe is slanted towards = 30°.	Beamforming techniques.	AI techniques.	1

TABLE 6.6
Analysis of Green Internet of Things Technology

Ref. No.	Title	Objective	Component	Drawback	Operational Factor	Methodology	Upcoming Plans	ROC
36	G-IoT adaptable software – defined node placement.	For an IoT system to be green, Ht-CPBDS is necessary.	G-IoT.	Homogeneouss's case fails.	Ideal for network implementation to G-IoT.	Hm-GBDS, and Ht-CPBDS.	A sophisticated technique is utilized in homogenous networks.	1
37	A safety measures architecture integrating G-IoT and CIDM for SC is called M-SMDM.	Information security and CMMN object verification.	G-IoT.	To better handle the majority of traffic.	Decreases the computational overhead by an average of 29% and 27%.	CIDM.	Develop a framework based on machine learning.	7
38	ISEC concept for Smart Urban SUST Cities.	Energy consumption was reduced from 21% to 15%.	G-IoT.	Insufficient to defend against Denial of Service (DoS) risks for the IoT network.	12% end-to-end latency, 36% route interruption, and 52% network overhead.	ISEC model.	Energy harvesting TECH in SUST developments.	12
39	Green energy harvesting strategies on EBIUC in SUST IoT.	Enhance the energy consumption of smart applications.	Green energy harvesting IoT.	Expensive.	Efficiency enhancement by 8%.	EBIUC.	AI-based techniques.	7

40	Energy analysis of IoT DMA for smart green CMMN networks.	Improve the efficiency of clustering.	Green CMMN network.	Less accurate.	Enhanced efficiency of clustering by 6%.	DMA, and DBSCAN.	GA.	5
41	EBDS framework using IoT for G-ENVT.	16% more energy was consumed, and 33% more packets were dropped.	IoT for green environment.	Less accurate.	Non-uniform data traffic results in an end-to-end delay of 15.5% and a 16% decrease in route stability.	EBDS framework.	By utilizing mobile sensors, the G-ENVT can use less power.	5
42	IoT green virtualization: energy-focused route and message planning strategy.	Less energy is consumed per unit of time than cutting-edge technology.	Green computing for IoT.	Slow and less accurate.	Lower average delay time as 2 ms.	M/M/1 queuing analysis and SPLL.	Utilizing heuristic-based methods for sensor-relied ENVT.	11
43	MGR: Multi-parameter Green Reliable CMMN for IoT in 5G network.	MIN energy requirements.	Green CMMN, IoT, and 5G.	Fails in big data.	Increase the data rate by 10%.	CA.	Improve our G-IoT for 5G proposal.	4

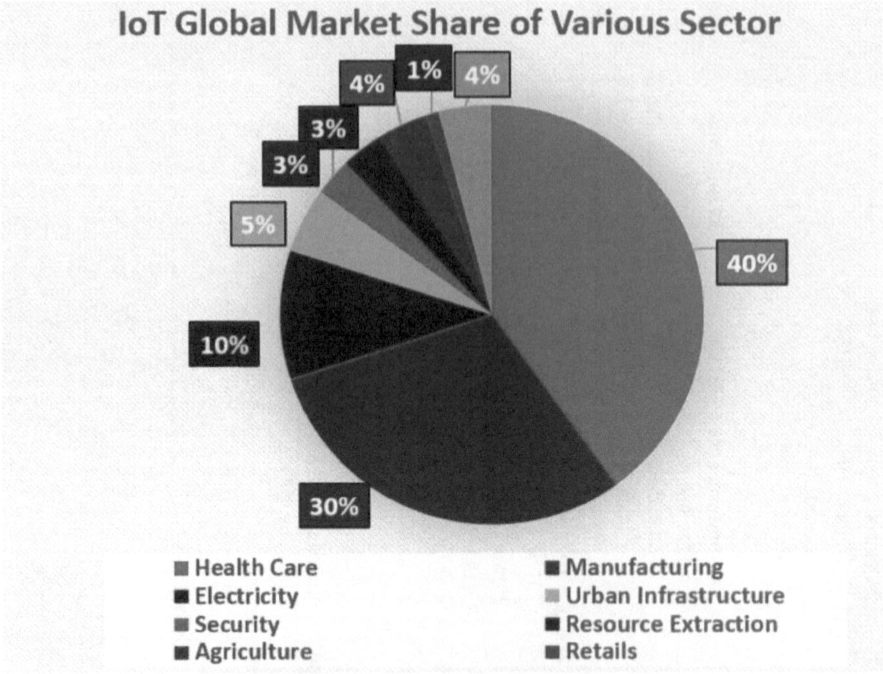

FIGURE 6.4 Various sector shares in G-IoT global market.

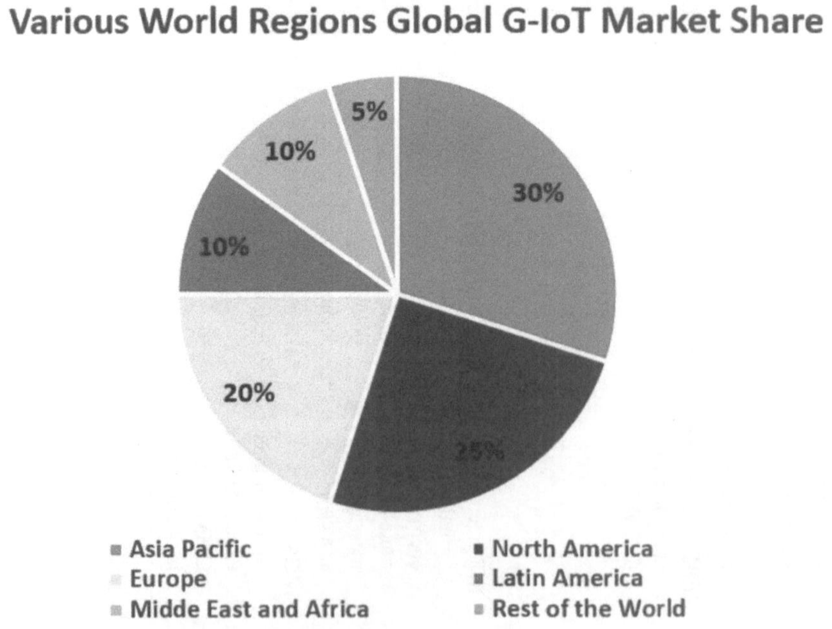

FIGURE 6.5 Various world region share in G-IoT global market.

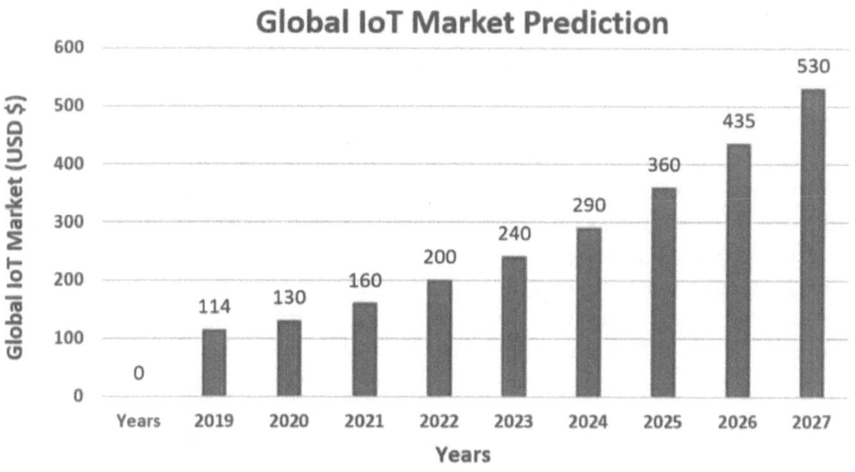

FIGURE 6.6 Global IoT market prediction from 2019 to 2027.

Business share in the global IoT market has been displayed in Figure 6.5 for several global regions, including Asia Pacific, Europe, Middle East & Africa, North America, South America, and the Rest of the World [45]. The largest G-IoT market in the world is in Asia Pacific.

Figure 6.6 displays the global IoT market's annual revenue in US dollars from 2019 to 2027. From this, it can be observed that the market will grow from 2019 to 2027, with valuations ranging from 114 billion to 530 billion dollars [46].

6.3.3 Case Study: Current and Future Trends in the Effect of Cell Phones on the Environment

Examining how mobile technology, including PDA (personal digital assistant) laptops, cell phones, and smartphones, affects the environment is the primary goal of green computing [47]. In this subsection, we examine a case study involving cell phones, a crucial element of the Internet of Things (IoT). We look into it and predict how it will be used, impacted, and recycled.

Lessen the influence on the environment: The study proposed using eco-friendly substances, modifying contract terms, minimizing packaging and peripherals, and engineering for disassembling and energy-saving batteries to lessen the ecological consequences of smartphone emissions. Several steps to support the environment are defined below:

- The US EPA has stated that trash reduction is hazardous and bio-accumulative synthetic compounds and cell devices are among the more harmful products. Mobile phone printed circuit boards (PCBs) comprise 63% metals, 24%

ceramics, and 13% polymers. The cellular phone component Cu (copper) is to blame for the water's toxicities.

- Hardware level indicators, such as the TEER (Telecommunications Energy Efficiency Ratio) proposed by the ATIS (Alliance for Telecommunications Industry Solutions), the TEEER (telecommunications equipment energy efficiency rating) by the Verizon NEBS Compliance, the CCR (consumer consumption rating), and the ECR (energy consumption rating), review for a pretty low productive rating of a single piece of mobile networking devices with specific functions in microforms.
- Making cellphones easier to disassemble, repair, or swap out individual parts would be a game-changer. Comparable to polylactic acid plastic (PLA), which is entirely manufactured from glucose or corn starch and is sustainable, disposable, and uses fewer components, recyclable plastic and common materials like bamboo are also available.
- The organic and ecological radical battery (ORB) charged in about 30 seconds without the need for any heavy metals that might be harmful to people.

Evaluation of the cell phone life cycle: The findings demonstrated that, among the three reuse approaches, the refurbishment has the most adverse environmental effects, except ODP (ozone depletion potential). Desktops cut power use and CO_2 emissions by 20%–55% and 18%–74%, respectively (VD). When the features of the phone from 2008 and the smartphone are compared, the life cycle assessment (LCA) result reveals an increase of 34 kg CO_2. For the cell devices and TVs, it is more constant than for PCs.

The emission and selling valuation of smartphones: The Japan Environmental Monitoring Organization for Industry approved Samsung Galaxy S4 emissions (JEMAI). The iPhone 6S, Galaxy S4, and Nokia Lumia 1520 are among them. As mentioned earlier, the cell's estimated revenues are shown in Table 6.7. We get the growth per year by factoring in the purchasing rates for 2014 and 2015.

Business Insider analysts estimated that 24 to 34 billion Internet-connected gadgets in 2020 will be IoT-based, and about $6 trillion will be invested in IoT-based services over the following 5 years. As per the IEEE Green ICT Action Plan, the ICT industry currently accounts for 2% of global CO_2 emissions, expected to double over the next 5 years. For every 85 kWh of electricity ingested, one tree is required to offset the ecological footprint already precipitated by that electricity, and in 2007, the ICT industry absorbed 18PWh of energy.

TABLE 6.7
Emission of Smartphones

Smartphones	Production (%)	Customer Use (%)	Transport	Recycling	Total CO_2 (kg)	Units Sell 2014	Units Sell 2020
iPhone 6S	83	11	65%	1.5%	81	190K	397K
Galaxy S4	3	17	2%	0.5%	22	306K	17,675K
Nokia Lumia 1520	73	14	10%	1.8%	38	185K	298K

6.4 OPPORTUNITIES AND CHALLENGES OF THE GREEN INTERNET OF THINGS

Concerning the transition from IoT to G-IoT, Green IoT has numerous problems. These may be based on various elements, including equipment, programming, routing formulas, repurposing of policies, and so forth as represented in Figure 6.2. These should be constructed with energy efficiency in mind [2]. Processors, sensors, servers, ICs, RFID, etc. are examples of hardware-based devices. Cloud computing, configuration MNGT, data centres, etc., are examples of software-based TECH. Smart metering systems, energy consumption forecasting, and other tools can be policy-based. Numerous issues and hurdles need to be resolved. We have listed the main difficulties below [4]:

- Power effectiveness must be incorporated into the IoT ecosystem to reach an acceptable pace.
- Implementations should be environmentally friendly to reduce their impact, and G-IoT must be reliable with energy usage algorithms.
- The frame of reference consciousness with a low-energy IoT network.
- Both the communicational tools and the protocols should use less power and be more energy-efficient.
- Green IoT models should have a minimally complicated pattern.
- A trade-off between effective spectrum management and effective dynamic spectrum sensing.
- IoT is promising because of effective energy mechanisms such as wind, solar, vibration, and thermal.
- Efficient cloud management in terms of energy usage.

Providing a greener environment, lowering energy costs, lowering latency and power fluctuations, manufacturing smart grids, and more prospects are provided by the E G-IoT. Voice-Powered Technology, retail data analysis, security, netphone, real estate, seasonal retail, health surveillance, and telemedicine are examples of G-IoT opportunities.

6.5 CONCLUSION AND FUTURE SCOPE

The G-IoT concept was discussed in this chapter. The pros and cons of G-IoT are explained in this section. It also examined the G-IoT life cycle and the TECH required to build a G-IoT system. Future energy consumption will undoubtedly be improved by the G-IoT models' fusion with a variety of cutting-edge technology. Finally, we may infer that the G-IoT application can offer a comprehensive approach that incorporates energy-saving methods, G-IoT TECH, and power-harvesting ideas. The fusion or combination of these methods produces an energy-efficient, long-lasting, and environmentally friendly IoT system. The finest G-IoT approaches use 6G. The entire ICT sector is expected to transform due to IoT, with the economic impact being paramount. While IoT is primarily employed in hospitals, North America is a significant market for EVs.

The following open issues and future research directions are noted regarding green IoT:

- Designing a green Internet of Things should be done while considering how much energy the entire system consumes, as long as service goals are met and performance standards are met (QoE).
- Acknowledging the features of distinct IoT uses and the facilities rendered for them is essential.
- Real power usage modelling of IoT system elements is required (such as WSN, core networking, embedded circuit, and CC).
- Well, with the ubiquitous use of sensors, it is possible to envision a standardised sensor as a service (SNaaS), wherein users get entry and management over their own personal IoT.
- Within the sense of SNaaS, looking into energy-saving system architecture, energy-saving service composition techniques, circumstance and situational consciousness relating to consumers and devices (such as learning and predicting), power-efficient WSN governance, and power-saving cloud monitoring is crucial.

REFERENCES

1. M. Bathre, and P. K. Das, "Review on an Energy Efficient, Sustainable and Green Internet of Things," 2nd International Conference on Data, Engineering, and Applications (IDEA), Bhopal, India, 2020, pp. 1–6, doi: 10.1109/IDEA49133.2020.9170736.
2. M. A. M. Albreem et al., "Green Internet of Things (IoT): An Overview," 2017 IEEE 4th International Conference on Smart Instrumentation, Measurement, and Application (ICSIMA), Putrajaya, Malaysia, 2017, pp. 1–6, doi: 10.1109/ICSIMA.2017.8312021.
3. Airtel Business, "IoT: Understanding the Logical and Physical Design of IoT," *Business Blogs*, 2020, www.airtel.in/blog/business/physical-and-logical-design-of-iot/.
4. G. Verma, and S. Prakash, "A Comparative Study Based on Different Energy Saving Mechanisms Based on Green Internet of Things (GIoT)," 2020 8th International Conference on Reliability, Infocom Technologies and Optimization (Trends and Future Directions) (ICRITO), Noida, India, 2020, pp. 659–666, doi: 10.1109/ICRITO48877.2020.9197848.
5. S. Sendra, L. García, J. Lloret, I. Bosch, and R. Vega-Rodríguez, "LoRaWAN Network for Fire Monitoring in Rural Environments," *Electronics*, vol. 9, no. 3, p. 531, 2020, doi: 10.3390/electronics9030531.
6. G. S. Karthick, and P. B. Pankajavalli, "A Review on Human Healthcare Internet of Things: A Technical Perspective," *SN Computer Science*, vol. 1, no. 198, 2020, doi: 10.1007/s42979-020-00205-z.
7. X. Zhu, S. Wen, A. Jolfaei, S. Camtepe, and Y. Xiang, "Synthesized Corpora to Evaluate Fuzzing for Green Internet of Things Programs," in *IEEE Transactions on Green Communications and Networking*, vol. 5, no. 3, pp. 1041–1050, Sept. 2021, ISNN: 2473-2400, doi: 10.1109/TGCN.2021.3068466.
8. D. Zhang, Y. Qiao, L. She, R. Shen, J. Ren, and Y. Zhang, "Two Time-Scale Resource Management for Green Internet of Things Networks," in *IEEE Internet of Things Journal*, vol. 6, no. 1, pp. 545–556, Feb. 2019, ISNN: 2327-4662-2327-2541, doi: 10.1109/JIOT.2018.2842766.

9. C. Zhu, V. C. M. Leung, L. Shu, and E. C. H. Ngai, "Green Internet of Things for Smart World," in *IEEE Access*, vol. 3, pp. 2151–2162, 2015, ISNN: 2169-3536, doi: 10.1109/ACCESS.2015.2497312.
10. X. Zhai, X. Guan, C. Zhu, L. Shu, and J. Yuan, "Optimization Algorithms for Multiaccess Green Communications in the Internet of Things," in *IEEE Internet of Things Journal*, vol. 5, no. 3, pp. 1739–1748, June 2018, ISNN: 2327-4662-2372-2541, doi: 10.1109/JIOT.2018.2792300.
11. Y. Chen et al., "Time-Reversal Wireless Paradigm for Green Internet of Things: An Overview," in IEEE Internet of Things Journal, vol. 1, no. 1, pp. 81–98, Feb. 2014, ISNN: 2327-4662:2372-2541, doi: 10.1109/JIOT.2014.2308838.
12. F. K. Shaikh, S. Zeadally, and E. Exposito, "Enabling Technologies for Green Internet of Things," in *IEEE Systems Journal*, vol. 11, no. 2, pp. 983–994, June 2017, ISNN: 1932-8184-1937-9234, doi: 10.1109/JSYST.2015.2415194.
13. L. Nie et al., "Intrusion Detection in Green Internet of Things: A Deep Deterministic Policy Gradient-Based Algorithm," in *IEEE Transactions on Green Communications and Networking*, vol. 5, no. 2, pp. 778–788, June 2021, ISNN: 2473-2400, doi: 10.1109/TGCN.2021.3073714.
14. T. Guo, K. Yu, G. Srivastava, W. Wei, L. Guo, and N. N. Xiong, "Latent Discriminative Low-Rank Projection for Visual Dimension Reduction in Green Internet of Things," in *IEEE Transactions on Green Communications and Networking*, vol. 5, no. 2, pp. 737–749, June 2021, ISNN: 2473-2400, doi: 10.1109/TGCN.2021.3062972.
15. Y. Dai, and L. Lyu, "NOMA-Enabled CoMP Clustering and Power Control for Green Internet of Things Networks," in *IEEE Access*, vol. 8, pp. 90109–90117, 2020, ISNN: 2169-3536, doi: 10.1109/ACCESS.2020.2992648.
16. Y. Ren, X. Zhang, T. Wu, and Y. Tan, "In-Network Caching for the Green Internet of Things," in *IEEE Access*, vol. 9, pp. 76413–76422, 2021, ISNN: 2169-3536, doi: 10.1109/ACCESS.2021.3082532.
17. X. Li et al., "Physical Layer Security of Cognitive Ambient Backscatter Communications for Green Internet-of-Things," in *IEEE Transactions on Green Communications and Networking*, vol. 5, no. 3, pp. 1066–1076, Sept. 2021, ISSN: 2473-2400, doi: 10.1109/TGCN.2021.3062060.
18. N. Hu, Z. Tian, X. Du, N. Guizani, and Z. Zhu, "Deep-Green: A Dispersed Energy-Efficiency Computing Paradigm for Green Industrial IoT," in *IEEE Transactions on Green Communications and Networking*, vol. 5, no. 2, pp. 750–764, June 2021, ISSN: 2473-2400, doi: 10.1109/TGCN.2021.3064683.
19. A. Ranjha, and G. Kaddoum, "Quasi-Optimization of Uplink Power for Enabling Green URLLC in Mobile UAV-Assisted IoT Networks: A Perturbation-Based Approach," in *IEEE Internet of Things Journal*, vol. 8, no. 3, pp. 1674–1686, 1 Feb. 2021, ISNN: 2327-4662-2372-2541, doi: 10.1109/JIOT.2020.3014039.
20. H. A. B. Salameh, M. Bani Irshaid, A. Al Ajlouni, and M. Aloqaily, "Energy-Efficient Cross-Layer Spectrum Sharing in CR Green IoT Networks," in *IEEE Transactions on Green Communications and Networking*, vol. 5, no. 3, pp. 1091–1100, Sept. 2021, ISNN: 2473-2400, doi: 10.1109/TGCN.2021.3076695.
21. K. E. Jeon, J. She, J. Xue, S. -H. Kim, and S. Park, "luXbeacon—A Batteryless Beacon for Green IoT: Design, Modeling, and Field Tests," in *IEEE Internet of Things Journal*, vol. 6, no. 3, pp. 5001–5012, June 2019, ISNN: 2327-4662-2372-2541, doi: 10.1109/JIOT.2019.2894798.
22. A. Kaur, and S. K. Sood, "Cloud-Centric IoT-Based Green Framework for Smart Drought Prediction," in *IEEE Internet of Things Journal*, vol. 7, no. 2, pp. 1111–1121, Feb. 2020, ISSN: 2327-4662-2372-2541, doi: 10.1109/JIOT.2019.2951610.

23. Z. He, Y. Zhang, B. Tak, and L. Peng, "Green Fog Planning for Optimal Internet-of-Thing Task Scheduling," in *IEEE Access*, vol. 8, pp. 1224–1234, 2020, ISSN: 2169-3536, doi: 10.1109/ACCESS.2019.2961952.
24. A. Tan, S. Wang, N. Xin, Y. Shi, and Y. Peng, "A Multi-Channel Transmission Scheme in Green Internet of Things for Underground Mining Safety Warning," in *IEEE Access*, vol. 8, pp. 775–788, 2020, ISNN: 2169-3536, doi: 10.1109/ACCESS.2019.2959659.
25. M. Kaur, D. Singh, V. Kumar, B. B. Gupta, and A. A. Abd El-Latif, "Secure and Energy Efficient-Based E-Health Care Framework for Green Internet of Things," in *IEEE Transactions on Green Communications and Networking*, vol. 5, no. 3, pp. 1223–1231, Sept. 2021, ISSN: 2473-2400, doi: 10.1109/TGCN.2021.3081616.
26. X. Xiang, J. Gui, and N. N. Xiong, "An Integral Data Gathering Framework for Supervisory Control and Data Acquisition Systems in Green IoT," in *IEEE Transactions on Green Communications and Networking*, vol. 5, no. 2, pp. 714–726, June 2021, ISNN: 2473-2400, doi: 10.1109/TGCN.2021.3068257.
27. Q. Zhou, Y. Sun, H. Lu, and K. Wang, "Learning-Based Green Workload Placement for Energy Internet in Smart Cities," in *Journal of Modern Power Systems and Clean Energy*, vol. 10, no. 1, pp. 91–99, Jan. 2022, ISBN: 2196-5625-2196-5420, doi: 10.35833/MPCE.2020.000271.
28. M. Wen, S. Lin, K. J. Kim, and F. Ji, "Cyclic Delay Diversity with Index Modulation for Green Internet of Things," in *IEEE Transactions on Green Communications and Networking*, vol. 5, no. 2, pp. 600–610, June 2021, ISBN: 273-2400, doi: 10.1109/TGCN.2021.3067705.
29. W. Yang, H. Huang, X. Jing, Z. Li, and C. Zhu, "Social Interaction Assisted Resource Sharing Scheme for Device-to-Device Communication Towards Green Internet of Things," in *IEEE Access*, vol. 8, pp. 71652–71661, 2020, ISSN: 2169-3536, doi: 10.1109/ACCESS.2020.2986785.
30. Z. Li, Y. Liu, A. Liu, S. Wang, and H. Liu, "Minimizing Converge-Cast Time and Energy Consumption in Green Internet of Things," in *IEEE Transactions on Emerging Topics in Computing*, vol. 8, no. 3, pp. 797–813, 1 July–Sept. 2020, ISBN: 2168-6750, doi: 10.1109/TETC.2018.2844282.
31. J. Sharma, and P. S. Mehra, "Secure Communication in IOT-Based UAV Networks: A Systematic Survey," in *Internet of Things*, vol. 23, p. 100883, Oct. 2023, doi: 10.1016/J.IOT.2023.100883.
32. M. R. Khosravi, and S. Samadi, "BL-ALM: A Blind Scalable Edge-Guided Reconstruction Filter for Smart Environmental Monitoring Through Green IoMT-UAV Networks," in *IEEE Transactions on Green Communications and Networking*, vol. 5, no. 2, pp. 727–736, June 2021, ISBN: 2473-2400, doi: 10.1109/TGCN.2021.3067555.
33. J. Ma, S. Shi, S. Gu, N. Zhang, and X. Gu, "Age-Optimal Mobile Elements Scheduling for Recharging and Data Collection in Green IoT," in *IEEE Access*, vol. 8, pp. 81765–81775, 2020, ISBN: 2169-3536, doi: 10.1109/ACCESS.2020.2990931.
34. M. Muniswamaiah, T. Agerwala, and C. C. Tappert, "Green Computing for Internet of Things," 2020 7th IEEE International Conference on Cyber Security and Cloud Computing (CSCloud)/2020 6th IEEE International Conference on Edge Computing and Scalable Cloud (EdgeCom), New York, NY, 2020, pp. 182–185, ISBN: 978-1-7281-6550-9, doi: 10.1109/CSCloud-EdgeCom49738.2020.00039.
35. R. Caso et al., "Antennas and Photovoltaic Panels: Toward a Green Internet of Things," 2015 IEEE 2nd World Forum on Internet of Things (WF-IoT), Milan, Italy, 2015, pp. 128–131, ISBN: 978-1-5090-0366-2, doi: 10.1109/WF-IoT.2015.7389039.
36. R. C. Shit, S. Sharma, M. S. Obaidat, and D. Puthal, "Adaptive Software Defined Node Deployment for Green Internet of Things," ICC 2020 – 2020 IEEE International

Conference on Communications (ICC), Dublin, Ireland, 2020, pp. 1–6, ISBN: 978-1-7281-5089-5, doi: 10.1109/ICC40277.2020.9149396.

37. A. Rehman, K. Haseeb, T. Saba, and H. Kolivand, "M-SMDM: A Model of Security Measures Using Green Internet of Things with Cloud Integrated Data Management for Smart Cities," *Environmental Technology & Innovation*, vol. 24, p. 101802, 2021, ISSN 2352-1864, doi: 10.1016/j.eti.2021.101802.
38. K. Haseeb, I. Ud Din, A. Almogren, I. Ahmed, and M. Guizani, "Intelligent and Secure Edge-Enabled Computing Model for Sustainable Cities Using Green Internet of Things," *Sustainable Cities and Society*, vol. 68, p. 102779, 2021, ISSN 2210-6707, doi: 10.1016/j.scs.2021.102779.
39. M. Lu, G. Fu, N. Beshir Osman, and U. Konbu, "Green Energy Harvesting Strategies on Edge-Based Urban Computing in Sustainable Internet of Things," *Sustainable Cities and Society*, vol. 75, p. 103349, 2021, ISSN: 2210-6707, doi: 10.1016/j.scs.2021.103349.
40. Z. Du, "Energy Analysis of Internet of Things Data Mining Algorithm for Smart Green Communication Networks," *Computer Communications*, vol. 152, pp. 223–231, 2020, ISSN: 0140-3664, doi: 10.1016/j.comcom.2020.01.046.
41. K. Haseeb, S. Lee, and G. Jeon, "EBDS: An Energy-Efficient Big Data-Based Secure Framework Using Internet of Things for Green Environment," *Environmental Technology & Innovation*, vol. 20, p. 101129, 2020, ISSN: 2352-1864, doi: 10.1016/j.eti.2020.101129.
42. L. Farhan, R. Kharel, O. Kaiwartya, M. Hammoudeh, and B. Adebisi, "Towards Green Computing for Internet of Things: Energy Oriented Path and Message Scheduling Approach," *Sustainable Cities and Society*, vol. 38, pp. 195–204, 2018, ISSN: 2210-6707, doi: 10.1016/j.scs.2017.12.018.
43. S. Din, A. Ahmad, A. Paul, and S. Rho, "MGR: Multi-Parameter Green Reliable Communication for Internet of Things in 5G Network," *Journal of Parallel and Distributed Computing*, vol. 118, no. 1, pp. 34–45, 2018, ISSN: 0743-7315, doi: 10.1016/j.jpdc.2017.12.012.
44. S. Muthuswamy, and P. Ganapathi, "Internet of Things – An Overview," Conference: UGC Sponsored Two Day National Conference on Internet of Things, Avinashilingam Institute for Home Science and Higher Education for Women University, 2016, www.researchgate.net/publication/301887107_Internet_of_Things_-_An_Overview
45. M. Alba, "IoT Trends: Manufacturing and Automation to Lead IoT Growth into 2017 and Beyond," 2017, engineering.com, Canada, www.engineering.com/story/iot-trends-manufacturing-and-automation-to-lead-iot-growth-into-2017-and-beyond.
46. BMJ Editorial Team, "Enterprise IoT Market 2019–2027," *Bitcoin Market Journal*, 2022, www.bitcoinmarketjournal.com/report-iot-tokens/enterprise-iot-market-2019-2027/.
47. R. Arshad, S. Zahoor, M. Ali Shah, A. Wahid, and H. Yu, "Green IoT: An Investigation on Energy Saving Practices for 2020 and Beyond," *IEEE Access*, vol. 5, pp. 15667–15681, 2017, doi: 10.1109/ACCESS.2017.2686092.

7 Network Optimization for Internet of Things (IoT)

Techniques and Algorithms

Aryaman Sharma

7.1 INTRODUCTION

7.1.1 Overview

To connect consumers and data providers, data packets with unique Internet Protocol (IP) addresses were initially sent over the Internet. Technology improvements have fueled the proliferation of Internet of things (IoT) devices and their numerous uses, resulting in a huge network of resource-limited gadgets. The IoT network faces a substantial difficulty in handling the significant amount of data these devices create. It is essential to create solutions that solve several network-related problems, such as energy efficiency, routing, heterogeneity, congestion, scalability, quality of service (QoS), dependability, and security, to fully utilize the network's potential. Within the context of IoT devices, optimization is essential to improving the performance and effectiveness of IoT systems. In order to minimize data transfer, cut down on power use, improve network connection, boost security, and make use of edge computing, several techniques and algorithms must be used. The chapter will examine various optimization strategies for IoT systems, including resource optimization, network optimization, and data optimization. Each kind of optimization has unique needs that call for particular algorithms and methods. We examine the various forms of IoT system optimizations in this chapter, including resource, network, data, security, and performance optimization. We look at the optimization strategies that can raise the effectiveness and performance of IoT systems. Comparing the numerous algorithms of IoT optimization approaches on the basis of their strengths, weaknesses, performances, and types, we can design more efficient, dependable, and secure IoT networks that can meet the increasing demands of IoT applications and devices [1, 2].

 DOI: 10.1201/9781003405535-9

7.2 TYPES OF OPTIMIZATION TECHNIQUES IMPLEMENTED TO OPTIMIZE IOT DEVICES

7.2.1 Resource Optimization

Efficiently managing resources is crucial for optimizing IoT systems and maximizing their utilization. In an IoT environment, numerous interconnected devices generate vast amounts of data that require processing, storage, and transmission. By effectively managing resources, we can enhance the efficiency, reliability, and scalability of IoT networks, ensuring the successful deployment of IoT systems. To optimize resources in IoT devices, we can focus on two levels: the Edge and Cloud tiers. This method reduces latency and bandwidth requirements by sending less data to the cloud for processing. Moreover, edge computing enhances reliability and availability by allowing IoT devices to function even when the network connection is lost. Researchers like M. Peyman et al., for example, have looked into the advantages of Edge, Fog, and Cloud computing, particularly in the context of growing data generation and data-driven projects in smart cities [3].

On the other hand, cloud computing offers a flexible and scalable platform for data processing, analysis, and storage. We can efficiently store, process, and grow huge volumes of data with cloud computing. Advanced security and compliance features, as well as streamlined management and monitoring capabilities, are all provided by cloud computing. Kaium et al.'s thorough analysis concentrates on various cloud data compression and storage optimization strategies, examining their impacts and possible future paths while emphasizing their importance for IoT data management, compression, and mining for intelligent decision support systems [4].

To optimize resources in IoT networks, federated learning develops as a machine learning technique. In contrast to centralized training on a single server, federated learning enables edge devices to perform training using local data, ensuring user privacy. The resulting models are then aggregated on a central server. By aggregating the models, achieving a balance between local and global model accuracy is possible. There is growing interest in data-driven deep learning (DL) systems that use Deep Reinforcement Learning (DRL) to optimize resource allocation due to the rising costs and scalability problems associated with deploying IoT sensors. In this approach, DL is used to learn network condition representations and interactions with the dynamic environment to optimize decision-making.

7.2.2 Network Optimization

IoT network optimization is essential to ensure the effective use of network resources and to minimize the impact of IoT-generated traffic on other services as the number of IoT devices grows quickly. IoT-produced traffic is heterogeneous regarding applications and device kinds, in contrast to traffic created by cellular networks. It is crucial to control IoT traffic and monitor how well IoT products and services are working. Although IoT applications only produce a modest quantity of data, adding devices increases traffic because of control plane messages, straining the network significantly. Providing a reliable system to handle and enhance control plane messages from IoT devices is crucial. Given the anticipated connectivity of billions of IoT devices to worldwide networks in the next years, academics and operators are paying more attention to the need for network optimization in IoT [5].

7.2.3 Data Optimization

To ensure efficient resource usage, lessen network congestion, and improve system performance, efficient data optimization is essential in IoT networks. The exponential expansion of data creation brought on by the spread of IoT devices presents management, processing, and analytical issues for network operators. Therefore, to reduce the amount of data carried over the network and enhance overall performance, effective data optimization techniques are needed. Recent years have substantially increased research interest in energy-efficient sensory data collecting. Various strategies have been suggested to do this. One such strategy is a two-step procedure that entails cluster construction and data collection [6]. IoT nodes take part in selecting a cluster head in the first step. A cluster's nodes gather data in the second stage and send it to the cluster head, which then gets in touch with the base station. However, the cluster creation process becomes resource- and time-intensive in scenarios with frequent IoT node movement or large-scale IoT networks. Data compression is the focus of a different strategy to cut down on data transmission requirements. However, these techniques could cause data distortion and might not be appropriate when the network has a high data flow [7].

7.3 ALGORITHMS USED FOR OPTIMIZATION

List of Optimization Algorithms	Particle Swarm Optimization (PSO)	Genetic Algorithms	Ant Colony Optimization (ACO)	Neural Network (NN)
Overview	The Particle Swarm Optimization (PSO) is a nature-inspired algorithm designed to solve complex optimization problems. PSO employs "particles" representing potential solutions that navigate through multidimensional search space seeking an optimal solution. Each particle's position adjustments depend on its best-known solution obtained and the swarm's overall.	Genetic algorithms are a series of algorithms inspired by natural evolutionary and selection processes. Working iteratively, they evolve groups or populations of individual solutions. Each iteration involves selecting segments from this population to generate offspring for upcoming generations,	ACO is a population-based stochastic optimization approach in which the optimization problem has been modified to find an optimal path on a weighted graph. These "ants" gradually develop solutions by moving down the graph.	Neural networks, at the core of deep learning, employ interconnected artificial neurons to optimize their parameters and minimize a loss function. The network modifies its weights and biases via methods such as gradient descent,

List of Optimization Algorithms	**Particle Swarm Optimization (PSO)**	**Genetic Algorithms**	**Ant Colony Optimization (ACO)**	**Neural Network (NN)**
	known best solution The repeated updating of particles' positions aims at converging toward global optimum, which makes PSO remarkably beneficial in problems involving optimization.	gradually resulting in increasingly optimized solution development over multiple iterations as part of systematic "evolution."		enabling it to solve various optimization problems effectively.
Application in IoT	PSO can be used to optimize the distribution of network resources in the IoT. It can assist in optimally allocating bandwidth, power, and computational resources to fulfill the needs of IoT applications and devices.	Genetic algorithms help study feasible/ optimal IoT device placement position and resource allocation. Their ability to find globally optimal solutions makes them well-suited for tackling optimization problems often found in IoT networks with numerous parameters and constraints.	This technique is an excellent approach to increase the scalability and efficiency of Internet of things systems; it takes inspiration from how ants find their way to food sources. It aids in the optimal deployment and routing of devices and data flows in the dynamic and variable environment of Internet of Things networks, reducing congestion, improving network performance, and ensuring dependable operations.	NNs can be utilized for various tasks in IoT network optimization. They can learn patterns in data traffic, predict network congestion, and optimize resource allocation based on real-time conditions.

(*continued*)

List of Optimization Algorithms	Particle Swarm Optimization (PSO)	Genetic Algorithms	Ant Colony Optimization (ACO)	Neural Network (NN)
Advantages	PSO can investigate various options because it is simple to build and provides a global search function. It has good convergence capabilities and is appropriate for continuous optimization issues.	Suitability for parallel computing, high reliability, adaptability, and versatility to handle different domain-specific topics all make Genetic algorithms suitable for IoT network optimization.	ACO is adaptive and decentralized, making it suitable for dynamic and distributed environments like IoT networks. It can handle complex routing problems and find near-optimal solutions.	NNs can learn from data, adapt to changing network conditions, and make intelligent decisions. They can capture complex patterns and relationships in IoT data, leading to efficient network optimization.
Limitations	PSO may suffer from premature convergence, where the algorithm gets stuck in a suboptimal solution. It requires careful parameter tuning and may struggle with high-dimensional problems.	In order to make the Genetic algorithm effective, tuning process involving the parameter tuning should be thorough for greater algorithm performance. This tuning process can be time consuming and may require pre-requisite knowledge on the problem.	ACO relies on the evaporation of pheromone trails, which can lead to slow convergence and difficulties in dealing with dynamic network conditions. It requires balancing exploration and exploitation.	NNs require a large amount of labeled training data and extensive computational resources for training. They may also face challenges in interpretability and vulnerability to adversarial attacks.

7.3.1 Particle Swarm Optimization (PSO)

PSO can be implemented in IoT networks to optimize data routing, energy usage, and resource allocation. For instance, the path selection in data routing can be optimized using PSO. This is important for reducing energy consumption and maximising network lifespan. PSO can also be employed to optimize the energy usage of IoT devices by dynamically altering the duty cycle of the devices based on the network traffic and the device's battery level. With PSO, IoT networks can achieve higher performance, scalability, and efficiency. PSO is a flexible and successful optimization method for the IoT network since it permits the optimization of various parameters. It can assist overcome the problems faced by IoT devices, such as limited processing power, memory, and storage capacity. PSO has the potential to greatly enhance IoT network performance, making them more dependable and long-lasting [8].

A swarm of particles traveling through a multidimensional search space is what the Particle Swarm Optimization (PSO) algorithm mimics. Each particle stands for a possible answer to the optimization issue. The algorithm iteratively adjusts the particles' velocities and positions based on the particles' individual and collective best positions at that point. The influence of the particle's optimum positions on its movement is governed by the cognitive and social learning components (c_1 and c_2). Exploration and exploitation are balanced according to the inertia weight (w). The particles eventually arrive at the best outcome as they explore and update the search space. When a predefined termination condition is met, usually after a predetermined number of iterations, the algorithm ends.

Algorithm: Particle Swarm Optimization (PSO)

1. Initialize swarm of particles with arbitrary positions and velocities as $X_i(t)$ and $V_i(t)$, respectively.
2. Position and Velocity of the particle is given by $X_i(t) = (x_i(t), y_i(t))$ and $V_i(t) = (v_{ix}(t), v_{iy}(t))$, respectively
3. Repeat until the termination condition is met:

Update each particle's velocity:

$$V_i(t+1) = \omega \cdot V_i(t) + c_1 \cdot u_1(t) \cdot (p_{best,i}(t) - X_i(t)) + c_2 \cdot u_2(t) \cdot (p_{best\,swarm}(t) - X_i(t))$$

- where: $u_1(t)$ is the random number which takes a value between 0 and 1.
  $u_2(t)$ is the random number which takes a value between 0 and 1.
- Update each particle's position: $Xi(t+1) = X_i(t) + V_i(t+1)$, where $X_i(t)$ denote the position of particle i in the search space at time step t; unless otherwise stated, t denotes discrete time steps. The position of the particle is changed by adding a velocity, $V_i(t)$, to the current position.
- Update the best position and fitness value for each particle if necessary.

4. Return the best solution found.

An innovative compute offloading mechanism for the IoT has been developed by Aichuan Li and colleagues using an upgraded PS O method in edge computing. This method addresses the problems of high energy consumption and time delay caused by traditional offloading techniques. Work offloading systems are optimized via the PSO method, which computes fitness values and updates particle positions through encoding. This updates particle locations and significantly reduces energy consumption during job allocation. The optimization objective function, which considers constraints from several perspectives, is based on the multiuser personalized compute job load model and the uplink transmission communication model. Simulation findings demonstrate that the proposed technique works better than the ARIMA-BP [9] and cross-server computation algorithms [10] when 80 mobile devices are utilized. The recommended method's average time delay and energy consumption are 0.205 seconds and 0.2 J, respectively. Consequently, this approach can reduce job execution latency while consuming less energy [11].

7.3.2 Genetic Algorithms

Genetic algorithms' main objective is to mimic the evolutionary process of natural selection by applying a set of rules to generate successive generations of solutions better suited to the current problem. When genetic algorithms are used for IoT optimization, a fitness function is used to evaluate the worth of a solution. The fitness function considers some characteristics, including throughput, network latency, energy usage, and other performance metrics. Finding a set of solutions that meet the optimization requirements and lower total operating expenses is the target. Due to their ability to handle challenging optimization problems, numerous studies have shown genetic algorithms to be crucial tools for IoT network optimization. Genetic algorithms can swiftly find the best answers while exploring large solution areas efficiently. They can also alter their parameters as the network environment evolves for better solutions. Genetic algorithms are capable of managing several objectives. The fitness function can be designed to simultaneously maximize a number of objectives, such as cutting down on network latency and energy usage. Evolutionary algorithms can be used to optimize IoT networks with many competing objectives because of their adaptability. Since genetic algorithms require a lot of work, it may take some time before the optimization process converges to the optimal outcomes. This is a major issue for IoT networks with limited processing power and battery life [12, 13]. The algorithm's performance is significantly influenced by the choices regarding factors like population size, mutation rate, and selection criteria. Selecting the ideal parameters can take a while and involve a lot of trial and error [14].

Algorithm: Genetic Algorithm (GA)

1. Initialize a population of individuals randomly:
   Population = {Individual$_1$, Individual$_2$, …, Individual$n$}
2. Fitness evaluation of each individual:
   Fitness(Individual$_x$) for $x = 1, 2, \ldots, n$
3. Repeat until termination condition is met:
   - Select parents based on fitness:
     Parents = Selection(Population)
   - Create offspring through crossover and mutation:
     Offspring = Crossover(Parents)
     Offspring = Mutation(Offspring)
   - Evaluate the fitness of the offspring:
     Fitness(Offspring$_x$) for $x = 1, 2, \ldots,$ OffspringSize
   - Replace some individuals in the population with the offspring:
     Population = Replacement(Population, Offspring)
4. Return the best individual found.

Research was done by A.V. Dhumane et al. to develop an efficient routing algorithm that would allow precise data transfer in an IoT-based smart object network. The suggested algorithm is a context-based routing algorithm that strongly emphasizes improving network performance. The program chooses the cluster head and calculates its energy cost using Energy-based Based Clustering and the K-means approach. The optimal path is then determined by considering the energy cost at the cluster head, the sensor nodes' dependability, and the path's length. The suggested method also uses Genetic algorithm-based optimization to increase the dependability of the selected path [15].

7.3.3 Ant Colony Optimization

An artificial ant colony is employed in an ACO algorithm to search the solution space and determine the best route for data packets through the network. By leaving behind pheromone trails for other ants to follow, each ant in the colony acts as a possible solution and moves through the network. Stronger pheromone trails indicate a better road, and they serve as a representation of a path's desirability. Ants walk through a network, leaving behind pheromones that enhance the attractiveness of the trails they travel. Ants can find new, possibly more efficient paths by following the pheromone trails created by other ants [16] (Figure 7.1).

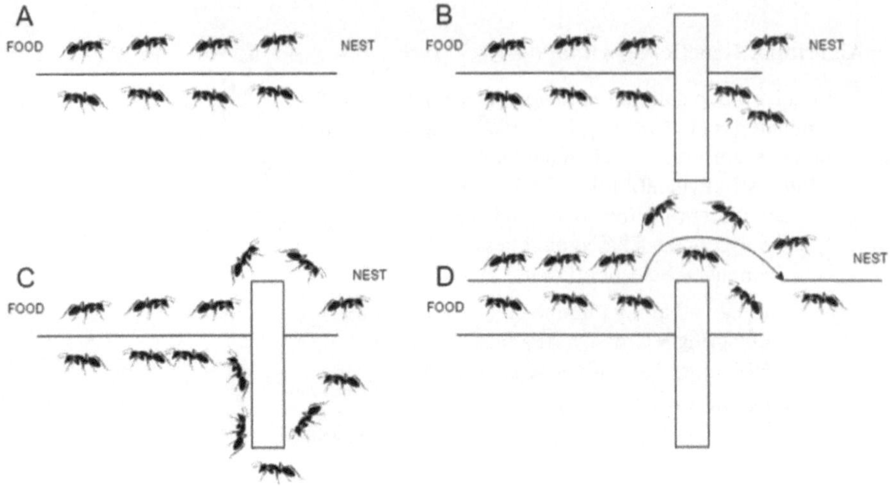

FIGURE 7.1 Ant colony optimization.

Algorithm: Ant Colony Optimization (ACO)

1. Initialization:
   - Set the pheromone trails on each edge (x, y) to an initial value τ_{xy}. τ_{xy} is the amount of pheromone deposited for transition from state.
2. Repeat until the termination condition is met:
   a. Create a set of ants, each starting from a random vertex.
   b. For each ant:
      - Construct a solution by moving the ant along the graph.
      - At each vertex, select the next vertex based on the probability function $P(x, y)$:

$$P(x,y) = \frac{\left(\tau_{xy}^{\alpha}\right)\cdot\left(\eta_{xy}^{\beta}\right)}{\Sigma\left(\tau_{xy}^{\alpha}\right)\cdot\left(\eta_{xy}^{\beta}\right)}$$

      - where α and β are parameters.
   c. Update the pheromone trails based on the ants' solutions:
      - Evaporate the existing pheromone trails: $\tau_{xy} \leftarrow (1-\rho)\cdot\tau_{xy}$, where ρ is the evaporation rate.
      - Deposit new pheromone trails: For each ant's solution, deposit an amount of pheromone $\Delta\tau_{xy}$.
   d. Check termination condition.

The pheromone trails on each edge of the graph are given an initial value at the beginning of the algorithm. These paths serve as the ants' means of communication, directing their travel and exploration. A collection of ants is formed in each cycle, and each ant begins at a different random vertex in the graph. The ants move iteratively through the graph to create their solutions. The next vertex at each point is decided probabilistically based on the edge's pheromone level and a heuristic value. The parameters control the pheromone and heuristic information's respective weights in decision-making. Pheromone trails are updated once all ants have built their solutions. While fresh pheromone trails are deposited dependent on the caliber of the solutions discovered by the ants, the current pheromone trails progressively disappear to prevent premature convergence. The evaporation rate determines the amount of pheromone decay, and the potency of each ant's solution affects how much pheromone is deposited. The algorithm iterates until a termination condition is satisfied, which could be reached after a predetermined number of iterations or by arriving at a satisfactory solution. The optimal answer to the optimization problem is the one that was discovered after most iterations.

ACO can be used in the context of IoT devices to optimize the routing of data packets between devices in a reliable and energy-efficient way. ACO algorithms produce synthetic pheromone trails between IoT devices to simulate the attractiveness of various communication channels, much as how ants leave pheromone trails to communicate with one another. The ACO algorithm deems a path more desirable the stronger the pheromone trail. Ants navigate the network by following the pheromone trails left by preceding ants as the ACO algorithm runs. They do this to investigate several paths between the devices, and the algorithm eventually converges on the best route for data transmission between the devices. The path for data transfer between devices in the IoT is critical to optimize for some reasons. Because IoT devices have limited resources, energy must be managed wisely to increase device longevity. ACO can assist in cutting down on energy use by sending data over the most energy-efficient routes [17].

7.3.4 Neural Networks

There are several ways that neural networks can be applied to enhance the functionality of IoT systems. First, predictive maintenance of IoT devices using neural networks can help to decrease downtime and increase overall system reliability. Neural networks can forecast when a device is likely to fail and plan repair before the failure occurs by analyzing sensor data. This can lower repair costs and eliminate unplanned downtime. In IoT systems, neural networks can also be utilized to optimize energy utilization. Neural networks can find chances to minimize energy usage without impacting system performance by looking for patterns in energy consumption data. This can aid in lowering running expenses and enhancing the long-term viability of IoT systems. The detection of anomalies is another use for neural networks in IoT optimization. Neural networks can find unusual patterns that might point to a security breach or other system issue by examining data from various sensors. Data breaches and other security events may be avoided as a result. In IoT networks, neural networks can be utilized for traffic prediction and congestion control. Neural networks can forecast future traffic flows and manage network resources by evaluating past traffic patterns, helping to avoid congestion and guarantee the network's smooth operation. Neural networks are an effective tool for enhancing the performance of IoT systems since they can help to increase system reliability, lower energy consumption, avoid security breaches, and increase network efficiency by evaluating data and spotting trends.

Algorithm: Neural Network Training

1. Initialize the neural network weights and biases:
   - Randomly initialize the weights $W^{(0)}$ and biases $b^{(0)}$ of the neural network.
2. Repeat until termination condition is met:
   a. Forward Propagation:
   Compute the predicted outputs y_{pred} of the neural network with activation function $g(z)$ for a given input x using the following equations:

$$z^{(1)} = W^{(1)} \cdot a^{(0)} + b^{(1)}$$

$$a^{(1)} = g\left(z^{(1)}\right)$$

   Where l denotes the layer index, $a^{(0)}$ is the input layer, and $g()$ is the activation function.

   b. Calculate Loss:
   - Compute the loss L between the predicted outputs y_pred and the actual outputs y using a suitable loss function.

   c. Backpropagation:
   Calculate the gradients of the loss with respect to the weights and biases using the following equations:

$$\frac{\partial L}{\partial z^{(L)}} = \left(y_{pred} - y\right) \cdot g'\left(z^{(L)}\right)$$

$$\frac{\partial L}{\partial W^{(l)}} = \frac{\partial L}{\partial z^{(l)}} \cdot (a^{(l-1)})^T$$

$$\frac{\partial L}{\partial b^{(l)}} = \frac{\partial L}{\partial z^{(l)}}$$

   d. Update Weights and Biases:
   Update the weights and biases using the following equations:

$$W^{(l)} = W^{(l)} - \eta \cdot \frac{\partial L}{\partial W^{(l)}} b^{(l)} = b^{(l)} - \eta \cdot \frac{\partial L}{\partial b^{(l)}}$$

   where η is the learning rate.
3. Return the trained neural network.

The algorithm is first used to initialize the weights and biases of the neural network. The initial values of these parameters are selected at random. Every iteration of the method employs forward propagation to determine the predicted outputs of the neural network. The activation function is used after the bias term has been included and the weighted sum of the outputs from the previous layer has been calculated. The appropriate loss function is then used to calculate the output difference between the expected and actual outputs. Next, backpropagation is used to compute the gradients of the loss concerning the weights and biases. This requires that the error be computed at the output layer, propagated backward through the network, and considered when calculating the activation function's derivatives. The weights and biases are then updated via gradient descent, with the updated step size determined by the learning rate. The algorithm doesn't stop iterating unless a termination condition is met, like reaching a maximum number of iterations or achieving a satisfactory degree of accuracy. The final result of the procedure is the trained neural network.

7.4 LIMITATIONS IN OPTIMIZATION TECHNIQUES

7.4.1 Lack of Standardization in IoT Devices

IoT device standardization or compatibility problems result from the fact that different companies frequently employ various hardware and software technologies when designing and producing IoT devices. IoT devices might not be interoperable with one another due to this non-concurrent manufacturing, making it difficult to optimize them for a particular application. An optimization algorithm might not function with other IoT devices that utilize other communication protocols or data formats, for instance, if created to work with a particular type of IoT device. Because they might not be portable between various devices, this can limit the scalability of optimization strategies for IoT devices. Different IoT devices can connect and share data effectively by adhering to standards for communication protocols, data formats, and security procedures. Without these standards, the interoperability of IoT devices may be limited, making it difficult to tailor them for particular purposes. Standardization of communication protocols, data formats, and security protocols is essential to addressing the compatibility issue in IoT device optimization. To increase interoperability and compatibility, open-source software platforms and libraries that support a variety of IoT devices are also being developed. Additionally, by utilizing widely accepted standardized data formats and communication protocols in the IoT sector, developers can create optimization algorithms with compatibility in mind. This strategy can boost the scalability of optimization techniques for IoT devices and ensure that optimization algorithms can be used across various IoT devices.

7.4.2 Limitations in Processing Power of IoT Devices

One of the major obstacles to implementing optimization techniques is the processing power limitation in IoT devices. The processing, memory, and storage capabilities of many IoT devices are constrained, making it difficult to run sophisticated optimization algorithms that need a lot of computing power. The necessity for IoT devices to be lightweight, portable, and energy-efficient restricts the resources and power available for computational operations. One of the key concerns with processing power constraints in IoT devices is that optimization techniques that demand vast amounts of data and computational resources to interpret information effectively may not function well. For instance, in order to accurately train models, neural networks or DL algorithms need a lot of data and processing capacity. Results from these techniques may be less precise or take longer to compute when used on IoT devices with constrained processing resources.

Some approaches to solving this issue have been discussed. A number of IoT nodes should share the computational workload, according to a suggestion by J. Coelho et al. This fosters a cooperative atmosphere that enables the delivery of more complex and resource-intensive services that would otherwise be impractical or ineffective [18].

7.4.3 Scalability of IoT Networks

The number of devices that need to be connected to the network is growing quickly as IoT devices expand. However, the network's performance may deteriorate as the number of connected devices rises, resulting in bottlenecks, lag times, and greater energy usage. The fact that many IoT devices have constrained processing, memory, and storage capabilities, making it challenging to handle massive amounts of data, can compound this limitation.

A novel scheduling method dubbed the Long Hop (LH) first algorithm has been created to prioritize data that travels great distances and must visit a greater number of devices in order to address the scalability issue. For wireless networks, a brand-new scheduling technique known as the Long Hop (LH) first algorithm was developed to increase the effectiveness of data transmission. Data packets that travel farther and must link to more devices in order to reach their destination are given higher priority using this strategy. The LH algorithm prioritizes packets that must make more or longer hops to reach the target node. This prioritizing is carried out by calculating a metric based on the distance, the number of intermediate nodes, and the number of devices to be accessed for each packet. The transmission order is then configured to send the packet with the highest metric first. It outperforms two competing scheduling algorithms, nearest job next (NJN) and first-come, first-served (FCFS), regarding power consumption and network speed. The LH algorithm reduces power consumption by up to 56 while increasing network throughput by up to 62.50%. Additionally, it lessens packet delay and loss, speeds up transmission, and extends the network's lifespan. These simulation-based results demonstrate the effectiveness of the LH method [19].

7.4.4 Dynamic Nature of IoT Environments

Through IoT, which is connecting more and more of the globe, optimization approaches have become crucial for ensuring effective operations and optimal resource usage. The dynamic nature of IoT networks, however, represents one of the most important limits in the optimization strategies of IoT devices and applications. It can be difficult to optimize the performance of the entire system because IoT devices are frequently deployed in erratic and continuously changing environments. This is due to how quickly and unexpectedly the network topology, device connectivity, and data flow can change, necessitating regular revisions to the optimization approaches used. For instance, moving an IoT device to a different place or adding a new device to the network may disturb the current network topology and need to be adjusted to improve performance. Additionally, controlling and optimizing IoT networks is more difficult due to the heterogeneity of IoT devices and applications. It is challenging to establish optimal resource usage and to balance the workload across devices in the network since IoT devices originate from many vendors, employ various communication protocols, and have diverse processing capabilities. For instance, whereas some IoT devices may have low processing capability, others may have more sophisticated processing capabilities, necessitating the use of various optimization approaches [20, 21].

REFERENCES

1. P. N. Beuchat, H. Hesse, A. Domahidi, and J. Lygeros, "Enabling Optimization Based Localization for IoT Devices," *IEEE Internet of Things Journal*, pp. 1–1, 2019, https://doi.org/10.1109/jiot.2019.2904559.
2. K. Hossain, M. Rahman, and S. Roy, "IoT Data Compression and Optimization Techniques in Cloud Storage," *International Journal of Cloud Applications and Computing*, vol. 9, no. 2, pp. 43–59, Apr. 2019, https://doi.org/10.4018/ijcac.2019040103.
3. M. Peyman, P. J. Copado, R. D. Tordecilla, L. do C. Martins, F. Xhafa, and A. A. Juan, "Edge Computing and IoT Analytics for Agile Optimization in Intelligent Transportation Systems," *Energies*, vol. 14, no. 19, p. 6309, Oct. 2021, https://doi.org/10.3390/en14196309.
4. K. Hossain, M. Rahman, and S. Roy, "IoT Data Compression and Optimization Techniques in Cloud Storage: Current Prospects and Future Directions," *International Journal of Cloud Applications and Computing (IJCAC)*, vol. 9, no. 2, pp. 43–59, 2019 Available: https://ideas.repec.org/a/igg/jcac00/v9y2019i2p43-59.html
5. N. N. Srinidhi, S. M. Dilip Kumar, and K. R. Venugopal, "Network Optimizations in the Internet of Things: A Review," *Engineering Science and Technology, an International Journal*, vol. 22, no. 1, pp. 1–21, Feb. 2019, https://doi.org/10.1016/j.jestch.2018.09.003.
6. A. V. Dhumane, R. S. Prasad, and J. R. Prasad, "An Optimal Routing Algorithm for Internet of Things Enabling Technologies," *International Journal of Rough Sets and Data Analysis*, vol. 4, no. 3, pp. 1–16, Jul. 2017, https://doi.org/10.4018/ijrsda.2017070101.
7. A. Salim, W. Osamy, A. Aziz, and A. Khedr, "SEEDGT: Secure and Energy Efficient Data Gathering Technique for IoT Applications based WSNs," *Journal of Network and Computer Applications*, 2022, Accessed: Feb. 25, 2023. [Online]. Available: www.semanticscholar.org/paper/SEEDGT%3A-Secure-and-energy-efficient-data-gathering-SalimOsamy/a3db1392127c9b1a5ee8ab8288f3b8a908c5fd7c
8. Manju, S. Singh, A. K. Dwivedi, A. K. Sharma, and P. S. Mehra, "Learning Automata Based Heuristics for Target Q-Coverage," in *ICRITO 2020 – IEEE 8th International Conference on Reliability, Infocom Technologies and Optimization (Trends and Future Directions)*, Institute of Electrical and Electronics Engineers Inc., Jun. 2020, pp. 170–173. https://doi.org/10.1109/ICRITO48877.2020.9197923.
9. M. Zhao and K. Zhou, "Selective Offloading by Exploiting ARIMA-BP for Energy Optimization in Mobile Edge Computing Networks," *Algorithms*, vol. 12, no. 2, p. 48, Feb. 2019, https://doi.org/10.3390/a12020048.

10. Y. Shi, Y. Xia, and Y. Gao, "Cross-Server Computation Offloading for Multi-Task Mobile Edge Computing," *Information*, vol. 11, no. 2, p. 96, Feb. 2020, https://doi.org/10.3390/info11020096.
11. A. Li, L. Li, and S. Yi, "Computation Offloading Strategy for IoT Using Improved Particle Swarm Algorithm in Edge Computing," *Wireless Communications and Mobile Computing*, vol. 2022, pp. 1–9, Feb. 2022, https://doi.org/10.1155/2022/9319136.
12. J. Sharma and P. S. Mehra, "Secure Communication in IOT-based UAV Networks: A Systematic Survey," *Internet of Things*, vol. 23, p. 100883, Oct. 2023, https://doi.org/10.1016/J.IOT.2023.100883.
13. D. Chawla and P. S. Mehra, "A Roadmap from Classical Cryptography to Post-Quantum Resistant Cryptography for 5G-Enabled IoT: Challenges, Opportunities and Solutions," *Internet of Things*, vol. 24, p. 100950, Dec. 2023, https://doi.org/10.1016/J.IOT.2023.100950.
14. A. Tam, "A Gentle Introduction to Particle Swarm Optimization," *Machine Learning Mastery*, Sep. 15, 2021. https://machinelearningmastery.com/a-gentle-introduction-to-particle-swarm-optimization/
15. A. K. Dwivedi, P. S. Mehra, O. Pal, M. N. Doja, and B. Alam, "EETSP: Energy-Efficient Two-Stage Routing Protocol for Wireless Sensor Network-Assisted Internet of Things," *International Journal of Communication Systems*, vol. 34, no. 17, p. e4965, Nov. 2021, https://doi.org/10.1002/DAC.4965.
16. A. Datta and S. Nandakumar, "A survey on bio inspired meta heuristic based clustering protocols for wireless sensor networks," IOP Conference Series: Materials Science and Engineering, 2017. [Online]. Available: www.semanticscholar.org/paper/A-survey-on-bio-inspired-meta-heuristic-based-for-DattaNandakumar/b787a0f04c14979bd67fab746354a7173818c785
17. B. Shi and Y. Zhang, "A Novel Algorithm to Optimize the Energy Consumption Using IoT and Based on Ant Colony Algorithm," *Energies*, vol. 14, no. 6, p. 1709, Mar. 2021, https://doi.org/10.3390/en14061709.
18. J. Coelho and L. Nogueira, "Enabling Processing Power Scalability with Internet of Things (IoT) Clusters," *Electronics*, vol. 11, no. 1, p. 81, Dec. 2021, https://doi.org/10.3390/electronics11010081.
19. L. Farhan, A. E. Alissa, S. T. Shukur, M. Hammoudeh, and R. Kharel, "An energy efficient long hop (LH) first scheduling algorithm for scalable Internet of Things (IoT) networks," 2017 Eleventh International Conference on Sensing Technology (ICST), Dec. 2017, https://doi.org/10.1109/icsenst.2017.8304511.
20. R. Mahmud, R. Kotagiri, and R. Buyya, "Fog Computing: A Taxonomy, Survey and Future Directions," *Internet of Things*, pp. 103–130, Oct. 2017, https://doi.org/10.1007/978-981-10-5861-5_5.
21. R. Kushwah, P. K. Batra, and A. Jain, "Internet of Things Architectural Elements, Challenges and Future Directions," *IEEE Xplore*, Mar. 1, 2020. https://ieeexplore.ieee.org/abstract/document/9182773.

Part III

Congestion Control and Heterogeneity in Intelligent IoT Network

8 Network Optimization on Internet of Medical Things

B. Santhosh and K. Viswanath

8.1 INTRODUCTION

IoMT (Internet of Medical Things) refers to embedding the medical devices and different connectivity devices with the internet to facilitate data collection, processing, analysis, and transmitting of the medical data, and to monitor constant patient health conditions for better telemedicine applications [1]. Real-time monitoring of vital signs, including blood pressure, heart rate, and oxygen saturation levels, is possible with the help of medical IoT devices and systems, as shown in Figure 8.1 [2] and notify healthcare professionals of potential health issues and aid in administering medical care.

This technology may lead to better patient outcomes, more effective healthcare, and lower healthcare costs [3]. However, it also poses substantial difficulties for data security, privacy, and regulatory compliance. It is crucial to overcome these issues to ensure the successful adoption of IoT in the healthcare industry. IoMT (Internet of Medical Things) utilizes a variety of networks, each with unique properties and benefits. The organization of a chapter on network optimization in IoMT is as outlined Brief overview of IoMT and its significance in healthcare; Introduction to network optimization in IoMT is discussed in the introduction.

The challenges in handling large volumes of medical data, data privacy and security concerns in IoMT, regulatory compliance and standards in healthcare data transmission, and network architecture and components in IoMT are explained in the scope of IoMT.

IoMT-specific networking technology, how hardware solutions affect data transmission efficiency, network optimization software solutions, and an overview of IoMT network optimization software tools and methods are discussed.

A description of tools for network monitoring and traffic management, software's role in organizing and prioritizing data traffic strategies for dynamic resource allocation, IoMT dynamic resource allocation, and how automatic allocation and real-time monitoring improve network performance is provided.

DOI: 10.1201/9781003405535-11

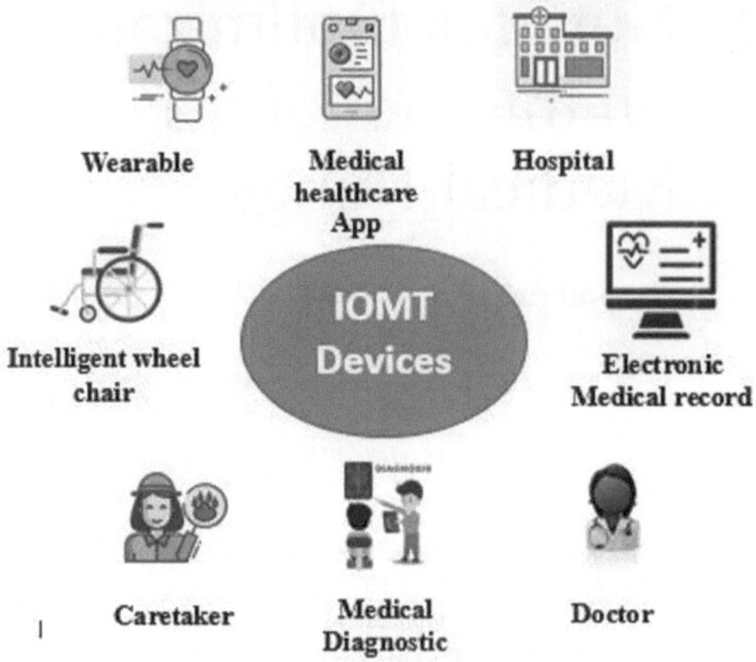

FIGURE 8.1 IoMT devices.

An introduction to machine learning and its function in improving IoMT networks, how machine learning improves network efficiency and minimizes congestion through predictive maintenance, traffic analysis, and anomaly detection is provided.

Display samples of IoMT network optimization in action, highlight effective tactics, obstacles overcome, results attained, and future trends and directions. Consideration of 5G, edge computing, and artificial intelligence (AI) advancements, Predictions for the evolution of network optimization strategies. Networks of the following popular types are frequently used in IoMT:

1. **Local Area Network (LAN):** A local area network, or LAN, links equipment in a confined space, like a hospital or clinic. Within a single location, it is frequently used to link computers, medical equipment, and other devices.
2. **Wide-Area Network (WAN):** A wide-area network, or WAN, connects devices across a large geographic area, such as a country or region. It is frequently used for remote patient monitoring, sending medical data over great distances between patients and healthcare providers.
3. **Personal Area Network (PAN):** A PAN is a network that links objects that are close together or within a short distance, as inside someone's body. It is frequently employed for sensors and wearable medical equipment that send data to a nearby smartphone or tablet.
4. **Wireless Sensor Network (WSN):** A WSN is a group of wireless devices working together to monitor and gather information about a certain area or

the surrounding environment [4]. In hospitals and clinics, it is frequently used for remote environmental monitoring of factors like temperature and humidity levels [5].

5. **Cloud-based Network:** A network architecture that is built on cloud computing technologies is known as a cloud-based network. It makes it possible for IoMT devices to send data to a centralized cloud-based server for analysis, processing, and storage. This kind of network is frequently used to store vast amounts of medical data and to analyze data in real time.

8.1.1 Need for Network Optimization in Medical IoT

The massive volumes of data created by medical devices and the requirement for this data to be delivered securely and effectively to healthcare professionals create the demand for network optimization in the medical IoT. Managing and configuring the network infrastructure to make it quick, dependable, and secure is known as network optimization. Making sure the network can manage the enormous amounts of data generated by medical devices and other IoT-related devices is one of the main goals of network optimization in the medical industry.

Wearable medical devices generate enormous amounts of data as in Figure 8.2, and timely patient care depends on the efficient transmission of this data [6]. Network

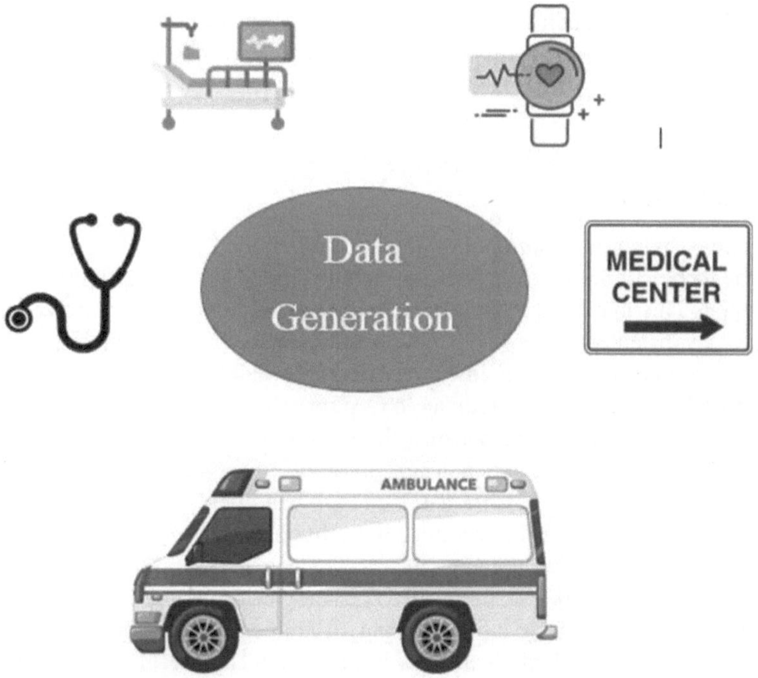

FIGURE 8.2 Data generation by medical devices.

optimization makes sure that the network can manage the load of data created and that data can be sent without any hiccups or delays.

In order to ensure that medical data is transmitted safely in the medical IoT, network optimization is also essential. Medical information is extremely private and sensitive, and any data breach could have serious repercussions for patient safety and privacy. Strong security protocols are put in place as part of network optimization to safeguard medical data from unauthorized access or interception.

Additionally, network optimization reduces maintenance and upkeep costs while decreasing patient care interruptions. This is crucial for remote patient monitoring, as patient treatment might be seriously delayed when data transmission is delayed.

8.1.2 Importance of Network Optimization in Medical IoT

IoT network optimization is essential for the medical industry because it guarantees that the network can handle the large amounts of data that wearables and medical devices produce. Managing and configuring the network infrastructure to make it quick, dependable, and secure is known as network optimization. In the context of medical IoT, the network must be capable of handling substantial amounts of data produced by several devices, frequently in real time. Network optimization makes certain that the data can be transmitted effectively and without any hiccups or delays. This is crucial for remote patient monitoring, in which severe healthcare delays can result from transmission delays. The secure transmission of medical data is another benefit of network optimization. Medical information is extremely private and sensitive, and any data breach could have serious repercussions for patient safety and privacy. Strong security protocols are put in place as part of network optimization to safeguard medical data from unauthorized access or interception.

8.2 UNDERSTANDING NETWORK CONGESTION IN MEDICAL IOT

When there is a significant amount of data created by medical devices and wearables that needs to be communicated over the network, medical IoT network congestion may happen. This may cause data transmission to be delayed or interrupted, which could have a substantial impact on patient care. To ensure that the network can handle the volume of data created by medical devices and wearables, it is essential to understand network congestion in medical IoT. When the network infrastructure is not functioning properly, it can be because the volume of data being transferred is too much for it [7]. This may happen when there are too many devices attempting to transmit data at once or when there is not enough bandwidth to support the data load. Data transmission delays or interruptions brought on by network congestion can have a serious negative impact on patient care.

Network congestion in the medical IoT sector can have serious repercussions, especially for remote patient monitoring. Healthcare professionals might not get timely notifications if medical data is not provided in real time, which could cause serious delays in patient care. For instance, if a patient's vital signs point to a medical emergency, a delay in data transmission could cause a delay in the healthcare providers' response, which could have a negative impact on the patient.

8.2.1 Indicators of Network Congestion

Medical IoT network congestion can have detrimental effects by causing disruptions or delays in the transmission of medical data, potentially leading to life-threatening delays in patient care [8]. Network administrators can recognize possible problems and take corrective action before they have an impact on patient care by being aware of the signs of network congestion. Some signs of medical IoT network congestion are shown in Figure 8.3 [9] in care for the patient.

One of the most frequent signs of network congestion in medical IoT is slow data delivery. Data transmission can become slow and cause delays in the transfer of medical data when network traffic exceeds the capacity that is available.

i. **Increased Latency:** Latency is the length of time that data must travel over the network between devices. Another sign of network congestion in medical IoT is increased latency. Data packets may be lost or delayed on a busy network, increasing latency.

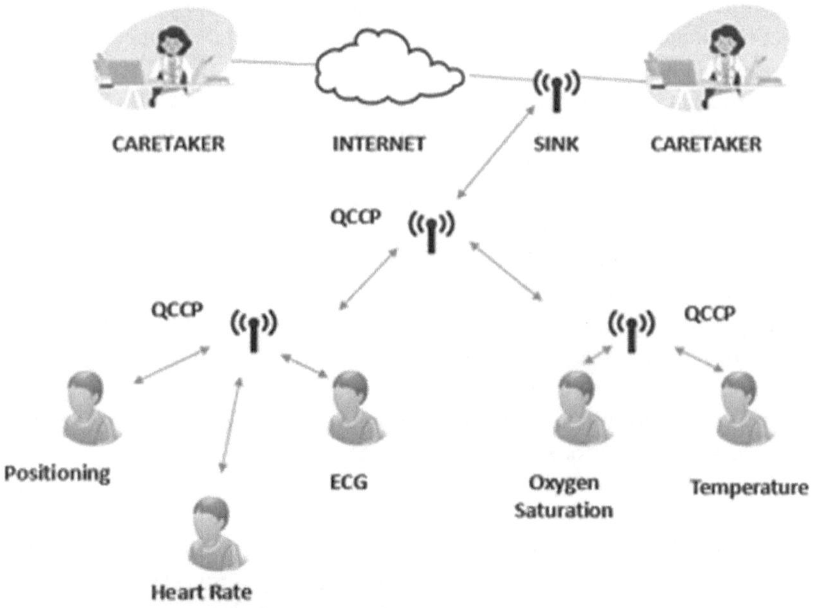

FIGURE 8.3 Signs of medical IoT network congestion.

Some signs of medical IoT network congestion

Data packets can be lost during transmission owing to network congestion or other problems, which is known as packet loss. The transmission of medical data may be delayed due to packet loss, which could have serious consequences for patient care.

ii. **Increased Network issues:** When the network is busy, issues like failed transmissions or damaged data packets can happen. Increased network errors may be a sign of congestion and may need to be fixed to avoid affecting patient care.
iii. **High CPU** (Central Processing Unit) **Consumption:** High CPU consumption on network hardware, such as switches and routers, maybe a sign of a congested network. Network devices may get overloaded and experience excessive CPU utilization when network traffic exceeds the bandwidth that is available.

8.2.2 Causes of Network Congestion

When there is a significant amount of data created by medical devices and wearables that needs to be communicated over the network, medical IoT network congestion may happen. The following categories can be used to classify the root causes of network congestion:

i. **High Network Traffic:** One of the most frequent reasons for network congestion in medical IoT is high network traffic. The network infrastructure may get overloaded if too many devices are transmitting data at once, which could cause delays or interruptions in data flow.
ii. **Insufficient Bandwidth:** Another typical reason for network congestion in medical IoT is insufficient bandwidth. Network congestion can happen when there is not enough bandwidth available to manage the data load produced by wearables and medical devices, which can cause delays or disruptions in the data transmission process.
iii. **Network Topology:** Network congestion in medical IoT can also be influenced by the network topology or the actual design of the network infrastructure. The effective transmission of data can be hampered by bottlenecks and congestion sites caused by improperly planned or configured network topologies.
iv. **Network Security**: Network congestion in medical IoT can also be caused by network security protocols. Security procedures can slow down data transfer and cause delays in medical treatment when they are too restrictive or improperly designed.
v. **Network Devices:** In medical IoT, network hardware like routers, switches, and firewalls can potentially cause network congestion. These devices may create a network bottleneck when they become old or overworked, causing data transmission to be delayed or interrupted.

8.2.3 Network Congestion on Medical Impact of IoT Systems

Medical IoT systems may be significantly impacted by network congestion, which could jeopardize patient outcomes and delay the delivery of medical care. The following are some effects of network congestion on medical IoT systems:

i. **Delayed Data Transmission:** Data transmission can be slow when the network is busy, which can cause delays in the provision of medical care. When making decisions about the care of patients in an emergency, rapid access to medical data is crucial. This delay can be very important.
ii. **Data Loss:** When medical data packets are dropped as a result of network congestion or other problems, this can also cause data loss. Important medical data may be lost as a result, which may have a negative impact on patient outcomes.
iii. **Erroneous Data:** Medical data packets may become corrupted as a result of network congestion, producing erroneous or incomplete data. Incorrect diagnoses and treatments can result from inaccurate data, potentially placing patients in danger.

    System downtime, in which Medical IoT systems are unavailable or offline, can occur when network congestion is significant. This could be disastrous in an emergency case where prompt access to medical information is essential to making wise patient care decisions.
iv. **Increased Costs:** By delaying or disrupting patient care, network congestion can significantly raise the price of healthcare. This may lead to longer hospital stays, more costly diagnostic procedures, and other related costs.

8.3 NETWORK OPTIMIZATION STRATEGIES

Network performance improvement techniques, notably in Medical IoT systems, are referred to as "network optimization strategies." These techniques aim to minimize the danger of network congestion and guarantee prompt patient treatment by guaranteeing that the network infrastructure can effectively handle the volume of data generated by wearables and medical devices. Among the important network optimization techniques applied in Medical IoT systems are:

i. **Bandwidth management:** It entails managing network resources to maximize performance and reduce congestion. This can be done via strategies like bandwidth allocation, which guarantees that vital applications have enough bandwidth, or traffic shaping, which gives priority to specific types of traffic.
ii. **Network Monitoring:** To monitor network performance and locate potential congestion locations, one must use specialist software tools [10]. Thus, IT staff may proactively deal with problems before they have an impact on patient care.
iii. **Network Segmentation:** It is the process of breaking the network infrastructure up into smaller, easier-to-manage portions. Restricting the volume of

traffic on any one segment and facilitating more effective traffic routing can help alleviate congestion [11].

iv. **Quality of Service (QoS):** QoS is a collection of methods used to give certain network traffic more priority than others [12]. In medical IoT systems, where vital medical data must be communicated in real time to guarantee prompt patient treatment, this can be very significant.

v. **Network Redundancy:** To ensure that crucial data can still be transmitted in the case of a network outage or other disturbance, network redundancy entails creating duplicate or backup network infrastructure.

8.3.1 Hardware Solutions

In Medical IoT systems, network optimization includes hardware solutions as a key component. These approaches frequently necessitate the employment of specialist networking hardware created to address the particular requirements of medical device data transmission.

Among the important hardware components utilized in Medical IoT systems are:

i. **Routers:** Routers are networking tools that link various network segments and ease data transfer between them. Routers can be used in Medical IoT systems to segment the network and prioritize traffic carrying vital medical data, lowering the likelihood of delays and congestion.

ii. **Switches:** Switches are networking tools that link various elements of a network and allow data to be sent back and forth between them. Switches can be employed in Medical IoT systems to guarantee that vital medical data is transmitted effectively and in real time, reducing the possibility of delays and data loss.

iii. **Wireless Access Points:** These are networking tools that allow wireless devices to connect to the network. Wireless access points can be used in Medical IoT systems to connect wearable medical devices and other wireless medical devices to the network, enabling real-time patient data monitoring.

iv. **Gateways:** Devices that connect several network types and allow data to be sent between them are known as gateways. Gateways can be used in Medical IoT systems to link various medical devices and wearables to the network, facilitating effective data management and transfer.

v. **Load Balancers:** Traffic in the network is split equally using these networking tools. Given the substantial amount of critical medical data stored in the cloud, a load balancer is essential in medical IoT systems to alleviate congestion and minimize delays.

8.3.1.1 Upgrading Network Infrastructure

Data from medical devices and wearables are increasing day by day, creating traffic in the network and resulting in difficulty handling the real-time data leading to congestion and delay. Delay plays a major role in medical IoT which needs to be addressed by upgrading network infrastructure.

The following are some of the main advantages of modernizing network infrastructure in Medical IoT systems:

i. **Increased Bandwidth:** Increasing the bandwidth lowers congestion and avoids delays. Providing security to the network against threats can also support efficient data transfer.
ii. **Enhanced Reliability**: In order to upgrade network infrastructure in Medical IoT systems, some important procedures must be taken. Prior to upgrading the network infrastructure, it is crucial to evaluate the current infrastructure and pinpoint any areas that require improvement. This may entail evaluating network security, potential congestion sites, and network performance analysis.
iii. **Making a Plan:** After evaluating the current infrastructure, a plan may be made to upgrade the network infrastructure. It can be necessary to upgrade current hardware and software, add new gear and software, or change the network architecture.

    Upgrading the network infrastructure is possible once the plan has been created. The network may need to be configured, new hardware and software installed, and the new infrastructure tested to make sure it is operating as planned.

8.3.1.2 Implementing Traffic Management Techniques

Another important tactic for network optimization in Medical IoT systems is the application of traffic management strategies. Utilizing traffic management techniques, it is possible to ensure that important medical data is prioritized and transferred quickly while blocking or deprioritizing non-important data.

Among the important traffic management strategies applied in Medical IoT systems is:

Service quality (QoS) – a networking protocol called Quality of Service enables network managers to give some traffic types priority over others. Critical medical data traffic can be prioritized in Medical IoT systems using QoS, ensuring that it is transmitted quickly and effectively [13].

Network administrators can manage the flow of traffic across a network by using a method called traffic shaping. Flow shaping can be used in Medical IoT systems to restrict the bandwidth utilized by non-critical data flow, lowering the likelihood of congestion and delays.

Packet filtering is a security method that involves preventing traffic based on predetermined standards. Packet filtering can be used in Medical IoT systems to restrict non-critical data traffic or communication from illegal devices, lowering network congestion risk and enhancing network security.

Administrators can divide a network into smaller virtual networks using virtual LANs (VLANs), a networking technology. The danger of congestion and delays can be decreased in Medical IoT systems by using VLANs to isolate vital medical data traffic from non-critical data traffic.

Allocating a particular amount of bandwidth to various types of traffic is a practice known as bandwidth allocation. Bandwidth allocation can be used in Medical IoT systems to guarantee that vital medical data traffic is given enough bandwidth, ensuring that it is transmitted effectively and in real time.

8.3.2 Software Solutions

Software solutions can be very important in optimizing networks in Medical IoT systems, in addition to hardware solutions and upgrading network infrastructure. The following are some essential software programs that can be utilized to enhance networks in the medical IoT:

Software for Network monitoring: Network administrators can keep an eye on network performance and traffic in real time, thanks to network monitoring software. This can assist in locating network bottlenecks and congestion areas, enabling administrators to address them before they worsen. Network Optimization Software: By lowering latency, increasing bandwidth utilization, and giving priority to important data, network optimization software is intended to optimize network traffic and performance [14]. By regulating the movement of traffic throughout the network, this software can also aid in the reduction of network congestion.

Software for Load Balancing: By distributing network traffic among several servers or devices, load-balancing software helps to improve network performance. As a result, Medical IoT solutions may operate more reliably and help avoid network congestion.

Software for Protocol Optimization: Software for protocol optimization is created to enhance the protocol for data transmission across the network [15]. Reducing latency and enhancing network speed can make it simpler to communicate vital medical data in real time.

Network Security Software: Network security software can help shield against cyber risks and stop unwanted access to Medical IoT systems. This software can assist prevent network congestion brought on by malicious assaults and increase the overall dependability of Medical IoT devices by enhancing network security.

8.3.2.1 Dynamic Resource Allocation

Another important network optimization technique in Medical IoT systems is dynamic resource allocation. It entails dynamically distributing computing resources according to the current network demands. By deprioritizing or blocking non-critical data, this strategy can ensure that important medical data is prioritized and transferred quickly. Key components of dynamic resource allocation in medical IoT include the following:

Real-time observation Dynamic resource allocation must be implemented while network traffic and resource usage are being monitored in real time. Administrators can pinpoint bottlenecks and congestion locations and allocate resources appropriately by continuously monitoring the network.

Automated Resource Allocation: Using software algorithms, automated resource allocation dynamically distributes computing resources in accordance with network demands. By blocking or deprioritizing non-essential data traffic, it can ensure that critical medical data traffic is prioritized and conveyed effectively.

Prioritizing Critical Data: To ensure that it is transmitted effectively and in real time, critical medical data traffic should be prioritized over non-critical data traffic. To detect vital data traffic and deploy resources appropriately, this necessitates extensive network monitoring and analysis.

Burden Balancing: To lessen the burden on any one device and enhance network performance, load balancing involves dividing network traffic among several servers or devices. By doing so, network congestion can be avoided and effective transmission of vital medical data traffic can be guaranteed.

Redundancy and Failover: In Medical IoT systems, redundancy and failover systems are essential elements of dynamic resource allocation. Administrators can guarantee that crucial medical data traffic is not lost in the case of a system breakdown or outage by establishing backup systems and failover techniques.

8.3.2.2 Machine Learning Algorithms

A sort of AI called machine learning algorithms can be employed in medical IoT systems to optimize networks. Administrators can optimize network performance and avoid network congestion by using these algorithms to analyze vast volumes of data, find trends, and make predictions.

Here are some examples of how machine learning techniques can be utilized for medical IoT network optimization:

Prevention-Based Maintenance: When network equipment is likely to malfunction or need maintenance, machine learning algorithms can be utilized to make predictions [16]. Administrators can prevent network congestion and guarantee that vital medical data is transmitted effectively by proactively identifying issues before they arise.

Traffic Analysis: Machine learning techniques can be used to examine patterns of network traffic and forecast when network congestion is most likely to happen. Administrators can prevent congestion and guarantee that traffic for vital medical data is prioritized by recognizing bottleneck locations in advance and taking appropriate action.

Machine learning techniques can be used to identify unusual network traffic that might point to a security breach or other network issue. Administrators can avoid network congestion and guarantee that vital medical data is transferred effectively by spotting and responding to irregularities in real time [17].

Machine learning techniques can be used to optimize QoS factors as bandwidth, latency, and packet loss. Administrators can guarantee that vital medical data traffic is prioritized and transported effectively by dynamically modifying QoS parameters based on network conditions [18].

Network traffic can be prioritized using machine learning algorithms according to its importance. Administrators can ensure that vital medical data traffic is transmitted quickly and effectively by giving it priority while deprioritizing non-critical data traffic to avoid network congestion.

8.3.3 Combination of Hardware and Software Solutions

An efficient method for network optimization in Medical IoT systems is the combination of hardware and software solutions. Administrators may improve network speed, avoid network congestion, and guarantee that vital medical data is transferred effectively by combining the best features of both hardware and software. The following are some strategies for combining hardware and software solutions for network optimization in medical IoT:

High-speed network interfaces are a useful tool for establishing quick and dependable connectivity between network devices. Administrators can lower latency and enhance network performance by utilizing hardware solutions like fiber-optic cable and high-speed switches.

Traffic Shaping: Using traffic-shaping software, network traffic can be prioritized according to its importance. Administrators can make sure that vital medical data traffic is prioritized and transmitted effectively while blocking or deprioritizing non-critical data traffic by implementing traffic-shaping techniques on network devices.

Stress Balancing: To lessen the stress on any one device and enhance network performance, load-balancing gear can be used to divide network traffic across numerous servers or devices. Network device managers can avoid Network latencies and guarantee the swift transmission of vital medical data by adopting load-balancing algorithms.

Virtualization: Virtual networks that are isolated from one another can be built using virtualization software. Administrators can avoid network congestion and guarantee that important medical data is transmitted effectively by separating critical from non-critical data traffic.

Machine learning techniques can be used to evaluate network traffic patterns and forecast when network congestion is likely to happen. Administrators can avoid network congestion and guarantee that traffic carrying crucial medical data is prioritized by dynamically modifying network parameters in response to network conditions.

8.4 CONCLUSIONS

Overall, ensuring these systems' effective and reliable operation depends on network optimization in IoT applications. The specific use case and the needs for data transfer, security, and dependability will determine the best network for IoMT.

In order to ensure that medical data is sent effectively, safely, and reliably, network optimization is crucial in medical IoT. To ensure that medical data is delivered effectively and in real time, enabling timely patient treatment and reducing the risk of negative outcomes, it is essential to understand network congestion in medical IoT.

Indicators of network congestion in the Medical IoT include slow data transmission, increased latency, packet loss, increased network faults, and excessive CPU utilization. Network optimization in Medical IoT can be handled by monitoring network traffic, real-time resource utilization, prioritizing critical data, and implementing load balancing, and redundancy. The tools are made up of high-speed network interfaces, traffic-shaping software, virtualization software load-balancing hardware, and machine learning algorithms.

REFERENCES

1. N. Anwar, Z. Khan, M. N. Alshahrani, and I. A. Tawalbeh, "A systematic review of network optimization techniques in the internet of medical things," *IEEE Access*, vol. 8, pp. 235–256, 2020.
2. N. Dilawar, M. Rizwan, S. Akram, and F. Ahamd, "Blockchain: securing Internet of Medical Things (IoMT)," *International Journal of Advanced Computer Science and Applications*, vol. 10, 2019. doi: 10.14569/IJACSA.2019.0100110.
3. Y. A. Sekercioglu, "The internet of things for medical applications," *IEEE Pulse*, vol. 9, no. 1, pp. 22–27, 2018.
4. S. S. Dhillon, S. S. Sohi, and M. S. Obaidat, "Energy-efficient routing algorithms for wireless sensor networks: a survey," *IEEE Communications Surveys and Tutorials*, vol. 18, no. 2, pp. 1229–1263, 2016.
5. H. K. Soni, N. P. Goyal, and M. Singh, "Performance analysis of a wireless sensor network for remote patient monitoring," *IEEE Access*, vol. 6, pp. 76890–76903, 2018.
6. www.altexsoft.com/blog/internet-of-medical-things/
7. P. Misra, V. K. Bali, and P. Kumar, "Impact of network congestion on telemedicine applications: a review," *Journal of Ambient Intelligence and Humanized Computing*, vol. 10, no. 9, pp. 3635–3655, 2019.
8. F. Al-Turjman, S. S. Al-Raweshidy, and A. Al-Abdullah, "Energy-efficient routing protocols for wireless body area networks: a survey," *IEEE Communications Surveys and Tutorials*, vol. 19, no. 2, pp. 921–952, 2017.
9. https://pub.mdpi-res.com/sensors/sensors-23-00923/article_deploy/html/images/sensors-23-00923-g001.png?1673591798
10. P. S. Aher, V. M. Phalle, and K. Wankhade, "A novel routing scheme for IoT-based healthcare systems," *Journal of Ambient Intelligence and Humanized Computing*, vol. 11, no. 7, pp. 2809–2823, 2020.
11. S. K. Das, S. S. K. Patra, and N. K. Suryadevara, "A survey on routing in wireless body area networks," *IEEE Communications Surveys and Tutorials*, vol. 19, no. 1, pp. 399–426, 2016.
12. C. Chang, T. H. Lin, Y. W. Lin, and T. H. Lin, "A machine learning-based approach to improving QoS in the internet of medical things," *IEEE Access*, vol. 8, pp. 22322–22333, 2020.
13. D. Chawla and P. S. Mehra, "QSMAH: a novel quantum-based secure cryptosystem using mutual authentication for healthcare in the internet of things," *Internet of Things*, vol. 24, p. 100949, Dec. 2023, doi: 10.1016/J.IOT.2023.100949.
14. S. Raza, S. A. Zahra, A. Khushnood, M. A. Naeem, and A. U. Rehman, "Performance analysis of deep learning models for predicting network traffic in healthcare IoT systems," *Journal of Ambient Intelligence and Humanized Computing*, vol. 12, no. 11, pp. 11965–11977, 2021.
15. F. Li, C. Liu, Y. Yang, H. Zhang, and S. Li, "A QoS-guaranteed data fusion scheme for IoMT in smart healthcare," *IEEE Access*, vol. 8, pp. 20420–20431, 2020.
16. Jiang, L. Gao, and X. Zhou, "A survey of machine learning for big data processing," *EURASIP Journal on Advances in Signal Processing*, vol. 2017, no. 1, pp. 1–15, 2017.

17. M. A. Alsanie and M. O. Farooq, "A survey on routing techniques for wireless sensor networks," *IEEE Communications Surveys and Tutorials*, vol. 21, no. 1, pp. 834–869, 2019.
18. Q. Zhang, Y. Sun, and Q. Liu, "A QoS-aware task scheduling method based on parallel evolutionary algorithm for IoT healthcare system," *IEEE Access*, vol. 8, pp. 157944–157958, 2020.

Part IV

Scalability, Reliability, and QoS in Intelligent IoT Network

9 Scalable IoT Architecture and Services

Challenges and Opportunities in IoT-Enabled Healthcare Monitoring Frameworks

Mohit Lalit, Gaurav Bathla, and Surender Singh

9.1 IOT: 4TH INDUSTRIAL REVOLUTION IN HEALTHCARE

Industry 4.0 aims to automate and streamline sensor data processing across most industries. Healthcare uses 22% of Internet of things (IoT) applications. Industry 4.0 is leading the healthcare industry with health and fitness tracking applications. Certain technology collects, processes, and acts on health data, and IoT is one of the leading technologies in this domain. Internet technology allows us to access data anytime, anywhere, and IoT has set a new bar for optimality by leveraging itself [1, 2]. As shown in Figure 9.1, IoT enables communication between items with integrated electronics devices to make human life more secure and comfortable.

Figure 9.2 describes the requirement for storage space has increased due to the rise of IoT in many applications, and cloud computing plays a significant role in data storage, analysis, and processing [3]. IoT applications are dynamic; therefore, their needs vary. Li et al. [4] put forward a dynamic solution for such IoT applications that revolve around data with respect to cloud platforms. IoT can succeed if the necessary hardware, software, and infrastructure are found. IoT requires a compatible configuration model; however, all configuration models might not be valid. IoT academics and developers look for flexible answers to IoT's diversity. Certain applications require real-time data processing with high definition and in the shortest amount of time, and cloud computing does not fit into these applications. Instead, these applications require developing concepts like fog computing (FC) for real-time data collection [5]. Due to the worldwide population surge, many people suffer from chronic diseases. The WHO predicts 643 million diabetics by 2030 and 783 million by 2045 [6]. Detecting and treating diseases has saved many lives, including those of TB patients, and from 2000 to 2020, death rates fell by 38% [7]. These services can only be possible with the help of smart healthcare, which can only be possible with the help of continuous IoT tracking. Healthcare 4.0 and Industry 4.0 are among the ideas covered

DOI: 10.1201/9781003405535-13

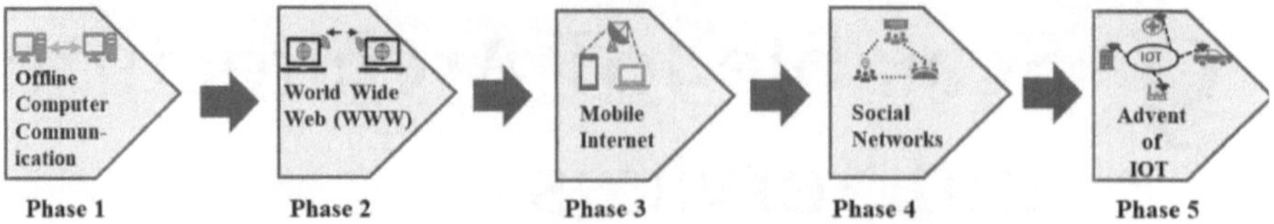

FIGURE 9.1 Development phase of IoT.

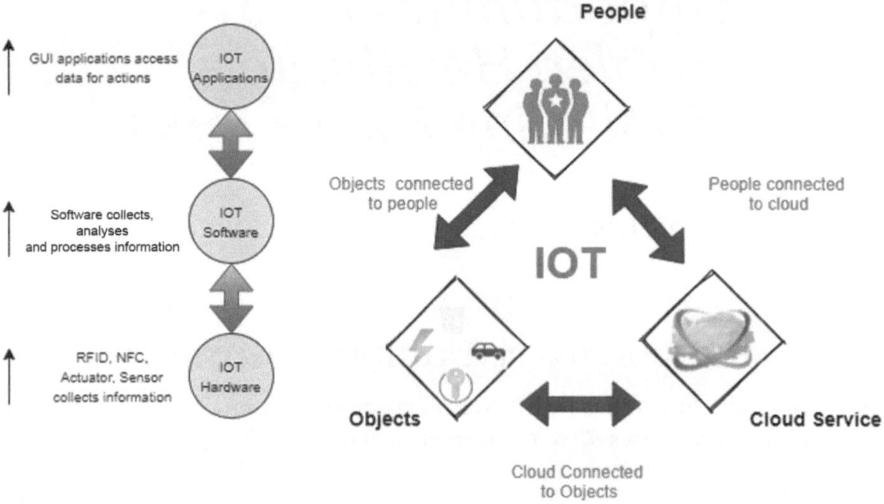

FIGURE 9.2 Working diagram of IoT with Cloud Computing.

in Section 9.1, which serves as an introduction. The conversation eventually turns to the Internet of Health Things (IoHT) and the need for FC in the healthcare industry. The issues facing the healthcare sector and the place of FC in Healthcare 4.0 are covered in Section 9.2. The part also analyses a recently suggested monitoring system made possible by IoT and FC. The analysis in Section 9.3 reveals the gaps in current knowledge, prompting the authors to provide a solution to address the present needs of the healthcare industry in Section 9.4. Conclusion and future application conclude Section 9.5.

9.1.1 Healthcare 4.0 and Industry 4.0

Healthcare 1.0–4.0 follows industry 1.0–4.0. Healthcare 1.0 increased quickly between 1970 and 1990 to increase productivity by automating and minimizing paperwork. Healthcare 2.0, introduced about 2000, overcomes Healthcare 1.0's limitations of standalone systems by integrating health IT systems and clinical imaging integration to produce health data. Between 2006 and 2015, Healthcare 3.0 emerged with electronic health records (EHRs), wearables, and implants, where EHRs enabled genetic data storage [8, 9]. In contrast, Healthcare 4.0 is backed by Industry 4.0 technologies,

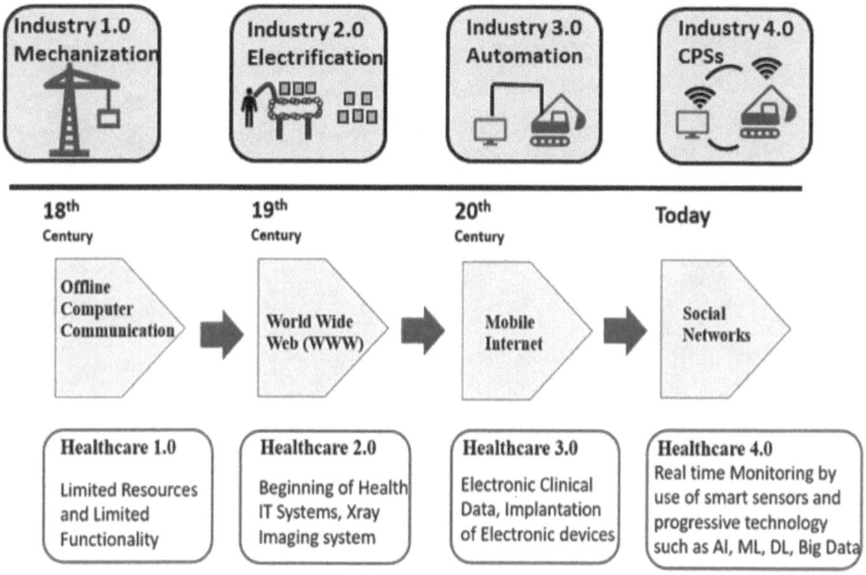

FIGURE 9.3 Healthcare 4.0 and Industry 4.0 Relevancy.

including artificial intelligence (AI), invisible user interfaces, cloud, fog, and edge computing, as seen in Figure 9.3. The ability to collect and share little data is more important than the quality of the information available to clinicians. Data portability allows patients and doctors to access information anytime and everywhere, and increased analytics allow for quick and imaginative diagnosis and treatment [10]. It can help in decision-making and development.

9.1.2 Internet of Health Things (IoHT)

IoT is utilized to control tangible objects via the Internet [11] remotely. The introduction of IoT in healthcare encompasses pharmaceuticals, microbiology, and biomedical engineering through the integration of emerging technologies such as AI, image processing, and big data analytics, among others [12]. Sensors are required to drive physical objects through the internet in smart health, smart home, smart industrial, smart building, smart electrical, and smart traffic management. As shown in Figure 9.4, [13] smart health has transformed the healthcare industry by making patients the primary source of information. Using wireless body sensors, smart health collects data remotely based on symptoms. Figure 9.4(a) depicts an IoT-based e-healthcare application for patients with chronic diseases. Most individuals carry cell phones that transmit body sensor data. Gateway transfers data to a server in the cloud for health record analysis and monitoring. The healthcare system comprises of sensors deployed on the human body to collect information about heart rate, blood pressure, mobility, and sugar levels and transmit the data as per availability of network [14]. Massive storage is required for analysing and automating continuous data [15]. This data is

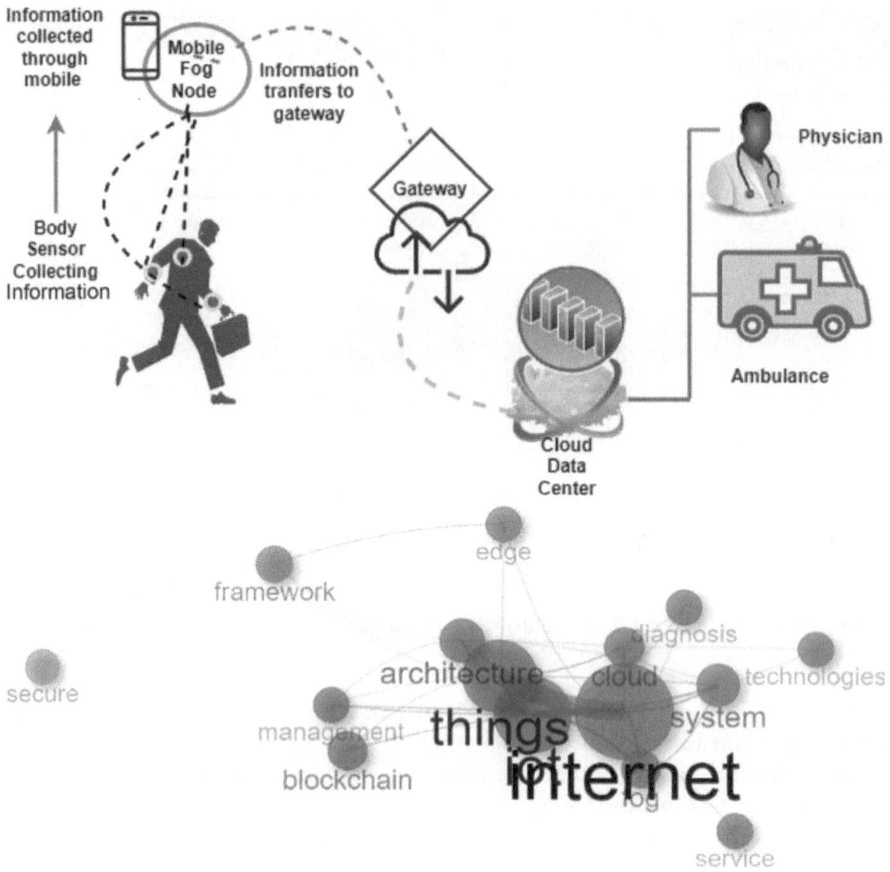

FIGURE 9.4 (a) IoT-enabled E-healthcare scenario. (b) Cooccurrence network.

highly required in the event of a severe medical emergency to acquire various medical services such as ambulance, hospital, physicians, or professional medical team. Such situations necessitate a quick, honest response [16]. Classical centralized cloud-computing-based IoT systems fail to maintain availability, continuity, and dependability when transmitting patient health records, and outsourcing data analytics may reduce accuracy and adaptability in constrained environments [17]. The concept of FC was proposed in 2012, for real-time data delivery with low latency. Figure 9.3 illustrates how FC brings storage closer to sensors in the field [18]. FC came into existence in 2012, while cloud computing is in extreme use in current technologies, and on the other side, edge computing is just a concept. However, in some cases, these concepts overlap each other, and to overcome these things, Table 9.1 differentiates these concepts based on different metrics. The execution and efficiency of any communication system is affected by response time, bandwidth, mobility, latency, dependability, deployment cost, maintenance scalability, network access, and security [19].

9.1.3 Fog Computing and Healthcare 4.0: WWW (Why, Where, and When)

Multiple FC applications influence daily life. This cooccurrence network demonstrates the urgency of FC. This database network diagram for the Web of Science database covers over 130 papers from 2021 to 2023, including 330 healthcare papers. The cooccurrence network in Figure 9.4(b) demonstrates that researchers are very interested in healthcare applications facilitated by FC.

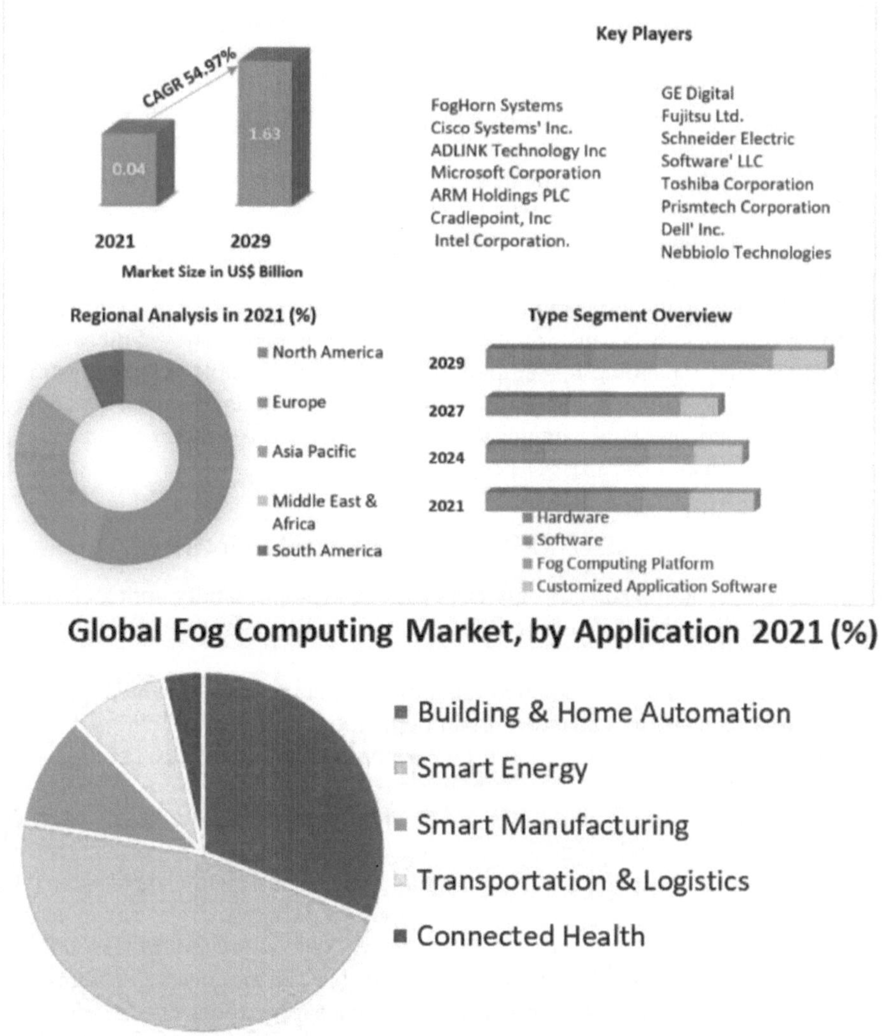

FIGURE 9.5 (a) Global fog computing data. (b) Global fog computing market by application [20].

In Figure 9.5(a), [20] global survey shows FC's enormous market share in diverse applications. CISCO introduced fog computing, which has a clear advantage over cloud computing in various applications.

It is expected that the global fog market will be 1.63 billion US dollars by 2029 which was 0.04 billion US dollars in 2021 as mentioned in Figure 9.5(b), and connected healthcare will be having the largest slice among all other fog-enabled applications. FC has the opportunity as cloud computing becomes more expensive in terms of latency, storage, security, and bandwidth for all Industry 4.0 applications.

9.1.4 Fog Computing Architecture

Access to existing healthcare infrastructure and amenities is hampered by the growing population. Healthcare is one of the most affected applications of FC, yet historically, the industry has been slow to implement technology [21]. IT has impacted inventory management, laboratory systems, and most recently, applications for Industry 4.0 [22]. The 4th industrial revolution encompasses intelligent industry, health, cities, buildings, energy management, and information technology. As time passes, Figure 9.6(a) demonstrates the architecture based on fog-to-fog communication and fog-to-cloud communication where enterprises can be considered as IoT-enabled application users.

FC is utilized in applications requiring real-time data transmission in environments with limited bandwidth because of its advantageous characteristics. Figure 9.6(b) illustrates why the expanding IoT empire utilized FC, while Table 9.1 demonstrates why fog is superior to cloud computing, whereas edge computing is just a concept so far [23]. Fog, cloud, and edge computing are compared based on their technical and financial characteristics in Table 9.1 [24].

9.2 HEALTHCARE INDUSTRY AND THEIR CHALLENGES

Because hospitals and physicians only treat patients who physically approach them, participation in treatment is impeded. Consequently, the patient must endure life-threatening conditions.

The major challenges are limited quality interaction with disabled patients, more adherent patients, an expanded urban population, costly treatment, and a lack of healthcare practitioners due to the growing population day by day mentioned in Figure 9.7 [25, 26].

IoT can acquire patient health information without interfering with their normal routine. It saves time and predicts the health of a patient after a period. Due to resource constraints, healthcare was referred to as Farahani HealthCare Version 1.0. In Healthcare 2.0, image processing and Internet data transmissions are included. EHRs maintain patient data charts in Healthcare 3.0 and data stored on clouds for further processing. The cloud computing stores, analyses, and processes voluminous data generated by number of patients, which consumes more time. The FC and edge computing facilitate real-time data streaming with minimal latency and bandwidth, thereby enhancing Healthcare 4.0 [27].

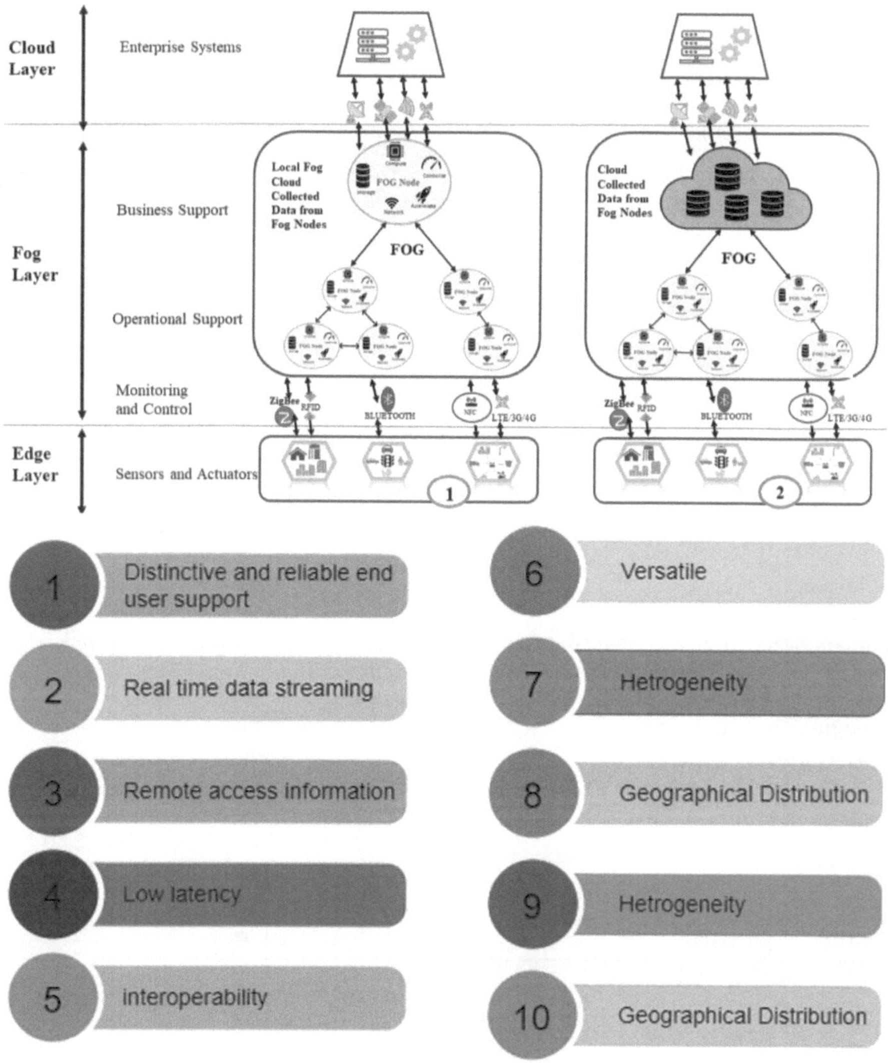

FIGURE 9.6 (a) Fog computing-enabled architecture. (b) Fog computing characteristics.

Health industry is affected by robustness, reliability, privacy, and security. IoT-enabled healthcare applications have issues with data integrity, dependability, and inadequate data security. These applications collect sensitive personal information; therefore, numerous data trust, privacy, and security issues were highlighted and addressed [28]. WIFI, LoRa, and other cloud data collection methods collect and promptly transmit health data to users. In addition, they voiced concerns about data validity, privacy, and secure data transfer [29]. If data leaks or path information is compromised and updated, the repercussions will be severe, and without patient-specific information, physicians may prescribe inappropriate medications. The issues

TABLE 9.1
Metrics Comparison of Edge, Fog, and Cloud Computing

Metrics	Edge Computing	Fog Computing	Cloud Computing
Bandwidth	Low bandwidth required for communication	More bandwidth required than edge for communication	High bandwidth required for communication
Mobility	High	High	Low due to various technical reasons
Latency	Very low	Medium security algorithms	Higher than edge and fog computing
Reliability	Yes	Yes, but requires security procedures to validate information	To attain reliability, various authentication, authorization, trust management, and data security algorithms required
Deployment cost	Low	Medium	High
Maintenance	Required multiskilled technicians	Require expert technicians from diverse areas	Require expert technicians from diverse areas
Scalability	High	Low	High
Network access	RAN, LAN	Wireless modes: wifi, Bluetooth,antenna, satellite	Wireless modes: wifi, Bluetooth, antenna, satellite
Operational cost	Low	High	High
Security	High	Medium	Low

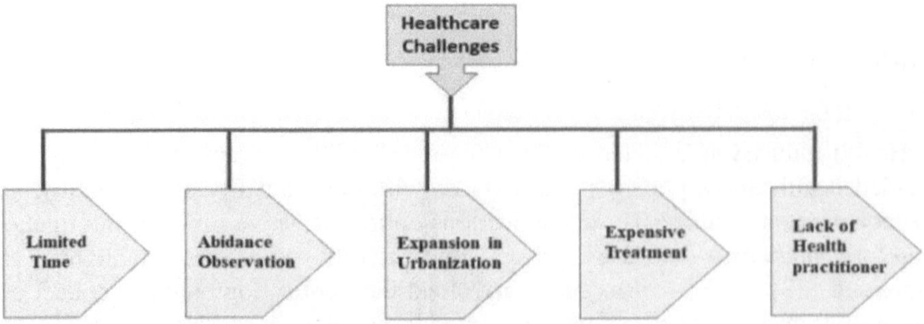

FIGURE 9.7 Healthcare industry challenges.

of latency, low bandwidth, and low power consumption are solved by FC up to some extent.

9.2.1 Fog Computing-Enabled Healthcare 4.0

Numerous healthcare frameworks have been developed with the assistance of FC. Since architectures and frameworks have been classified in a regressive manner [30], and a distinct viewpoint is required to identify voids and evaluate other proposed designs. This chapter analyses gaps in existing frameworks and architectures and proposes a new framework to address the issue at hand.

9.2.2 Recent State of Frameworks in Healthcare

Frameworks for an FC environment have been suggested and built by several academics in healthcare applications. These frameworks fall within the following category:

- Data management
- Data processing and analytics
- Fog node configuration
- Monitoring systems
- Security and privacy

9.2.3 Why to Analyse Healthcare Monitoring Architectures and Frameworks?

In recent years, the research community has moved towards healthcare, though most of the work is in monitoring. These details support our claim. Table 9.2 displays Web of Science statistics for the years 2019–2023 that employ "Healthcare + Framework + Node Configuration + Fog Computing." Other search terms, including "Healthcare + Framework + Data Management + Fog Computing," did not yield any results. Additional keyword combinations returned 14 results, including "Healthcare + Framework + Data Processing and Analytics + Fog Computing." We received 43 results for "Healthcare + Framework + Security and Privacy + Fog Computing" and 93 results for "Healthcare + Framework + Monitoring + Fog Computing." From 2019 to 2023, all frameworks will be indexed by Web of Science data. In the past four years, research has changed, as demonstrated by the differences.

9.3 RELATED WORK

Sensitive healthcare applications alert users of heart attacks and strokes in real time. Applications in healthcare that are time-sensitive necessitate continuous surveillance and tracking of health indicators. This conversation prompted us to investigate the recently proposed health monitoring tools and architectures. In 2021, Ali et al. [31] created a data mining-based architecture for evaluating and storing healthcare data.

TABLE 9.2
Keyword-Based Analysis of Frameworks

Framework Type	Keywords Used				Year	Articles Found
Node configuration	Healthcare	Framework	Node configuration	Fog computing	2019–2023	1
Data management	Healthcare	Framework	Data management	Fog computing	2019–2023	41
Data processing and analytics	Healthcare	Framework	Data processing and analytics	Fog computing	2019–2023	14
Security and privacy	Healthcare	Framework	Security and privacy	Fog computing	2019–2023	43
Monitoring	Healthcare	Framework	Monitoring	Fog computing	2019–2023	93

The healthcare architecture minimized latency and network usage [32]. The program iFogSim simulation [33] presented another remote monitoring method for early detection of diabetes patients. The Pima Indian Diabetes Database is used to construct support vector machines for variables sensitive to diabetes. The patients with Parkinson's disease cost society $29,695 and $21,482 in healthcare; therefore, Raza et al. [34] proposed a remote monitoring strategy based on IoT for this condition, which affects society. In a typical interior setting, the intelligent framework analyses patients' health and informs administration. Using fog-enabled smart wearables and IoT sensors, Hussain et al. [35] designed a healthcare monitoring system for gym fitness analysis. Vital statistics, body activity, and health information comprised the fitness industry's foundation.

The mosquito-borne disease dengue fever is well known and Sood et al. [36] proposed in 2021 an intelligent system for real-time dengue disease monitoring, prediction, and alerting.

Hu et al.'s 5G-enabled healthcare system [37] aims for fast, accurate health data readings and blockchain-secure data exchange. Elhadad et al. [38] proposed real-time patient body monitoring in healthcare in which wearable or improved body sensors measure the patient's heart rate, body temperature, and blood pressure. Lakhan et al. [39] created a medical enterprise blockchain socket Remote Call Procedure (RPC) system. The measurements minimize blockchain security, service, and data storage costs in distributed mobile cloud networks. Yildirim et al. [40] proposed an IoT-powered healthcare monitoring architecture employing wireless body area networks (WBANs) and technologies such as big data, cloud, and FC to evaluate health data. Support vector machines based on cloud integration of FC and fuzzy logic are utilized for fast, accurate decision-making.

For remote patient monitoring in the gym, healthcare facilities, disease-specific monitoring, etc., the literature's frameworks and architectures rely heavily on FC and cloud computing. These frameworks incorporate neural networks, machine learning, deep learning, support vector machines, and large amounts of data. Several framework performance measures have not yet been evaluated, as shown in Table 9.3.

9.3.1 Research Challenges of Present State of Proposed Monitoring Frameworks

Every application or service begins with a framework, and tools are necessary for applications that necessitate a specific environment. Frameworks for healthcare surveillance assist in tracking patients' health in a variety of conditions and meeting business requirements. Various authors employ industry 4.0 components such as machine learning, blockchain, big data, FC, cloud computing, and AI to improve framework execution. The frameworks were evaluated according to several performance indicators, including (1) reliability, (2) scalability, (3) latency, (4) privacy, (5) resource management, (6) mobility, (7) energy and power efficient, and (8) cost effective. These performance areas demand deep dive investigation in various areas such as:

TABLE 9.3
Analysis of Present State of Proposed Monitoring Frameworks

Author and Year	Proposed Framework	Focus	1	2	3	4	5	6	7	8
Raza et al. 2020 [34]	Remote monitoring of disease-specific patients'	Parkinson's disease remote monitoring	□	□	✓	□	✓	□	□	✓
Sodhro et al. 2021 [41]	Proposed an artificial intelligence-enabled framework	E-healthcare applications	□	□	□	□	□	□	✓	□
Ali et al. 2021 [42]	Proposed an architecture for extracting information from large amount of data	Use of data mining for information extraction	✓	✓	□	□	✓	✓	✓	□
Hussain et al. 2021 [43]	The framework used to track fitness data in a gym environment	To track the exercise patterns of trainers, players, and physicians	□	□	✓	□	□	✓	□	□
Sood et al. 2021 [44]	The proposed approach diagnoses using a Nave Bayesian Network (NBN)	Framework for detection, monitoring, and alerting prediction and control of Dengue	□	□	✓	□	□	□	□	✓
Ramesh et al. 2021 [33]	Remote monitoring framework for diabetic patients for early identification	Pima Indian Diabetes Database is used to build the support vector machine idea for diabetes-sensitive variables	□	□	□	✓	□	□	□	□
Asghar et al. 2021 [32]	Developed an architecture for monitoring health with minimal latency and network use	Load Balancing Scheme for balancing the load between fog nodes	□	□	✓	□	□	✓	□	□
Elhadad et al. 2022 [38]	Framework for real-time monitoring of a patient's body in healthcare	Data such as heart rate, body temperature, and blood pressure are collected from body via advanced body sensors	□	□	✓	□	□	□	✓	□

TABLE 9.3 (Continued)
Analysis of Present State of Proposed Monitoring Frameworks

Author and Year	Proposed Framework	Focus	1	2	3	4	5	6	7	8
Lakhan et al. 2022 [39]	Developed a system for medical enterprises based on the use of blockchain in socket Remote Call Procedure (RPC)	The metrics are designed to reduce the cost of blockchain security, service, and data storage in distributed mobile cloud networks	□	□	✓	✓	□	□	□	□
Yildirim et al. 2023 [40]	Proposed architecture, wireless body area networks (WBANs), and health big data are analysed utilizing cloud and fog computing technologies	Proposed method considers the diabetes prediction use cases with the help of support vector machine	□	□	✓	□	□	□	□	✓

- Factors like privacy and security have not been considered by the previously analysed frameworks.
- Another essential point is that dependability cannot be evaluated under diverse conditions, such as when transmitting data to a healthcare facility with multiple patients in single, parallel, or mixed modes. These instances do not account for reliability.
- Although the scalability of the framework is described in certain works, it cannot be realized unless the dependability is attained in settings that use parallel, mix, and serial modes.
- Just a few frameworks are available for indoor healthcare applications, which the modern healthcare industry desperately needs for remote patient monitoring.
- Most of the frameworks suggested do not consider the impact of interoperability and energy consumption, even though these parameters are crucial in environments with limited device capabilities.

A new framework is proposed to fill the gaps identified above for more effective monitoring of patients.

9.4 PROPOSED FRAMEWORK

The objective of this research and analysis is to develop a fog computing-enabled energy-efficient framework for healthcare 4.0 that is secure, scalable, interoperable

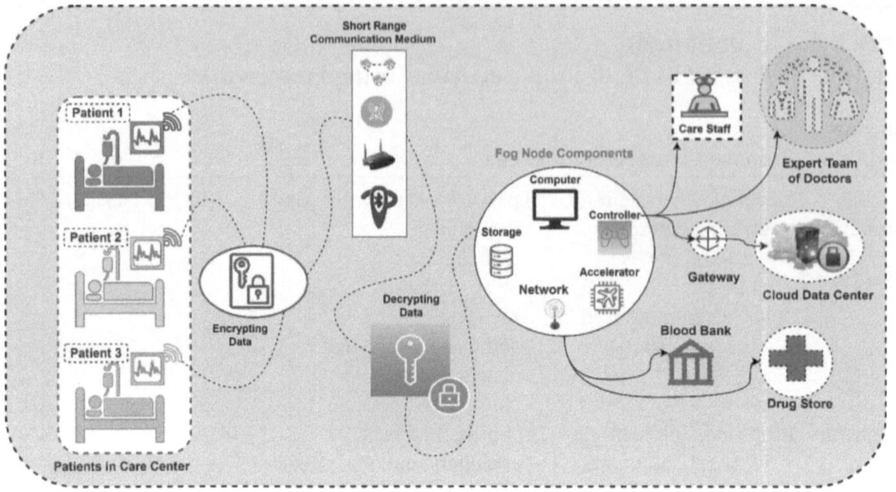

FIGURE 9.8 Proposed framework for healthcare 4.0.

too. The implanted body sensors will continually produce data, and they will relay that data to the closest fog node (i.e., local server within the healthcare centre).

If a threshold value is detected, the system will notify doctors, care staff, blood banks, and drug store. A gateway will transfer the data for future use on a cloud server. Cloud servers enable professionals to independently analyse and recommend treatment. Figure 9.8 depicts the whole working structure of proposed framework. The implanted sensors send the data to fog nodes for local processing. The fog node comes with all the necessary parts, including a controller and network capabilities to accept data, data storage and analysis, calculations, and data processing acceleration. The latency is decreased as a result of local data processing, and since sensors are closer to the fog data centre, communication reliability has risen proportionately. The transmission method, i.e., whether data is sent in serial, mixed, or parallel form, often determines the communication's reliability. In order to determine what sort of distribution the data will be sent in, the medium of communication must be further investigated.

9.5 CONCLUSION AND FUTURE SCOPE

The chapter discusses the use of technology in automated healthcare. This chapter covered IoT applications in a variety of industries. There are many practical applications of technology, and each has its own pros and cons. The cloud provides shared processing and storage but shared access is plagued by increased latency, which many applications cannot tolerate. Healthcare is a real-time application in which the smallest delay can be fatal. A trustworthy, energy-efficient, and secure architecture based on FC reduces data transmission latency. Cloud computing is inferior to FC in terms of dependability and latency. It can also be used for immediate healthcare indoors. Consequently, the authors proposed an FC framework for real-time patient monitoring in a healthcare facility and alerting authorities if any health-related threshold value exceeds the limits and endangers patients. The proposed architecture may be problematic due to the limited storage capacity of fog nodes and the potential Random Access Memory (RAM) exhaustion caused by the continuous generation of monitoring data. The system will lose crucial information if fog devices do not save the generated data. In this case, storing data in the cloud is necessary, which may cause delays and compromise security.

REFERENCES

1. M. Yun and B. Yuxin, "Research on the architecture and key technology of Internet of Things (IoT) applied on smart grid," *ieeexplore.ieee.org*, Accessed: Feb. 27, 2023. [Online]. Available: https://ieeexplore.ieee.org/abstract/document/5557611/?casa_token=gQVn3dmT_0wAAAAA:c3loPz7frSiJfz01oikjaP4YzAbZFUJsd4DlenSZVAQbePLNlIFQC9ktR8bJK-6D2wT9TA
2. N. Goyal, S. Sharma, A. Rana, and S. Tripathi, *Internet of Things: Robotic and Drone Technology*. 2022. Accessed: Mar. 06, 2023. [Online]. Available: https://books.google.com/books?hl=en&lr=&id=x3dYEAAAQBAJ&oi=fnd&pg=PP1&dq=Internet+of+Things:+Robotic+and+Drone+Technology&ots=ZId0ZI5DVI&sig=r2NB-RzP5wrNBO2I1CZhP5q_68M
3. D. Bandyopadhyay and J. Sen, "Internet of things: Applications and challenges in technology and standardization," *Wirel Pers Commun*, vol. 58, no. 1, pp. 49–69, May 2011, doi: 10.1007/S11277-011-0288-5.
4. F. Li, M. Voegler, M. Claessens, and S. Dustdar, "Efficient and scalable IoT service delivery on cloud," in 2013 IEEE Sixth International Conference on Cloud Computing, Jun. 2013, pp. 740–747. doi: 10.1109/CLOUD.2013.64
5. M. Dave, V. Rastogi, M. Miglani, P. Saharan, and N. Goyal, "Smart fog-based video surveillance with privacy preservation based on blockchain," *Wirel Pers Commun*, vol. 124, no. 2, pp. 1677–1694, May 2022, doi: 10.1007/S11277-021-09426-8.
6. "Health, Pharma & Medtech Diseases – Statistics & Facts," *Diseases – Statistics & Facts*. www.statista.com/topics/2070/diseases/#topicOverview (accessed Mar. 01, 2023).
7. Organização Mundial de Saúde, *World health statistics 2022 (Monitoring health of the SDGs)*. 2022. [Online]. Available: http://apps.who.int/bookorders.
8. D. Chawla and P. S. Mehra, "QSMAH: A novel quantum-based secure cryptosystem using mutual authentication for healthcare in the internet of things," *Internet of Things*, vol. 24, p. 100949, Dec. 2023, doi: 10.1016/J.IOT.2023.100949.
9. Z. Pang *et al.*, "Introduction to the special section: Convergence of automation technology, biomedical engineering, and health informatics toward the healthcare 4.0," *ieeexplore.ieee.org*, Accessed: Feb. 24, 2023. [Online]. Available: https://ieeexplore.ieee.org/abstract/document/8421122/
10. S. Malik, K. Gupta, D. Gupta, A. Singh, and M. Ibrahim, "Intelligent Load-Balancing Framework for Fog-Enabled Communication in Healthcare. Electronics 2022, 11, 566," 2022, Accessed: Mar. 06, 2023. [Online]. Available: www.academia.edu/download/87577838/pdf.pdf
11. L. Atzori, A. Iera, and G. Morabito, "The internet of things: A survey," *Elsevier*, Accessed: Nov. 1, 2022. [Online]. Available: www.sciencedirect.com/science/article/pii/S1389128610001568

12. G. Karthick and P. B. Pankajavalli, "A review on human healthcare Internet of things: a technical perspective," *Springer*, vol. 1, no. 4, p. 198, Jul. 2020, doi: 10.1007/s42979-020-00205-z.
13. Z. Pang, "Technologies and Architectures of the Internet-of-Things (IoT) for Health and Well-being," 2013, Accessed: Jan. 24, 2023. [Online]. Available: www.diva-portal.org/smash/record.jsf?pid=diva2:621384
14. D. Chawla and P. S. Mehra, "A roadmap from classical cryptography to post-quantum resistant cryptography for 5G-enabled IoT: Challenges, opportunities and solutions," *Internet of Things*, vol. 24, p. 100950, Dec. 2023, doi: 10.1016/J.IOT.2023.100950.
15. Y. Luo *et al.*, "Deep learning-based anomaly detection in cyber-physical systems: Progress and opportunities," *dl.acm.org*, vol. 54, no. 5, p. 106, Jun. 2021, doi: 10.1145/3453155.
16. V. K. Quy, N. Van Hau, D. Van Anh, and L. A. Ngoc, "Smart healthcare IoT applications based on fog computing: architecture, applications and challenges," *Complex and Intelligent Systems*, 2021, doi: 10.1007/S40747-021-00582-9.
17. I. Azimi *et al.*, "HiCH: Hierarchical fog-assisted computing architecture for healthcare IoT," *dl.acm.org*, vol. 16, no. 174, Sep. 2017, doi: 10.1145/3126501.
18. K. Tange, M. De Donno, X. Fafoutis, and N. Dragoni, "A Systematic Survey of Industrial Internet of Things Security: Requirements and Fog Computing Opportunities," *IEEE Communications Surveys and Tutorials*, vol. 22, no. 4, pp. 2489–2520, Oct. 2020, doi: 10.1109/COMST.2020.3011208.
19. J. Sharma and P. S. Mehra, "A survey of security challenges and existing prevention methods in FANET," in *Intelligent Data Analytics, IoT, and Blockchain*. Auerbach Publications eBooks, 2023, pp. 252–262. doi: 10.1201/9781003371380-24.
20. "Polyurethane Market – Global Industry Analysis and Forecast (2022-2029)," *Fog Computing Market: Global Industry Analysis and Forecast (2022-2029)*, 2022. www.maximizemarketresearch.com/market-report/global-polyurethane-market/70607/
21. A. K. Sarangi, A. G. Mohapatra, T. C. Mishra, and B. Keswani, "Healthcare 4.0: A voyage of fog computing with iot, cloud computing, big data, and machine learning," *Signals Commun Technol*, pp. 177–210, 2021, doi: 10.1007/978-3-030-46197-3_8.
22. S. Rani, A. Kataria, and M. Chauhan, "Fog Computing in Industry 4.0: Applications and Challenges—A Research Roadmap," *Lecture Notes on Data Engineering and Communications Technologies*, vol. 74, pp. 173–190, 2022, doi: 10.1007/978-981-16-3448-2_9.
23. A. Kumar, S. Sharma, N. Goyal, … S. G.-I. J., and undefined 2022, "Energy-efficient fog computing in Internet of Things based on Routing Protocol for Low-Power and lossy network with Contiki," *Wiley Online Library*, vol. 35, no. 4, Mar. 2022, doi: 10.1002/dac.5049.
24. A. M. Alwakeel, "An overview of fog computing and edge computing security and privacy issues," *mdpi.com*, Accessed: Feb. 01, 2023. [Online]. Available: www.mdpi.com/1396884
25. B. Farahani, F. Firouzi, V. Chang, M. Badaroglu, N. Constant, and K. Mankodiya, "Towards fog-driven IoT eHealth: Promises and challenges of IoT in medicine and healthcare," *Elsevier*, Accessed: Jan. 23, 2023. [Online]. Available: www.sciencedirect.com/science/article/pii/S0167739X17307677
26. F. Firouzi, K. Chakrabarty, and S. Nassif, *Intelligent internet of things: From device to fog and cloud*. 2020. Accessed: Jan. 23, 2023. [Online]. Available: https://link.springer.com/content/pdf/10.1007/978-3-030-30367-9.pdf
27. A. Sarangi, A. Mohapatra, T. C. Mishra, and B. Keswani, "Healthcare 4.0: A voyage of fog computing with iot, cloud computing, big data, and machine learning," *Springer*,

Accessed: Jan. 30, 2023. [Online]. Available: https://link.springer.com/chapter/10.1007/978-3-030-46197-3_8
28. Y. Shaikh, V. Parvati, and S. R. Biradar, "Survey of smart healthcare systems using internet of things (IoT)," *ieeexplore.ieee.org*, Accessed: Mar. 11, 2022. [Online]. Available: https://ieeexplore.ieee.org/abstract/document/8668128/
29. J. Leng, Z. Lin, and P. Wang, "An implementation of an internet of things system for smart hospitals," *ieeexplore.ieee.org*, Accessed: Mar. 11, 2022. [Online]. Available: https://ieeexplore.ieee.org/abstract/document/9097614/
30. A. R. Nair and S. Tanwar, "Fog computing architectures and frameworks for healthcare 4.0," in *Signals and Communication Technology*, pp. 55–78, 2021, doi: 10.1007/978-3-030-46197-3_3.
31. F. Ali *et al.*, "An intelligent healthcare monitoring framework using wearable sensors and social networking data," *Future Generation Computer Systems*, vol. 114, pp. 23–43, 2020, doi: 10.1016/j.future.2020.07.047.
32. A. Asghar, A. Abbas, H. A. Khattak, and S. U. Khan, "Fog based architecture and load balancing methodology for health monitoring systems," *IEEE Access*, vol. 9, pp. 96189–96200, 2021, doi: 10.1109/ACCESS.2021.3094033.
33. J. Ramesh, R. Aburukba, and A. Sagahyroon, "A remote healthcare monitoring framework for diabetes prediction using machine learning," *Healthcare Technol Lett*, vol. 8, no. 3, pp. 45–57, Jun. 2021, doi: 10.1049/htl2.12010.
34. M. Raza *et al.*, "Intelligent IoT framework for indoor healthcare monitoring of Parkinson's disease patient," *ieeexplore.ieee.org*, Accessed: Feb. 18, 2023. [Online]. Available: https://ieeexplore.ieee.org/abstract/document/9186157/?casa_token=DZzyjwV1Hw8AAAAA:VMf41TDvGJD_BKY-19dJLP-zjzbrhEInTk6Ewje0xvbph1um6--9Wa-k5A2vDBrrFTVBcg7nDKM_AFo
35. A. Hussain, K. Zafar, and A. R. Baig, "Fog-Centric IoT based framework for healthcare monitoring, management and early warning system," *IEEE Access*, vol. 9, pp. 74168–74179, 2021, doi: 10.1109/ACCESS.2021.3080237.
36. S. K. Sood, V. Sood, I. Mahajan, and Sahil, "An intelligent healthcare system for predicting and preventing dengue virus infection," *Computing*, 2021, doi: 10.1007/s00607-020-00877-8.
37. J. Hu, W. Liang, O. Hosam, M. Y. Hsieh, and X. Su, "5GSS: a framework for 5G-secure-smart healthcare monitoring," *Conn Sci*, vol. 34, no. 1, pp. 139–161, 2022, doi: 10.1080/09540091.2021.1977243.
38. A. Elhadad, F. Alanazi, A. I. Taloba, and A. Abozeid, "Fog Computing Service in the Healthcare Monitoring System for Managing the Real-Time Notification," *J Healthc Eng*, vol. 2022, 2022, doi: 10.1155/2022/5337733.
39. A. Lakhan, T. Morten Groenli, A. Majumdar, P. Khuwuthyakorn, F. Hussain Khoso, and O. Thinnukool, "Potent Blockchain-Enabled Socket RPC Internet of Healthcare Things (IoHT) Framework for Medical Enterprises," *Sensors*, vol. 22, no. 12, pp. 1–16, 2022, doi: 10.3390/s22124346.
40. E. Yıldırım, M. Cicioğlu, and A. Çalhan, "Fog-cloud architecture-driven Internet of Medical Things framework for healthcare monitoring," *Med Biol Eng Comput*, 2023, doi: 10.1007/S11517-023-02776-4.
41. A. H. Sodhro and N. Zahid, "AI-enabled framework for fog computing driven E-healthcare applications," *Sensors*, vol. 21, no. 23, 2021, doi: 10.3390/s21238039.
42. F. Ali *et al.*, "An intelligent healthcare monitoring framework using wearable sensors and social networking data," *Future Gener Comput Syst*, vol. 114, pp. 23–43, 2021, doi: 10.1016/j.future.2020.07.047.

43. A. Hussain, K. Zafar, and A. R. Baig, "Fog-centric IoT based framework for healthcare monitoring, management and early warning system," *ieeexplore.ieee.org*, Accessed: Feb. 18, 2023. [Online]. Available: https://ieeexplore.ieee.org/abstract/document/9430526/
44. S. K. Sood, V. Sood, I. Mahajan, and Sahil, "An intelligent healthcare system for predicting and preventing dengue virus infection," *Computing*, 2021, doi: 10.1007/S00607-020-00877-8.

10 A Study on Intrusion Detection System for RPL Protocol Attacks in Internet of Things

Kajol Mittal and Payal Khurana Batra

10.1 INTRODUCTION

IoT (Internet of Things) represents a technological revolution which serves as a global network for connecting physical things to the Internet and facilitates the exchange of information and services.The future Internet will comprise heterogeneously linked devices that will expand the world's borders with physical and virtual entities. It is predicted that in comparison to 2019, the USD 5 billion IoT investment would triple to reach USD 15 billion. The IoT investments in India are expected to triple to reach ~USD 15 billion by 2021 compared to USD 5 billion in 2019 [1, 2]. Gartner estimated there would be a 21% hike from 2019 and around 5.8 billion endpoints in 2020 in the enterprise and automotive IoT market [3]. The industries are adopting IoT to increase profit and become the leading industry in the market.

IoT empowers the development of various industrial applications and user-oriented IoT applications. IoT facilitates users' living more smartly with the help of its applications, such as smart agriculture, smart cities, smart environments, and many more. IoT is vulnerable to security threats and multiple other vulnerabilities inherited from the standard communication protocols and the Internet's foundations. A report revealed by Hewlett Packard [4] states that serious vulnerabilities occur in 70% of the most frequently operated IoT devices. The drastic increase in the number of IoT devices attracts attackers' attention and creates high security and privacy risks.

Maximum IoT applications are enabled by Low power and Lossy Networks (LLNs) that use IPV6 low-power wireless personal area networks (6LoWPAN). Low throughput and lossy communication links are provided by the LLNs [5]. Due to the limited resources, LLNs constrain the usage of conventional computers and communication technology. Furthermore, such networks employ resource-constrained devices (nodes) that are low power, low energy, and have limited onboard memory and processing capabilities [6]. The Internet Engineering Task Force (IETF) suggested the 6LoWPAN stack to enable interoperability and IPv6 functionality for IoT devices [7].

DOI: 10.1201/9781003405535-14

The peculiarities and hardware limits of networks linked with IoT devices provide several issues, particularly for routing protocols [8]. "The Routing over Low-power and Lossy Networks (ROLL) group created the RPL (IPv6 Routing Protocol for Low-Power and Lossy Networks) routing protocol for Low-Power and Lossy networks in March 2012" [5]. RPL is a distance-based routing protocol presented in RFC6550 [9]. RPL is a proactive protocol that creates a destination-oriented Directed Acyclic Graph (DODAG) that connects leaf nodes and the sink node (border router). RPL has a low energy consumption mechanism and provides self-organizing and self-healing functionality to handle node failures. A route is constructed by sending information from node to node until the target destination is reached. Malicious nodes carry out their activities during route discovery or forwarding so that information routing might be vulnerable to various attacks [10]. RPL is vulnerable to various threats or attacks, and to provide seamless connectivity in IoT, it is required that communication should not be disruptive.

This chapter will present a survey on attacks that can happen on RPL. This chapter focuses on the available solutions to handle the attacks using various machine learning methods. This chapter will help readers to know about the various attacks and solutions available in the literature. The remaining chapter is arranged as follows: Section 10.2 presents an outline of the RPL protocol. In Section 10.3, security attacks on RPL protocol are discussed. Section 10.4 presents various security solutions based on an Intrusion Detection System (IDS). In Section 10.5, performance metrics are discussed. Section 10.6 discusses limitations with IDS based on machine learning (ML) and Deep Learning (DL) models. In Section 10.7, the conclusion is presented.

10.2 OUTLINE OF RPL PROTOCOL

Implementing a routing protocol for 6LoWPAN is challenging with respect to resource-constraint environment and low-power IoT devices. RPL facilitates managing data routing of nodes in a restricted resource environment efficiently and successfully [11]. RPL protocol is a network layer protocol.

"RPL generates the network topology using a DAG (Destination-Oriented Graph), consisting of one or more DODAG." The DODAG is oriented towards the root node, the LBR (LLN border router), or the sink node. The objective function (OF) builds the DODAG in RPL and selects the preferred parent among the neighbouring parents. There are two main types of OF: (1) OF zero and (2) minimum rank OF, which considers the routing metrics such as ETX (Expected Transmission Count), and hop count to select the optimum path between child nodes and the root node of DODAG. The route selection depends on parameters such as available node power, throughput, DODAG link, latency, or reliability. Each DODAG in a Directed Acyclic Graph (DAG) is differentiated by the four main factors: DODAG ID version number, RPL Instance ID, and Rank. Furthermore, the sink nodes of every DODAG in a DAG are linked with each other. Every node selects the best suitable parent from the parent list and sets the nodes to maintain. Rank helps detect loops, differentiate parent nodes from sibling and child nodes, and build parent-child relationships. To avoid the formation of loops, RPL uses the rank metrics rule. RPL uses control messages to form

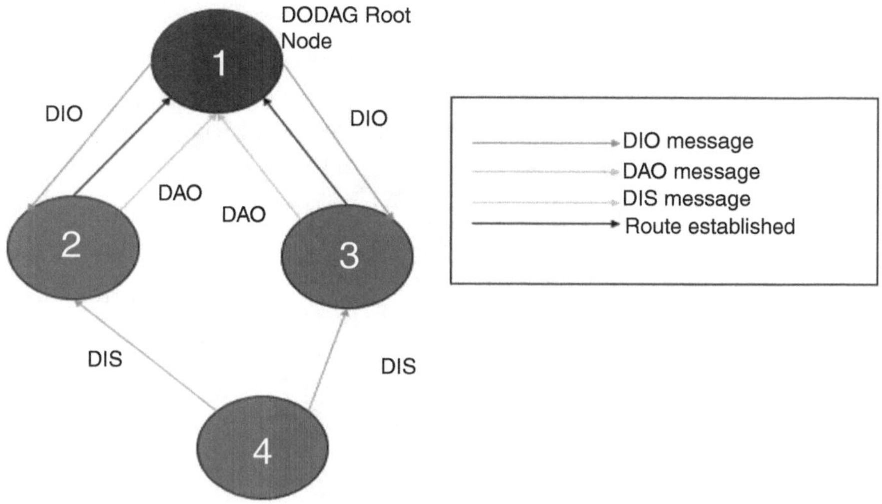

FIGURE 10.1 Construction of DODAG.

and maintain the routing in the networks. Figure 10.1 illustrates the DODAG formation with the help of control messages. The control messages are as follows: DODAG Information Object (DIO), DODAG Information Solicitation (DIS), and destination advertisement object acknowledgement (DAO).

10.3 RPL PROTOCOL SECURITY ATTACKS

RPL is vulnerable to various threats, some of which come from the traditional wireless sensor networks (WSNs). The RPL protocol is susceptible to both inside and external attacks. It is challenging to identify and defend against these attacks due to the vulnerable nature of nodes, LLN networks, and wireless networks; a lack of infrastructure; inadequate physical security; changing topology; unreliable connectivity; mobility of nodes; and resource limitations. The RPL protocol provides multiple configuration settings to meet various mobility, heterogeneity, and topology change needs. Resource-constrained nodes, changing topologies, and the passive wireless medium encourage malicious acts [12]. Examples of this information include the DODAG version, node rank values, and IDs, although this list is not limited. A combination of attacks may have severe impacts like resource exhaustion of nodes, leading to a significant rise in communication packets and major degradation of the protocol's data transfer efficiency. Figure 10.2 represents the classification of RPL attacks in various categories and sub-categories.

10.3.1 Attacks on Resources

Attacks on resources usually comprise causing legitimate nodes to conduct extra processing to drain their resources by exhausting node energy, memory, or computational

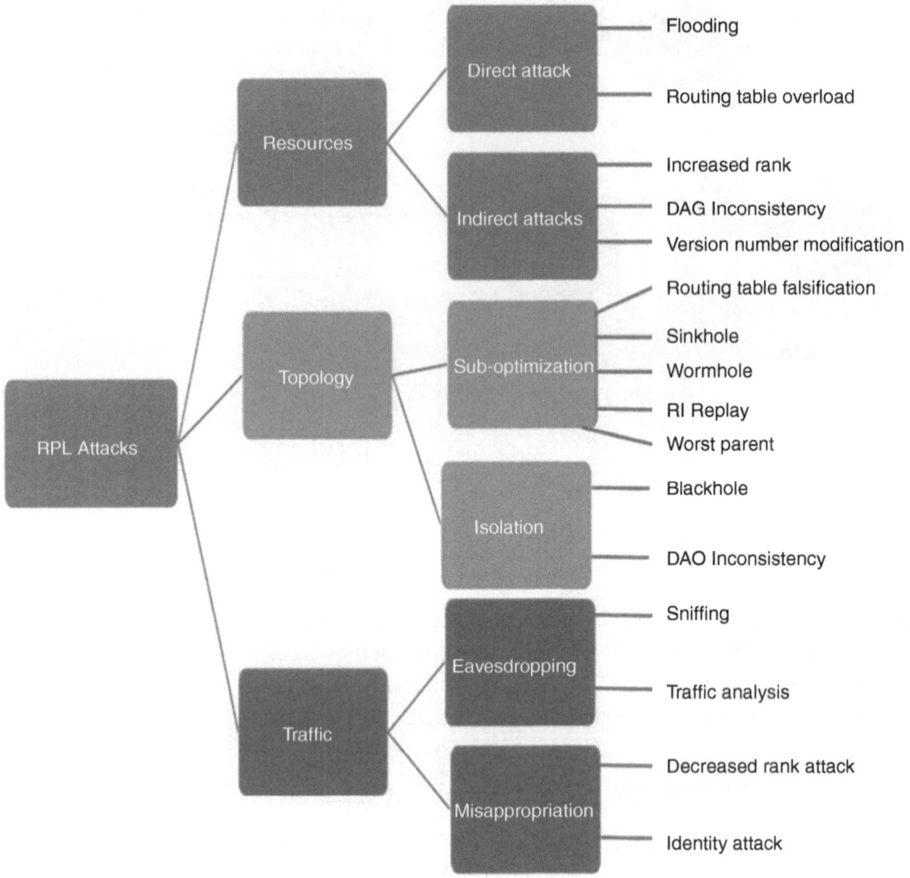

FIGURE 10.2 RPL attacks classification.

power. It could influence network availability by saturating available connections and, hence, network lifespan, which can be greatly reduced. Direct and indirect attacks on resources are the two types of attacks.

10.3.1.1 Direct Attacks: A malicious node produces the overload directly to degrade the network performance. The attacker does the resource depletion. This is typically accomplished when the storage mode is enabled by launching flooding attacks or overloading attacks on routing tables. The direct attack can be categorized into two attacks:

Flooding Attack: This attack is carried out by flooding DIO and DIS messages. To join the network nodes, send the first message as a HELLO. DIO messages are called Hello messages in RPL and are used to send information about DODAG [13]. This attack creates network congestion and may lead to node saturation.

Routing Table Overhead: The proactive nature of RPL implies that when the storing mode is activated, they generate and preserve routing tables. Routing table overflow involves publishing false routes utilizing DAO messages to overload a node's routing table. This limits legitimate route building, affects network performance, and might cause memory overflow.

10.3.1.2 Indirect Attacks: These attacks are those in which the hostile node causes another node to cause an overloaded network. It consists of three types of attacks:

Increasing Rank Attack: It involves intentionally increasing the RPL node's rank to cause network loops. Whenever the node wishes to modify its rank, it requires updating its parent's list by removing nodes with a higher rank with a low-rank value. When sending a packet to the root node, a node in a DODAG selects a parent from this list to reduce routing costs. A malicious node promotes a greater rank than it really has, which may lead to an attack if no loop avoidance method is used.

DAG Inconsistency Attack: An DAG node identifies a RPL inconsistency when the packet arrives with a Down 'O' bit set from a node with a higher rank and vice versa [14], whenever the node does not maintain the rank relationship. This could lead to loop formation in the graph. This problem is managed via the Rank-Error 'R' flag bit. One of two things may occur whenever a node detects an inconsistency: (i) the node sets the Rank-Error flag if it is not set and transmits the packet; (ii) the node rejects the packet and restarts the timer. Consequently, control messages can be issued frequently. The malicious node only needs to change the flags or add new flags to the header to compromise the integrity of the network. The direct result of this attack is a reset of the targeted node's DIO trickle timer. This also depletes the batteries of the nodes and influences the availability of connections. The attack affects the node's whole neighbourhood because it needs to filter excessive traffic. Furthermore, tampering with genuine traffic means a target node can discard all packets. This can lead to a black hole attack.

Version Number Modification: The version number is sent down the DODAG graph without being altered in any way, and the root node is the sole node that increases to it whenever there is a need for the rebuilding of the DODAG (also known as "global repair"). An earlier value implies that the node has not been updated to the new DODAG graph, which prevents it from being used as a parent node. Whenever a DIO message is sent to the neighbours, an attacker may modify the version number by maliciously increasing this field. An attack of this kind necessitates the needless rebuilding of the entire DODAG graph which can lead to exhaustion of node resources and congestion in the network. This attack has the possibility of causing several loops, leading to data packet loss.

10.3.2 Attacks on Topology

Additionally, the RPL protocol is susceptible to topological assaults, which may be further classified into the following two categories:

10.3.2.1 Sub-Optimization Attacks: These attacks affect the network's optimum convergence capability; that is, they inhibit the construction of optimal routes, hence impacting network traffic and degrading network services.

Routing Table Falsification: An adversary may attack by falsifying/forging the DAO control messages, forcing legitimate nodes to construct nonexistent false downstream routes. As a result, when authorised nodes attempt to send data to nonexistent nodes, consequences can lead to unnecessary packet latency, packet drop, DAO inconsistency, and high control overhead results. This attack is only possible if the storage mode is enabled.

Sinkhole: An attack of this kind is made in two parts (i) the malicious node pulls a large amount of traffic by advertising false information data; (ii) after receiving the traffic in an illegal way, it alters or drops it. In RPL networks, the attack may be simply executed by altering the rank value. Due to this forged advertising, other nodes wrongly choose a malicious node as an appropriate parent, while not delivering higher performance. The attack changes the network's topology or reduces its performance. Furthermore, the attacker can perform a blackhole attack if it decides to drop all communications [15].

Wormhole: Wormhole attack uses two RPL attacker nodes, considered as nodes 1 and 2, joined by a private network connection. Each packet received by node 1 is routed via the wormhole to node 2 and replayed afterwards. Wormhole attacks disrupt the routing route and are especially dangerous in RPL networks. When an adversary sends routing information through a tunnel to another portion of the network, nodes that are truly far apart perceive each other as if they were neighbours. Consequently, based on the objective function, they cannot construct any optimal routes.

Routing Information Reply: Valid control messages received from other nodes are saved by a node and sent later in the network. This attack is highly harmful because the topology and routing routes are frequently updated in dynamic networks. It will result in a deformed topology.

Worst Parent: This attack is the result of a "Rank attack," which entails selecting the least preferred parent based on the objective function. Consequently, the resultant route is not optimized, leading to performance degradation. This attack is difficult to deal with since child nodes depend on their parents to route packets, and neighbours cannot monitor it.

10.3.2.2 Isolation Attacks: These types of attacks on the topology isolate either a single node or a group of nodes in the RPL network. This implies that such types of nodes cannot interact with their parents or the root.

Black hole: A malicious node discards all packets meant to transmit in a black hole attack. In combination with a sinkhole attack, this attack may be extremely harmful, and consequently, result in the loss of a substantial portion of the traffic. There is also a grey hole (or selective forwarding attack) form of this technique in which the attacker only discards a specific subset of the network trace.

DAO Inconsistency: It arises when a node has a previously learned downward route from a DAO message, but this specific route is not valid in the child node's routing table, hence delivers the packet to another neighbour.

10.3.3 Attacks on Traffic

10.3.3.1 Eavesdropping Attacks: The pervasiveness of RPL networks may enable the placement of malicious nodes engaged in eavesdropping activity such as:

Sniffing: This attack involves listening to the network traffic. This attack seems quite frequent in wired and wireless networks, and it threatens communication confidentiality. The attacker could use a compromised device or a shared medium to directly capture packets to carry out this attack in wireless networks. Because of its passive nature, this attack is hard to detect. When the attacker is external, the only option to avoid sniffing is to secure communications.

10.3.3.2 Traffic Analysis: This attack is possible even when the packets are encrypted. Similar to sniffing attacks, by identifying parent-child links, the objective is to get information regarding the RPL network, such as a topology's limited view. A malicious node could be able to initiate further attacks using the information collected as a result of this attack. The consequences are dependent on the attacker's node rank.

10.3.3.3 Misappropriation Attacks: The identity of a valid node is stolen in misappropriation attacks, and performance is overstated. The RPL network is not significantly affected by these attacks. They enable the attacker to understand the network's topology better, gain high access, or intercept a huge proportion of the traffic. They are, however, often utilized as a preliminary to other types of attacks, like those found in the earlier two categories.

10.3.3.4 Decreased Rank: The lower the rank in a DODAG graph, the closer the node is to the root, and it is expected to manage more traffic. Misrepresenting one's performance occurs whenever a malicious node purposefully broadcasts a lower rank value than it possesses. Consequently, the attacker connects several valid nodes to the DODAG graph and brings in a significant portion of the traffic. An attacker may modify the rank value of the RPL protocol by falsifying DIO messages which can lead to sinkholes and eavesdropping attacks.

10.3.3.5 Identity Attack: Identity attacks are a combination of spoofing and Sybil attacks.

A spoofing attack, also known as a clone ID attack, happens when a malicious node masquerades as a legal existing node. The root node in an RPL network is essential in a DODAG graph. It constructs and maintains the topology by transmitting routing details. The network traffic is sniffed by the attacker to find the root node. Once this recognition is complete, it may spoof the DODAG root's address and gain control over the network. Identity attacks are used as a foundation for further operations.

10.4 SECURITY SOLUTIONS BASED ON INTRUSION DETECTION SYSTEM

An IDS is responsible for monitoring network traffic to identify potentially malicious behaviour and sends notifications when these activities are detected. This section explores several IDS-based security solutions to detect RPL protocol routing attacks. Contrary to WSNs, which do not presuppose continual access to the sink node, the IoT assumes that the sink node (6LoWPAN border router) is always available.

TABLE 10.1
Design Requirements [16]

Design Requirements	Description
RPL specification compliance	This covers the DODAG structure, control message exchange logic, and the Trickle timer mechanism.
Low overhead	Even security solutions for LLNs should include resource availability. A "low-budget" strategy should focus on exchanging control messages and use the standards to train the system and identify anomalies. Maintaining control overhead leads to conserving energy in transceivers which are the major consumers of constrained devices.
Scalability	The network size is sufficient for evaluating IDS in terms of its ability to safeguard the protocol, even if it is significantly expanded in terms of connected devices. Scalability shouldn't compromise the need for low overhead.
Robustness	Due to the variety of attacks, IDS must be able to detect them all.
Extendibility	IDSs can be expanded to handle more scenarios, such as it can be used as a "static" binary rationale to detect whether a threat pattern is recognized or not. Systems should use all existing technological assets to stay up-to-date and prepare for unexpected attacks.
Low false (negative/ positive) detections	The frequency of false positives/negatives affects a system's efficiency and detection accuracy. IDS should be accurate, not just robust and expandable. It should deliver alerts for targeted attacks while limiting false positives. If this is not met, the network operator must monitor the control overhead and the number of node parent changes, the packet delivery rate, and the number of RPL local repairs. Thus, it helps distinguish between normal but unexpected processes and attacks.
Mobility support	Mobile IoT applications have evolved over the past decade, and RPL functioning under mobility is the main challenging problem as it requires connection handovers and extra topology management. Basic security measures should address both stationary and mobile nodes, and the simple solutions are not available in the literature.

Table 10.1 presents seven design requirements that can be considered to design IDS for RPL [16].

10.4.1 Signature-Based IDS

Attacks are identified by signature-based intrusion detection systems depending on specific network traffic patterns. Signatures are the patterns that the IDS recognize. Signature-based intrusion detection systems can easily spot attacks for whom the pattern (signature) is already present on the system. Despite this, it is hard to identify new attacks whose pattern (signature) is undiscovered.

The [17] proposes an IDS to identify DIS and Virtual Network (VN) attacks. The authors suggest two types of extra nodes called IDS routers and sensors. IDS routers have detection and firewall modules. Sensors or IDS detectors monitor and report harmful traffic to network nodes. IDS routers examine passing traffic for malicious packets. The IDS detector monitors sensor traffic and calculates the Received Signal Strength Indicator (RSSI), packet drop rate, and packet sending rate. Based on each node's data, the 6LoWPAN Border Router's (6BR) detection module decides if a node is malicious.

IDS proposed in [18] uses distributed monitoring for version number attacks (VNAs). It focuses on the DODAG version, DIO messages, and nodes' rank. The method establishes various monitoring nodes that deliver a list of malicious nodes to the DODAG root by monitoring the RPL's specification parameters. When IDS modules are installed in constrained devices, more storage and processing capabilities are required [19]. Designing IDS with the purpose of protecting all incoming lists, the root informs network nodes to stop communicating with adversaries. Small and medium-sized networks function well, whereas large networks have excessive false positives/negatives. For large-scale networks, cross-monitoring each node with at least two other nodes can be a solution.

ELNIDS [20] uses artificial intelligence (AI) and ML on central premises to detect signatures. Ensemble learning detects attacks such as sinkholes, selective forwarding, Sybil, black holes, flooding, clone ID, and local repair attacks. The authors make use of an analysis engine, sniffer, sensor events/traffic repository, signature database, feature extraction module, and alarm/attack notification manager. The feature extraction module helps the analyser classify network traffic using ensemble models. The sniffer module detects and stores network traffic. ELNIDS has great accuracy. The limitations of the system are that it is challenging to identify new attacks, and the false-positive rate is excessively high.

10.4.2 Anomaly-Based IDS

The system or host's normal behaviour is observed, and any divergence from normal behaviour is reported as an anomaly. This kind of IDS can recognize novel attacks.

INTI [21] anomaly detection uses clustering (i.e., group of nodes). Each cluster has a leader node, associated node, and member nodes. Using node rankings and data, the system estimates trust. Dempster-Shafer evidence theory detects attacks and isolates malicious nodes. The system mitigates sinkhole attacks but requires a lot of computing. InDReS [22] is an enhancement of INTI [21] that maintains the core concepts of functioning while reducing computational overhead, hence conserving resources that are essential for LLNs. When malicious nodes are identified, the algorithm reconstructs the topology of the network by removing those nodes. However, the performance of InDReS is not examined in terms of false positives/negatives and mobility support compared to its predecessor. The IDS proposed in [23] utilizes the Sequential Probability Ratio Test (SPRT) and a dynamically adjustable threshold to detect selective forwarding attacks. It is composed of two modules. The first module makes decisions at the root node. The second module operates on a routing node that monitors incoming and outgoing packets. Monitoring nodes deliver data to the root through random routes. The root examines SPRT data and assigns each node a maliciousness probability. A threshold identifies malicious nodes. The root executes a DODAG global repair to isolate adversaries and alerts non-malicious nodes of their presence. The evaluation demonstrates the system's effectiveness and resource use. IDS isn't scalable due to high resource needs.

CoSec-RPL [24] addresses the combining of both flooding and replay attacks or "copycat attacks." Additionally, it can identify DIS flooding, DAO insider, and wormholes. The system employs a customized form of the Inter quartile Range (IQR) Outlier Detection (OD) method that utilizes the median rather than the mean value and is easier to implement. CoSec-RPL identifies nodes exhibiting abnormal behaviour. CoSec-RPL is activated every time a DIO message is received and checks the interval between messages. When measurements surpass specific criteria, the state of a node becomes "suspected." During this phase, contact with the node is indeed possible; if a second threshold is met, the node is judged malicious and "blocked"; no communication is allowed. The system's memory needs are not avoidable because each node requires a neighbouring table to hold relevant data. CoSec-RPL is beneficial in both mobile and static network environments. The main limitation of Anomaly IDS is that it also suffers from false positives.

10.4.3 Hybrid IDS

A hybrid IDS integrating both methods has been proposed to address the limitations of signature and anomaly-based IDS [25–26]. This combination allows the hybrid IDS to identify both misuse and anomaly-based attacks. The authors [27] proposed a Compression Header Analyzer Intrusion Detection System (CHA-IDS) that combines anomaly and signature-based IDS. CHA-IDS uses a centralized detection system. CHA-IDS uses a 6LoWPAN compression header as a feature for ML algorithms and classifies attacks [28–32]. The signature or rule generated by machine learning algorithms is inserted at 6BR. 6BR will be updated when a new routing attack signature becomes available. It accurately identifies Hello flooding, sinkhole, and wormhole attacks, separately or in combination [33, 34]. The selected characteristics using the correlation method are then analysed using six ML methods: Decision Trees (J48), Naive Bayes (NB), Logistic Regression (LR), Random Forest (RF), Multilayer Perceptron (MLP), and Support Vector

Machine (SVM). According to the authors, the system demonstrates a satisfactory balance between overhead and performance.

The authors [35] presented an Scalable and Versatile Lightweight Threat Evaluation (SVELTE) which is an anomaly and signature-based IDS. It uses distributed, hybrid, and centralized methods and places IDS modules in 6BR and restricted nodes. Mapper, Analyzer and Detector, and mini-firewall are centralized modules on 6BR. The system was evaluated against DODAG inconsistency, selective forwarding, sinkhole, and black hole attacks. SVELTE presents good performance considering the evaluation parameters as Packet Delivery Ratio (PDR) and control packet overhead. Given the fact that SVELTE employs a rank threshold to detect anomalies, it has a significant proportion of false positives and negatives [15, 36]. The limitation of SVELTE, which is a high false alarm rate, is reduced by the authors considering the timing inconsistency between the rank measurements [37]. To increase the security of 6LoWPAN networks, the authors [38] expand SVELTE with an ETX metrics-based intrusion detection module and also provide geographical information for identifying malicious nodes that perform attacks against ETX-based networks.

10.4.4 Machine Learning (ML)-Based IDS

ML is an emerging technology, and the researcher can use machine learning to provide an IDS. In recent times, there has emerged a significant rise in interest in developing intrusion detection methods based on ML and DL to improve security in IoT networks. Deep learning algorithms make use of Neural Networks (NNs). Recurrent Neural Network (RNN) is a kind of Artificial Neural network (ANN) that was first utilized in language models because of its capacity to remember long-term dependencies. However, when time lags grow longer, the gradient of RNNs may disappear as RNNs unfold into very deep feed-forward NNs. Several RNN topologies with forgetting units were developed to address the gradient vanishing issue. They were created to give memory cells the ability to decide when to forget specific pieces of information, hence determining the appropriate time lags [39].

In ML, the intrusion problem is divided into small sections, which are addressed one by one, and then the results are produced as a sum of all the obtained solutions. ML can effectively solve most of the challenges and help the network develop without producing unexpected difficulties. For example, security is a critical issue that must be maintained with each new technology. ML can enhance connected device security. ML uses less data to train and test; however, the accuracy it provides is less. On the other hand, deep learning takes a lot of time and a lot of data to train the system, but it often results in superior accuracy [29]. There are four main categories for ML algorithms [40]:

10.4.4.1 Supervised Learning: It is used when specified goals are given for a collection of inputs. Supervised models need pre-configuration to select acceptable data points (i.e., features). The data is first labelled and then trained using labelled data (inputs and outputs). It attempts to automatically find rules from available datasets, establish classes, and forecast element (object, person, and criterion) class membership. IoT devices have regular, basic behaviour due to their services and their sole purpose. As a result, they have a small quantity of meaningful data, hence reducing the requirement for techniques (like deep learning) designed for large datasets.

10.4.4.2 Unsupervised Learning: It can categorize unlabelled data into categories based on the analysis of similarities among unlabelled data. It helps to identify malicious from legitimate communications without telling the model how malicious traffic looks.

10.4.4.3 Semi-Supervised Learning: It bridges the gap of the above two kinds of ML categories as either no label is provided, or labels are present for all observations. Labelling is expensive in many cases because it needs to be experienced by human professionals. These methods are the best options for model generation when labels are only present in a small number of observations.

10.4.4.4 Reinforcement Learning (RL): In this learning method, the agent gains knowledge from feedback after having interaction with the environment. RL algorithms allow for the execution of numerous scenarios based on action and state variations. It performs actions and creates judgements based on the reward. It adopts "winning" techniques that are dependent on the surroundings. Implementing RL-based algorithms is costly at the network level since different environmental states and actions must be described for the whole network. The choice of reward function is critical since the agent's success or failure is determined by the overall reward.

This section describes the proposed or implemented ML and DL techniques for preventing or detecting RPL protocol attacks in IoT environments. There are various ML algorithms considering the main four ML categories, the most popular algorithms such as Random Forest (RF), K-Nearest Neighbour (KNN), Ensemble Learning, DT, SVM, K-means clustering, and PCA (principal component analysis) used in IoT IDS. Ensemble learning is used by various researchers to get better results. There are various evaluation/performance metrics; some of them are True Positive Rate (TPR), Root Means Squared error (RMSE), False Positive rate (FPR), PDR, energy consumption, accuracy, Mean Absolute Error (MAE), Precision, Area Under the Curve (AUC), Delay, Recall, F1-Score, Relative Absolute Error (RAE) and Receiver Operating Characteristics Curve (ROC). Table 10.2 presents the work done by researchers in the field of ML and DL-based IDS.

10.5 PERFORMANCE METRICS

ML provides a plethora of measures for evaluating the performance of a classifier. Selecting performance indicators based on an application's needs is essential. Certain classifiers might perform well on one parameter and not others. Various performance metrics are used to assess the quality of an ML model. This section discusses these performance metrics which are based on a confusion matrix generated by a ML model.

Where, T_RP denotes True Positive value, means a total number of cases in which predicted attack is same as actual attack.

TABLE 10.2
ML and DL-Based IDS Solutions

Reference	Publication Year	Contribution	Target Attack	Dataset	Tools/ Programming	Parameters
[41]	2023	A hybrid model using supervised deep ANN and semi-supervised auto encoder.	DIS attack, rank and wormhole attack	Simulated data set	Python (tensor flow and keras)	Precision, recall, F1-score, model loss and accuracy
[42]	2022	Proposed a DL-ESD model which first does feature extraction using LDA and then normalizes the features and does binary classification using different ML models.	Hello flooding (HF), decreased rank (DR) and version number attack (VNAs)	Simulated data set	Wire shark analyser and python	Precision, recall, F1-score, support, training and testing time
[43]	2022	Machine learning based IDS system (ML IDS) is proposed which uses feature scaling and principal component analysis for better results of ML models.	HF, DR and VNAs	Simulated data set	Not mentioned	Accuracy, precision, F1-score, feature importance, kappa, and Mathew correlation coefficient
[14]	2021	A light gradient boosting machine based lightweight VNA detection model has been introduced.	Version number attacks (VNAs)	Simulated data set	Cooja, python	Accuracy, precision, true negative rate, F-score, FPR

(continued)

TABLE 10.2 (Continued)
ML and DL-Based IDS Solutions

Reference	Publication Year	Contribution	Target Attack	Dataset	Tools/ Programming	Parameters
[44]	2021	A new secure RPL routing protocol based on machine learning is introduced. The classification of attacks is done with the help of SVM.	Hello flood, DODAG information solicitation (DIS), reduced rank, increase version	Simulated data set	Cooja, python	Packet delivery rate (PDR), energy consumption
[45]	2020	To counter HF attacks on the IoT RPL protocol, a gated recurrent unit network model based on DL was designed.	Hello flooding	Simulated data set	Cooja, python	Accuracy in terms of PDR and delay
[46]	2020	A novel data set was created, and machine learning algorithms were applied to consider the combined attack and vulnerabilities on two objective functions (OFs). Random forest and multilayer perceptron are most successful.	Rank, sybil, version, blackhole, decreased path metric, rank	Simulated data set	Cooja, weka	RSME, MSE, MAE, accuracy, ROC

[47]	2018	A novel data set is created, and deep learning-base models were proposed to detect the attacks.	Modification attacks, DR, hello flood, version number	Simulated dataset	Cooja	Performance based on F1 score and AUC test score
[20]	2019	Introduced ELNIDS uses ensemble learning to detect routing assaults.	HF and sinkhole, local repair, selective forwarding blackhole, sybil, clone ID	RPL-NIDDS17		Accuracy
[27]	2018	CHA-IDS is introduced.	HF, sinkhole, and wormhole attacks combined	Simulated dataset	Cooja	Precision, recall, accuracy, TPR, FPR, MAE, RMSE, RAE

T_RN signifies True Negative value means a total number of cases in which the predicted attack is not the same as the actual attack.

F_AP denotes False-Positive value, means the total number of cases in which prediction is attacked but actually it's not there.

F_AN signifies False-negative value, means the total number of cases in which prediction is not attacked but actually it's there.

10.5.1 Accuracy

Accuracy is measured as successfully identifying correct attack instances to the total attack instances. It can be defined as follows:

$$accuracy = \frac{T_RP + T_RN}{T_RP + F_AP + T_RN + F_AN} \tag{10.1}$$

10.5.2 Precision

It is defined as the ratio of true positive identified attack cases to total positive identified attack cases. It is calculated as given in Eq. (10.2):

$$precision = \frac{T_RP}{T_RP + F_AP} \tag{10.2}$$

10.5.3 Recall

It measures the actual attack cases which are predicted correctly, also called as sensitivity. It is the ratio of the number of attack cases that have been accurately classified to the total number of positive attack cases as shown in Eq. (10.3):

$$recall = \frac{T_RP}{T_RP + F_AN} \tag{10.3}$$

10.5.4 F1-Score

F1-measure considers accuracy and recall and is defined as the harmonic mean of the two measures. Its value lies between 0 and 1. It can be calculated as follows:

$$F1 - score = \frac{2 \times (precision \times recall)}{(precision + recall)} \tag{10.4}$$

10.6 LIMITATION WITH DL AND ML

Deep learning can identify large-scale attacks in IoT with computing efficiency in a multilayered manner. DL delivers cutting-edge approaches, efficiency, and performance gains not attainable with conventional ML algorithms. A proposed or trained model for one kind of problem may not function well when applied to another problem in the same area. Retraining of models is needed for applying models to the same kind of problem. ML-based techniques produce acceptable performance metrics but not without restrictions. Almost every ML algorithm contains hyperparameters whose values are specified before the learning process, and these parameters affect the learning system's behaviour. Due to the fixed data sets used in supervised learning, it is deemed stable, whereas RL and deep reinforcement learning (DRL) are not stable [48]. Training in a simulated environment utilizes an infinite amount of training data and deterministic system dynamics and hence does not accurately replicate real-world behaviour. Due to real-time data streaming in real-time applications, DRL and RL systems suffer from significant and unexpected delays in computing rewards and calculations. There are significant restrictions in DL techniques, such as no intra-layer connection, the network needs to be manually configured, and expensive models in terms of power usage and other computational resources [49]. Compiling complex ML algorithms is difficult since they need a lot of memory and energy when processing IoT systems. IoT devices have limited resources and extensive data collection [18, 50, 51].

ML and DL also suffer from the butterfly effect, which is a situation in which a little alteration in a system's intake results in turmoil in the output. Both ML and DL are sensitive to this impact in this situation. This problem exposes ML and DL algorithms used in IoT vulnerable to attacks in which adversaries alter input data on purpose to render the system unstable [52].

RL methods have a longer convergence time, which may be inappropriate for real-time applications. Furthermore, timing is crucial for safety-critical systems, and the system cannot tolerate delays. Therefore, additional research is required to increase the convergence rate. Training time is another barrier to deep learning models; as the training dataset grows, training time increases significantly [53].

There are issues associated with the lack of relevant and sufficient data sets for data-driven ML and deep learning approaches. Dataset availability and data security is also a major limitation. Learning techniques often examine and utilize various attributes of accessible data sets to generate training data sets. If a training data collection has low-quality data that incorporates noise, deploying a comprehensive and accurate learning method may not be perfect. As a result, authentication of training data sets is a significant difficulty in ML approaches for successful IoT network security [54]. Significant data sets are required to fully use ML algorithms in IoT systems, which are frequently difficult to collect depending on whether the system can recognize dangers and take appropriate steps. In this regard, data augmentation is a significant way to generate sufficient data collection based on existing actual data. However, there is a difficulty in that the generated new data samples must be correctly divided into different classes in order for ML algorithms to achieve optimum accuracy.

Identifying attacks is another important difficulty in IoT security to identify benign from malicious network states. A fundamental process in machine learning is feature selection, which affects the efficacy of ML IDS. To develop a successful IDS utilizing ML, researchers aim for the appropriate reduced feature set that does not compromise the attack categorization accuracy. There are no predefined rules for classifying the types and quantities of features. Researchers examine the dataset from a variety of viewpoints. If an intruder understands the attack type and can change the training data set used for ML approaches, they may easily adjust their attack kinds and network impacts.

10.7 CONCLUSION

The demand for IoT devices and the interest of both consumers and attackers are increasing daily. RPL is susceptible to numerous attacks that may be harmful to the user. Several approaches have been developed and proposed to identify and prevent these attacks. However, each method has its limitations. This chapter provides the researchers with insight into available intrusion detection mechanisms based on ML and other types of IDS. By analysing RPL attacks and the solutions related to those attacks based on IDSs, the potential for further research towards comprehensive solutions is presented.

REFERENCES

1. Al-Gburi, A., Al-Hasnawi, A., &Lilien, Differentiating security from privacy in Internet of Things: A survey of selected threats and controls. In *Computer and network security essentials* (pp.153–172). Springer, Cham, 2018.
2. GlobalData. (2018). IoT market forecast and growth opportunities. www.globaldata.com/store/report/gdtc0168ir--iot-market-forecast-and-growth-opportunities/.
3. Gartner. (2019). Enterprise and automative IoT market. www.gartner.com/en/newsroom/press-releases/2019-08-29-gartner-says-5-8-billion-enterprise-and-automotive-io/.
4. Hewlett Packard. (2014). HP study reveals 70 percent of Internet of Things devices vulnerable to attack. www8.hp.com/us/en/hp- news/press-release.html?id=1744676#.X_ibanYzbIV/.
5. Sobral, José V.V., Joel J.P.C. Rodrigues, Ricardo A.L. Rabêlo, Jalal Al-Muhtadi, and Valery Korotaev. "Routing protocols for low power and lossy networks in internet of things applications." *Sensors* 19, no. 9 (2019):2144.
6. Verma, Abhishek, and Virender Ranga. "The impact of copycat attack on RPL based 6LoWPAN networks in Internet of Things." *Computing* 103 (2020):1–22.
7. Verma, Abhishek, and Virender Ranga. "Mitigation of DIS flooding attacks in RPL-based 6LoWPAN networks." *Transactions on Emerging Telecommunications Technologies* 31, no. 2 (2020):e3802.
8. Kamble, Arvind, Virendra S. Malemath, and Deepika Patil. "Security attacks and secure routing protocols in RPL-based internet of things: Survey." In 2017 International Conference on Emerging Trends & Innovationin ICT (ICEI), pp. 33–39. IEEE, 2017.
9. Winter, Tim, Pascal Thubert, Anders Brandt, Jonathan W. Hui, Richard Kelsey, Philip Levis, Kris Pister, Rene Struik, Jean-Philippe Vasseur, and Roger K. Alexander. "RPL: IPv6 Routing Protocol for Low-Power and Lossy Networks." rfc6550 (2012):1–157.

10. Airehrour, David, Jairo Gutierrez, and Sayan Kumar Ray. "Secure routing for internet of things: A survey." *Journal of Network and Computer Applications* 66 (2016):198–213.
11. Kharrufa, Harith, Hayder A.A. Al-Kashoash, and Andrew H. Kemp. "RPL-based routing protocols in IoT applications: A review." *IEEE Sensors Journal* 19, no.15 (2019):5952–5967.
12. Musaddiq, Arslan, Yousaf Bin Zikria, Oliver Hahm, Heejung Yu, Ali Kashif Bashir, and Sung Won Kim. "A survey on resource management in IoToperating systems." *IEEE Access* 6 (2018):8459–8482.
13. Mayzaud, Anthea, Remi Badonnel, and Isabelle Chrisment. "A taxonomy of Attacks in RPL-based Internet of Things." *International Journal of Network Security* 18, no. 3 (2016):459–473.
14. Osman, Musa, Jingsha He, Fawaz Mahiuob Mohammed Mokbal, Nafei Zhu, and Sirajuddin Qureshi. "ML-LGBM: A Machine Learning Model based on light gradient boosting machine for the detection of version number attacks in RPL-basedNetworks." *IEEE Access* (2021) 9 83654–83665.
15. Vasseur, J., Navneet Agarwal, Jonathan Hui, Zach Shelby, Paul Bertrand, and Cedric Chauvenet. "RPL: The IP routing protocol designed for low power and lossy networks." *Internet Protocol for Smart Objects (IPSO) Alliance* 36 (2011) 1–20.
16. Simoglou, George, George Violettas, Sophia Petridou, and Lefteris Mamatas. "Intrusion detection systems for RPL security: a comparative analysis." *Computers & Security* 104 (2021):102219.
17. Ioulianou, Philokypros, Vasileios Vasilakis, Ioannis Moscholios, and Michael Logothetis. "A signature-based intrusion detection system for the internet of things." *Information and Communication Technology Form* (2018).
18. Mayzaud, Anthéa, Rémi Badonnel, and Isabelle Chrisment. "A distributed monitoring strategy for detecting version number attacks in RPL-based networks." *IEEE Transactions on network and service management* 14, no. 2 (2017):472–486.
19. Wallgren, Linus, Shahid Raza, and Thiemo Voigt. "Routing attacks and countermeasures in the RPL-based internet of things." *International Journal of Distributed Sensor Networks* 9, no.8 (2013):794326.
20. Verma, Abhishek, and Virender Ranga. "ELNIDS: Ensemble learning based network intrusion detection system for RPL based Internet of Things." In 2019 4th International conference on Internet of Things: Smartinnovation andusages (IoT-SIU), pp.1–6. IEEE, 2019.
21. Cervantes, Christian, Diego Poplade, Michele Nogueira, and Aldri Santos. "Detection of sinkhole attacks for supporting secure routing on 6LoWPAN for Internet of Things." In 2015 IFIP/IEEE International Symposium on Integrated Network Management (IM), pp. 606–611. IEEE, 2015.
22. Surendar, M., and A. Umamakeswari. "Indres: An intrusion detection and response system for internet of things with 6LoWPAN." In 2016 International Conference on Wireless Communications, Signal Processing and Networking (WiSPNET), pp. 1903–1908. IEEE, 2016.
23. Gara, Fatma, Leila Ben Saad, and Rahma Ben Ayed. "An intrusion detection system for selective forwarding attack in IPv6-based mobile WSNs." In 2017 13th International Wireless Communications and Mobile Computing Conference (IWCMC), pp. 276–281. IEEE, 2017.
24. Verma, Abhishek, and Virender Ranga. "CoSec-RPL: Detection of copycat attacks in RPL based 6LoWPANs using outlier analysis." *Telecommunication Systems* 75, no. 1 (2020):43–61.

25. Bostani, Hamid, and Mansour Sheikhan. "Hybrid of anomaly-based and specification-based IDS for Internet of Things using unsupervised OPF based on MapReduce approach." *Computer Communications* 98 (2017): 52–71.
26. Kasinathan, Prabhakaran, Gianfranco Costamagna, Hussein Khaleel, Claudio Pastrone, and Maurizio A. Spirito. "An IDS framework for internet of thing sem powered by 6LoWPAN." In Proceedings of the 2013 ACM SIGSAC Conference on Computer & Communications Security, pp. 1337–1340. ACM, 2013.
27. Napiah, Mohamad Nazrin, Mohd Yamani Idna Bin Idris, Roziana Ramli, and Ismail Ahmedy. "Compression header analyzer intrusion detection system (CHA-IDS) for 6LoWPAN communication protocol." *IEEE Access* 6 (2018):16623–16638.
28. Jhanjhi, N. Z., Sarfraz Nawaz Brohi, and Nazir A. Malik. "Proposing a rank and wormhole attack detection framework using machine learning." In 2019 13th International Conference on Mathematics, Actuarial Science, Computer Science and Statistics (MACS), pp. 1–9. IEEE, 2019.
29. A. Jamalipour, and S. Murali. "A Taxonomy of Machine Learning based Intrusion Detection Systems for the Internet of Things: A Survey." *IEEE Internet of Things Journal*, doi:10.1109/JIOT.2021.3126811.
30. Thakkar, Ankit, and Ritika Lohiya. "A review on machine learning and deeplearning perspectives of IDS for IoT: Recent updates, security issues, and challenges." *Archives of Computational Methods in Engineering* 28, no. 4 (2021):3211–3243.
31. Al Hajri, Mohamed I., Nazar T. Ali, and Raed M. Shubair. "Classification of indoor environments for IoT applications: A machine learning approach." *IEEE Antennas and Wireless Propagation Letters* 17, no. 12 (2018):2164–2168.
32. Breiman, Leo. "Randomforests." *Machinelearning* 45, no. 1 (2001):5–32.
33. Sharma, Mridula, Haytham Elmiligi, Fayez Gebali, and Abhishek Verma. "Simulating attacks for rpl and generating multi-class dataset for supervised machine learning." In 2019 IEEE 10th Annual Information Technology, Electronics and Mobile Communication Conference (IEMCON), pp. 0020–0026. IEEE, 2019.
34. Momand, Mohammad Dawood, and Mohabbat Khan Mohsin. "Machine Learning-based Multiple Attack Detection in RPL over IoT." In 2021 International Conference on Computer Communication and Informatics (ICCCI), pp. 1–8. IEEE, 2021.
35. Raza, Shahid, Linus Wallgren, and Thiemo Voigt. "SVELTE: Real-time intrusion detection in the Internet of Things." *Ad hoc networks* 11, no. 8 (2013):2661–2674.
36. Verma, Abhishek, and Virender Ranga. "Security of RPL based 6LoW-PAN Networks in the Internet of Things: A Review." *IEEE Sensors Journal* 20, no. 11 (2020):5666–5690.
37. Matsunaga, Takumi, Kentaroh Toyoda, and Iwao Sasase. "Low false alarm attackers detection in RPL by considering timing inconstancy between the rank measurements." *IEICE Communications Express* 4, no. 2 (2015):44–49.
38. Shreenivas, Dharmini, Shahid Raza, and Thiemo Voigt. "Intrusion detection in the RPL-connected 6LoWPAN networks." In Proceedings of the 3rd ACM international workshop on IoT privacy, trust, and security, pp. 31–38. ACM, 2017.
39. Fu, Rui, Zuo Zhang, and Li Li. "Using LSTM and GRU neural network methods for traffic flow prediction." 2016 31st Youth Academic Annual Conference of Chinese Association of Automation (YAC). IEEE, 2016.
40. Zeadally, Sherali, and Michail Tsikerdekis. "Securing Internet of Things (IoT) with machine learning." *International Journal of Communication Systems* 33, no. 1 (2020):e4169.

41. Al Sawafi, Yahya, Abderezak Touzene, and Rachid Hedjam. "Hybrid deep learning-based intrusion detection system for RPL IoT Networks." *Journal of Sensor and Actuator Networks* 12, no. 2 (2023):21.
42. Albishari, M., Li, M., Zhang, R., & Almosharea. "Deep learning-based early stage detection (DL-ESD) for routing attacks in Internet of Things networks." *The Journal of Supercomputing* 79, no. 3 (2023):2626–2653.
43. Saheed, Y. K., A. I. Abiodun, S. Misra, M. K. Holone, & Colomo-Palacios. "A machine learning-based intrusion detection for detecting internet of things network attacks." *Alexandria Engineering Journal* 61, no. 12 (2022):9395–9409.
44. Kasinathan, Prabhakaran, Gianfranco Costamagna, Hussein Khaleel, Claudio Pastrone, and Maurizio A. Spirito. "An IDS framework for internet of things empowered by 6LoWPAN." In Proceedings of the 2013 ACM SIGSAC conference on Computer & communications security,pp. 1337–1340. ACM, 2013.
45. Cakir, Semih, Sinan Toklu, and NesibeYalcin. "RPL attack detection and prevention in the Internet of Things networks using a GRU based deeplearning." *IEEE Access* 8 (2020):183678–183689.
46. Foley, John, Naghmeh Moradpoor, and Henry Ochenyi. "Employing a Machine Learning Approach to Detect Combined Internet of Things Attacks against Two Objective Functions Using a Novel Dataset." *Security and Communication Networks* 2020(2020).
47. Yavuz, Furkan Yusuf, DevrimÜnal, and Ensar Gül. "Deep learning for detection of routing attacks in the internet of things." *International Journal of Computational Intelligence Systems* 12, no. 1 (2018):39–58.
48. Dulac-Arnold, Gabriel, Daniel Mankowitz, and Todd Hester. "Challenges of real-world reinforcement learning." arXivpreprint arXiv:1904, no.12901 (2019).
49. Ahmad, Rasheed, and Izzat Alsmadi. "Machine learning approaches to IoT security: A systematic literature review." *Internet of Things* 14 (2021):100365.
50. Pongle, Pavan, and Gurunath Chavan. "Real time intrusion and worm-hole attack detection in internet of things." *International Journal of Computer Applications* 121, no. 9 (2015):1–9.
51. Verma, Abhishek, and Virender Ranga. "ELNIDS: Ensemble learning based network intrusion detection system for RPL based Internet of Things." In 2019 4th International conference on Internet of Things: Smart innovation and usages (IoT-SIU), pp.1–6. IEEE, 2019.
52. Hussain, Fatima, et al. "Machine learning in IoT security: Current solutions and future challenges." *IEEE Communications Surveys & Tutorials* 22, no. 3 (2020):1686–1721.
53. Roopak, Monika, Gui YunTian, and Jonathon Chambers. "Deep learning models for cyber security in IoT networks." In 2019 IEEE 9th annual computing and communication workshop and conference (CCWC), pp. 0452–0457. IEEE, 2019.
54. Sezer, Omer Berat, Erdogan Dogdu, and Ahmet Murat Ozbayoglu. "Context-aware computing, learning, and big data in internet of things: A survey." *IEEE Internet of Things Journal*5, no.1 (2017):1–27.

Part V

Security and Privacy in Intelligent IoT Network

11 Lightweight Cryptography Algorithms, Authorization, and Authentication Techniques in IoT

Diksha Chawla and Pawan Singh Mehra

11.1 INTRODUCTION

The Internet of Things (IoT) is going to be an integral part of our lives in the next few years. The IoT includes heterogeneous devices that gather all types of data using sensors, actuators, and gateways. IoT is envisioned to provide uninterrupted and accurate functioning of smart applications. The smart applications include automated vehicles, smart agriculture, healthcare, and surveillance systems. With the development and innovation of IoT communication, E-healthcare, automated cars, virtual reality, and Unmanned Aerial Vehicle (UAVs), security and surveillance use cases become possible. IoT applications are beneficial. However, these applications are vulnerable to attacks, such as an attacker can take control of automated vehicular system by malicious vehicles sending fraudulent messages. It can endanger the lives of any person in the network by blocking traffic light signals. The open access internet connectivity and physical systems increase the exposure to attacks such as data breaches, side-channel attacks, irregular/no updates of malware and ransomware, Man-in-the-Middle (MIMA), and impersonation attacks. These heterogeneous IoT applications generate, process, store, and transfer huge amounts of critical user data. The deployment of large-scale IoT applications is possible only when we can utilize these applications to make critical decisions. The challenging task is to assure the confidentiality, data security, and user authenticity.

The aforementioned issues motivated our discussions to analyze security concerns, requirements, solutions, and challenges for the IoT communication framework. The classical security mechanisms which are PC-based face many challenges due to the

DOI: 10.1201/9781003405535-16

resource-constrained nature of IoT such as low memory power, limited battery, limited computational power, and real-time data access. To address these issues, lighter cryptographic schemes need to be designed. The lightweight cryptographic schemes can be executed on IoT platforms like Radio Frequency Identification(RFID) and wireless sensor network(WSN) with ease. The lightweight schemes mainly considered small code measures, little clock sizes, limited block sizes, and smaller key sizes. The major part involving the lightweight schemes is the hash function, stream, and block cipher.

In this chapter, we provide an in-detail discussion of the security issues of IoT architecture applications and the different technologies involved. Concerning the above-mentioned issues in this work, we reviewed major lightweight algorithms for securing resource-constrained IoT. But for IoT, the major challenges are the implementation of these schemes such as (1) depending on the application identifying the best cryptographic solutions, (2) implementing a symmetric key onto the device, (3) handling generation and distribution of secure key, and (4) ensuring mutual authentication with variety and varsity of devices. Our research helps other researchers to find the best suitable lightweight cryptographic solutions for secure IoT communication frameworks.

11.2 LITERATURE SURVEY

With the continuous advancement in IoT, researchers have enlightened IoT architectures and IoT applications and requirements. As smart home applications are gaining popularity after it is accepted commercially. Shouran et al. [1] in his work discussed that the majority of IoT devices generate and process confidential data. It includes military applications, smart homes, and smart wearables and such devices have security and privacy concerns. The authors discussed that these devices lack authentication and confidentiality. These vulnerabilities lead to impersonation, repudiation, reply, message, modification, etc. Their schemes do not provide any cryptographic solutions for smart home data security [2]. IoT architecture and major applications are discussed by Hassija et al. [3]. The authors also discussed the major security concerns in the IoT communication framework. The authors provide blockchain and fog computing-based solutions for IoT security. However, in [4] IoT-based UAV applications are discussed and highlighted that blockchain methods are very time-consuming and not efficient. It is due to the time to mine the block being too high which is not suitable for real-time applications.

11.3 A SYSTEMATIC APPROACH TO IOT

Recently, tremendous growth and opportunities have been found in IoT-enabled communication. Therefore, security is one area that must be prioritized when implementing IoT-enabled communication. In this section, the IoT architectures, technologies, and applications concerning security issues are discussed.

11.3.1 IoT Architecture

In IoT, communication privacy and confidentiality needs to be considered for each layer because each layer involves data collection, processing, analyzing, and

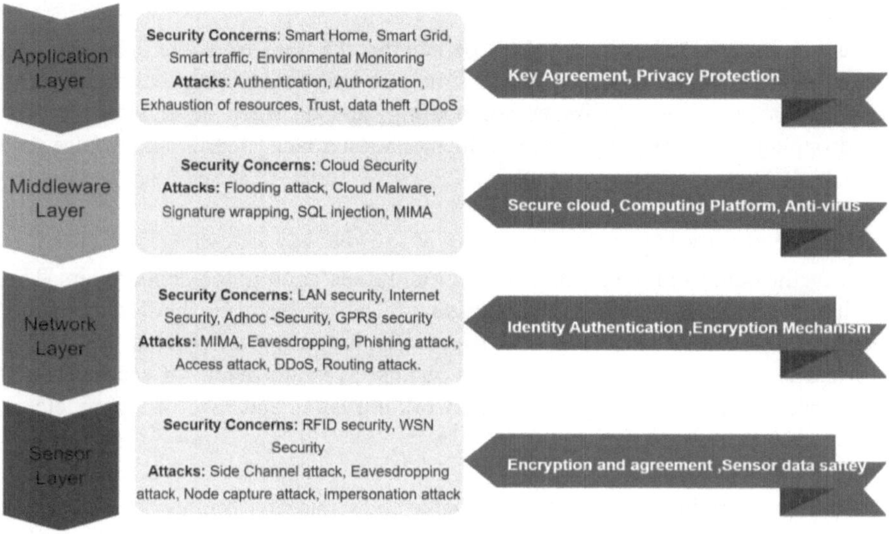

FIGURE 11.1 IoT layered architecture.

TABLE 11.1
IoT Architecture Security Issues and Countermeasures by other Researchers

IoT Layer	Security Challenge	Counter Measures	Ref.
Perception Layer	- Security from the malicious sensor node. - Cryptographic and key management. - Routing packets to different routes.	- Public key algorithm - Private key encryption algorithm	[5]
	- Denial of Service	- Routing packets to different routes - Intrusion detection	[6]
Network layer	- Internet security - Cloning protection	- Symmetric authentication - Asymmetric authentication - Hybrid approaches	[7]
Application layer	- Device authentication - User identification - Sensor node authentication	- Symmetric key algorithm with the capabilities of access control	[8]
		- Blockchain solutions	[9]
		- Public key encryption	[10]

exchanging information between the devices and communicating entities. The layered specific architecture of IoT is depicted in Figure 11.1. It represents security concerns, attacks, and solutions that need to be provided for secure communications. The major security concerns at each layer are represented in Table 11.1. The detailed description is as follows:

11.3.1.1 IoT Sensor Layer

The functioning of this layer involves perceiving the data from the surroundings. Such data is accessed by the user in real time or by using the cloud-based platform. Major sensor layer security issues are booting attacks, sleep deprivation attacks, eavesdropping and interference, node impersonation attack, flooding, and false data injection attack.

11.3.1.2 Network Layer

It collects the data from the perception layer and transfers it to the upper layers. This layer involves data collection and transmission by using various technologies such as WiFi, Sigfox, LoRa, ZigBee, NB-IoT, Bluetooth, and protocols such as IPv4, IPv6, and MQTT. The major security concerns at this layer are denial of service (DoS) attacks, access attacks, routing attacks, network intrusion, and eavesdropping, etc.

11.3.1.3 Middleware Layer

The functioning of this layer involves a reliable and secure IoT framework. The major use of this layer is to process and transport the data to the above application layer. The major threats at this layer include eavesdropping attacks, false data injection attacks, signature wrapping attacks, and cloud malware injection attacks.

11.3.1.4 Application Layer

Futuristic applications include intelligent homes, intelligent meters, smart parking, smart wearables, etc. Fingerprint and timing-based attacks (FATS) are the popular attacks at this layer. The major security concerns at this layer are mutual authentication, information privacy data thefts, and privacy issues.

11.3.2 IoT Technologies

As IoT connect heterogeneous applications, there have been several different technologies that provide support for long-range and short-range communication. For massive IoT connectivity, low-power wide area (LPWA) technologies are popular and provide wide area coverage, high energy efficiency, data rate, and low power consumption. LPWA supports long-range connectivity using technologies such as Sigfox, LoRa, WiFi, ZigBee, and NB-IoT. Such technologies are easily accessed through 5G, which provides unlicensed or unused spectrum bands. 4G will not be sufficient with the tremendous annual growth rate in IoT by 25% annually and is expected to reach 80 billion by 2030. 5G extensive communication support enhanced mobile broadband (eMBB), enhanced machine type communication (eMTC), and critical communications (URLLC). Such massive connectivity allows machine type communications (MTC) to be a smart city and smart transportation. However, the major concern is to ensure security as these applications generate diverse forms of data traffic.

11.3.3 IoT Applications

In IoT communication, the existing applications are smart metering and grids, automated cities, automated healthcare, IIoT, green IoT-based agriculture, autonomous vehicles, security, and emergencies. These applications need a high-security framework. Some of these applications with security concerns are explained in this section:

11.3.3.1 Smart Cities

When it comes to enhancing people's lives with technology, automated cities play a major role. Smart cities can monitor and control major sectors such as climate change, scarcity of resources, and parking and transportation. Many of these services can be utilized through the smart card, which threatens citizens' privacy. These applications need to be designed with a high level of security concerning threats such as information leakage, traffic analysis, user impersonation, and smart card theft. Blockchain technology enables communicating parties to exchange information securely while preserving privacy and threats. However, researchers need to pay much attention to threats to confidentiality, integrity, and authenticity.

11.3.3.2 Smart Health

Remote health monitoring systems monitor and analyze patients' health parameters remotely by using sensor devices. However, ensuring the safety and privacy of such applications is essential as it involves the life of patients. Such applications are vulnerable to threats such as:

- **Fingerprint and Timing-Based Snooping:** In this scenario, intruder tries to access information during data transient between sensors and from the sensor to a private user location.
- **Router Attack:** As routing allows data to be delivered remotely and encourages network versatility, e-healthcare systems require data to be delivered securely and protect user identity.

11.3.3.3 Smart Agriculture

With the use of modern technologies such as IoT, blockchain, cloud computing, and big data, agriculture information collected from various sensor nodes such as temperature, light, soil, and humidity can be stored and analyzed with ease. The information collected from various sources is processed using various artificial intelligence (AI) programs and stored on the central server. Cyberattacks in agriculture are continuously increasing due to social networking, phishing, and other malware, such as banking fraud. The entire UAV-based communication is based on wireless communication. Therefore, the communication channels are highly vulnerable to snooping, dispatch systems, Wi-Fi, and data modification attacks. Some of the recent attacks on the agriculture sector are:

- **Data attacks** include false data injection, insider and cloud data leakage, and missing information attacks.
- **Networking and equipment attacks contain** side-channel, botnet, malware injection, DoS, and radio frequency jamming.

TABLE 11.2
Comparative Analysis of other Researchers' Work on IoT Security Issues

Author & Ref.	Year	Vision	Security Concerns
Noor and Hassan [5]	2018	Analysis of IoT perception, network, and application layer.	RFID and WSN attacks such as jammers, injection, and side-channel attacks.
Rahimi et al. [11]	2018	Analysis of IoT security from a 5G perspective.	Flooding on the cloud, RF jamming, sleep deprivation attacks.
Sfar et al. [12]	2018	IoT components and their interaction with each other and the environment.	Privacy, trust, integrity, and authentication
Butt et al. [13]	2019	IoT applicability in the health sector and security issues.	Insider and external attack.
Shouran et al. [1]	2019	Smart home privacy issues	Eavesdropping, replay, user impersonation attack.
Song et al. [14]	2020	Security concerns in green IoT-based architecture are analyzed.	Man-in-middle attack, attack on cloud storage
Tsiknas et al. [15]	2021	Analysis of the industrial IoT sector.	Phishing attacks, supply chain attacks

11.4 IOT SECURITY ISSUES, REQUIREMENTS, AND THREATS

Table 11.2 represents the major security concerns, requirements, and threats concerning the IoT communication framework. The in-detail descriptions are discussed in this section:

11.4.1 Security Issues

To achieve a high-security IoT framework, the following issues need to be considered:

A. **Maintaining Data Integrity:** The IoT data needs to maintain the integrity of the network. Integrity ensures reliable communication between users. The IoT architectures must be capable of providing integrity. Integrity confirms that the data is safeguarded from unauthorized users.
B. **Maintaining Data Confidentiality:** It ensures that the user's confidential information cannot be leaked to any unauthorized user. The major concern of IoT smart objects is maintaining the confidentiality of data among communicating entities. In addition to that, important communication protocols such

as constrained application protocol (CoAP), IEEE 802.15.4 standard, and the IPv6 over low-power wireless personal area networks (6LoWPAN) are used to achieve data confidentiality.

C. **Maintaining Privacy of Communication:** The privacy of communication, data, and images are the main privacy and security concerns. Such issues arise due to multiple resource-constrained device connectivity. User credentials such as passwords, IDs, images, and other sensitive information such as location need to be maintained.
D. **Ensuring Mutual authentication:** Authentication is required at different levels in the IoT framework as it involves sensors, Gateway Node (GWN), and user devices. The user's authenticity should be verified before communication takes place [16].

11.4.2 Key Requirements for IoT Security

Therefore, an hour is required to secure IoT communication so that the majority of people can effectively use it. Confidentiality, authentication, and integrity are important requirements of IoT communication. The IoT devices are resource-constrained. Therefore, there must be a strong authentication method to identify the legitimate user of the system.

11.4.3 IoT Threats

All IoT devices are receiving data from sensors and transfer that data over the internet to communicating entities. Due to an open internet connection, such communication is susceptible to many attacks [17, 18]. Some attacks modify the user data while others can only get relevant information. Considering the above-mentioned cases, attacks on IoT systems are categorized as:

1. **Denial of service (DoS)**: Either by jamming or by flooding the network component. These attacks do not allow legitimate users to access the communication link.
2. **Impersonation attack**: The attacker by masquerading legitimate identity behaves like a legitimate user and accesses the communication link. An adversary can access the sensor node and send false information to legitimate users. An authentication scheme is required to address this issue.
3. **Reply attack**: A malicious node by copying the original message sends the same message to the receiver. An authentication scheme should avoid the use of the same messages each time for communication.
4. **Mutual authentication**: In an IoT communication scenario, mutual authentication is a must between users, sensor nodes, and base station.
5. **Man-in-middle attack**: Where the intruder may attempt to control the communication channel.

11.5 RECENT LIGHTWEIGHT CRYPTOGRAPHY FOR IOT SECURITY

11.5.1 Asymmetric Encryption

Public key (asymmetric) encryption is used to secure IoT communication by sharing secret keys with the legitimate user. The encryption is based on the mathematical framework, where public and private keys are generated randomly. Anyone with access to public keys can encrypt data, but only authorized users with private keys can decrypt the information. Recently, [19] analyzed that due to the complexity of integer factorization, the popularity of public key encryption techniques such as RSA is increasing continuously. The researchers [20] also focus on elliptic curves cryptography (ECC) due to its advantages in lightweight cryptography. In IoT, alternative approaches such as hyper-elliptic curve cryptography [21], NTRU [22], and Blue Ray [23] were proposed, which are based on the shorter key size.

11.5.2 Symmetric Key Encryption

The single key is used at both ends to send and receive messages. The major challenge is to share secret keys with legitimate users. It is difficult to identify eavesdroppers in the communication channel; therefore, ensuring that keys are shared with only legitimate users is a difficult task. The symmetric key cryptography algorithm, such as advanced encryption standard (AES), is popular for encrypting data without sacrificing keys [3].

11.5.3 Hashing

Hash-based authentication is widely used for ensuring integrity, verification, and secure digital signature and hash-based biometric authentication. The efficiently implemented hash functions are believed to be resistant to quantum adversaries. Hafizul and Prosanta [24] security issues in IoT communication have been prevailing for many years as IoT deals with the heterogeneous type of data sharing and generation. Therefore, the authors [25] work on securing multimedia communication. Their protocol uses authentication to secure user credentials based on the two-step process. Therefore, it allows only authenticated users to access the communication channel.

11.6 LIGHTWEIGHT SCHEMES FOR IOT

IoT devices have limited batteries and limited computational capabilities. Therefore, it is expedient to develop lighter yet stronger cryptographic solutions. Based on AES cryptographic primitive, the author [26] proposed a lightweight security enhancement scheme. Their work is based on a dynamic-key generation process based on time to provide authentication. Symmetric algorithms such as height, Tiny Encryption Algorithm (TEA), and RC5 provide solutions for possible attacks such as saturation-related and differential attacks [5]. A secured IoT communication framework is

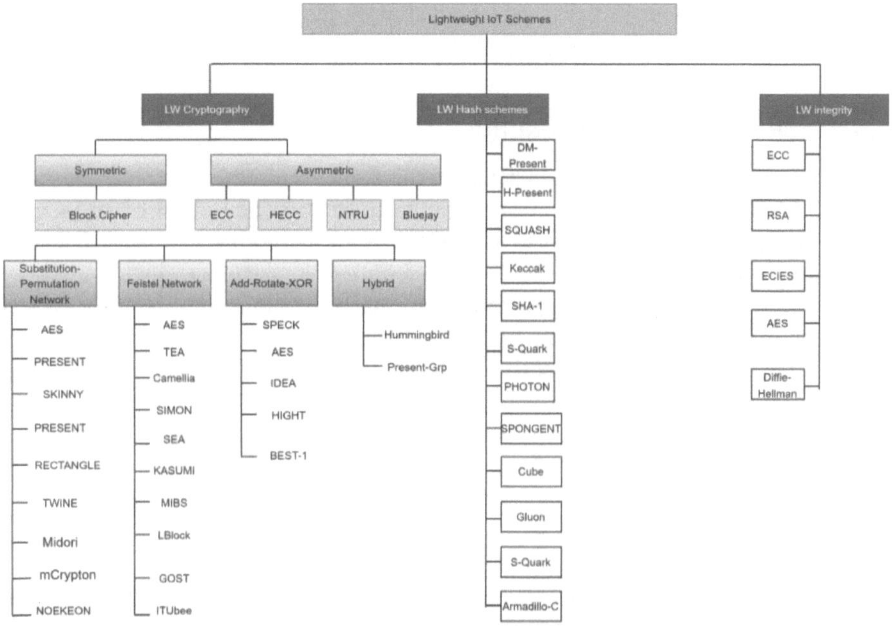

FIGURE 11.2 Classification of lightweight cryptographic schemes.

achieved using asymmetric AES. The lightweight encryption schemes based on hash and bit-wise X-OR operation are used for IoT devices. Lightweight cryptography is divided into three major categories: lightweight cryptography; lightweight hash functions; and lightweight integrity schemes. The main task of lightweight schemes is to set up an adequate control mechanism for data privacy that must consider the data collection, data transmission, and storage. The classification for these schemes is represented in Figure 11.2:

11.6.1 Lightweight Cryptography

The substitution-permutation network-based PRESENT [27] uses a 64-bit block based on 80-bit and 128-bit keys. However, this algorithm is hardware efficient and uses 4-bit S-boxes, whereas it takes large cycles in the block. Therefore, an upgraded version is required of this algorithm. Another work on security challenges and lightweight cryptography was discussed by Thakor et al. [28]. The authors discussed that while collecting data from the physical environment, they become an attractive target for many authors. They emphasize that conventional cryptography is not acceptable for resource-constrained devices. Therefore, the authors discussed lightweight cryptography algorithms based on block cipher such as (a) substitution-permutation network, (b) Generalized Feistel Network (GFN), (c) nonlinear feedback shift register (NLFSR), and (d) hybrid cipher. Reddy et al. [29] proposed a lightweight cryptographic scheme. Their protocol provides resistance against replay attacks, clock synchronization problems, smart card stolen attacks, insider attacks, and user impersonation attacks.

11.6.2 Lightweight Hash Function

These cryptographic hash functions generate the fixed representation of the message. These hash functions are based on one-way hash function and collision resistance. The one-way hash function is defined as if, H (m) is known by the adversary. It should be impossible to extract the message 'm'. The collision resistance ensures H (m) ≠ H (m'). The secure messages digest and secure hash functions are used in a variety of applications such as message authentication, intruder detection, and digital signature. SHA3 has four fixed-length hash algorithms SHA3-224, SHA3-256, SHA3-384, and SHA3-512 [30]. NIST has recommended SPONGENT, PHOTON, Quark, and Lesamnta-LW hash schemes. These hash functions are popular due to their capability to produce much smaller memory footprints [31].

11.6.3 Lightweight Integrity Schemes

Miller in 1985 defined the hard cryptographic algorithm known as ECC. ECC is considered a more suitable solution in comparison with RSA as it requires a small key size. Smart homes contain smart devices such as smart meters, smart doors, smart bulbs, and smart electricity systems. Any unauthorized users can take control of these systems and access data. Cryptographic approaches such as symmetric and asymmetric are used to maintain the integrity of network communication. Therefore, Kaur and Kumar [32] provide a two-parameter-based authentication method to secure user data. Their protocol is based on the ECC scheme. ECC is the most popular choice to ensure the integrity of IoT data. Recent approaches such as Software Defined Networks (SDN) and blockchain are also utilized to maintain integrity.

11.7 LIGHTWEIGHT AUTHENTICATION SCHEMES FOR IOT

11.7.1 Two-Factor Authentication Schemes

Authenticating user devices before joining the network is proposed by Yeh et al. [33]. The authors efficiently utilize the ECC- based schemes for secure communication. Additionally, to address security concerns, their protocol uses smart card-based authentication. Their work considers mutual authentication and provides resistance from insider and masquerade attacks. Later, it was found that passwords are susceptible to intercept [34]. Therefore, Althobaiti et al. [35] proposed the authentication scheme. In this scheme, the user iris is used to authenticate the user whenever he wants to login into the network. For authentication and key agreement, RSA and Diffie Hellman algorithms were proposed. Authors identified that previous schemes are not resistant to masquerading attacks. Therefore, the authors implemented the Tiny PK authentication protocol [36].

11.7.2 Three-Factor Authentication Schemes

For resource-constrained IoT communication, three parameters are used to ensure secure communication. The ECC-based scheme, which provides key agreement and user authentication, is proposed by the author [37] based on lightweight computation. The information must remain confidential while transferring to the legitimate user. Therefore, the authors proposed a protocol for securing multimedia information and preserving the data from external attacks [25]. The BAN logic is used to evaluate the security goal of the proposed protocol. The significant growth in IoT communication is due to the WSN; the nodes are spread in unattended targeted areas to provide real-time information. This issue has been addressed by authors [38] by using the MAC function. The results are analyzed and proved mathematically using BAN logic and the Random Oracle Model (ROM).

11.7.3 Multi-factor Authentication Schemes

The lightweight schemes are used for authentication such as multi-factor authentication based on ECC, XOR, and bit-wise operations [39]. Other than password schemes, graphical authentication schemes are also popular [40] for authenticating users. Similarly, many schemes based on three parameters [41] were proposed. These schemes are designed to overcome the drawbacks of previous schemes, such as stolen card attacks and offline password guessing attacks. The multi-parameter schemes are classified according to authentication factors such as (a) what you know, e.g. user credentials. (b) What you have, e.g. authorized smart card. (c) What are you, e.g. your biometric identity? Examples of multi-factor schemes are Google 2-step and FIDO U2F (universal two-factor) [42].

11.8 LIGHTWEIGHT CRYPTOGRAPHICALLY SECURE AUTHENTICATION MODEL

In the IoT communication framework, the data is collected from heterogeneous sensor nodes. These sensor nodes are responsible for data dissemination. Such nodes are susceptible to false data injection and node impersonation attacks. In Figure 11.3, we have illustrated the architecture for secure IoT communication by using lightweight schemes. In this architecture, by using internet communication, various IoT-based applications such as smart cities, smart agriculture, and smart automobiles are possible for the betterment of human life. The attacker can easily intercept the information communicated over the network. The attacker can capture, store, modify information, and send it to the unauthorized user. Therefore, lightweight authentication and lightweight cryptography preserve the privacy of user data. In the proposed architecture, we have assumed that all sensor nodes deployed in one particular area send the information to the cluster head node. These nodes are then connected to the base station and gateway node. In an IoT-enabled WSN, these nodes forward the sensitive data to nearby gateway nodes through dedicated communication channels by using cryptographically secure schemes.

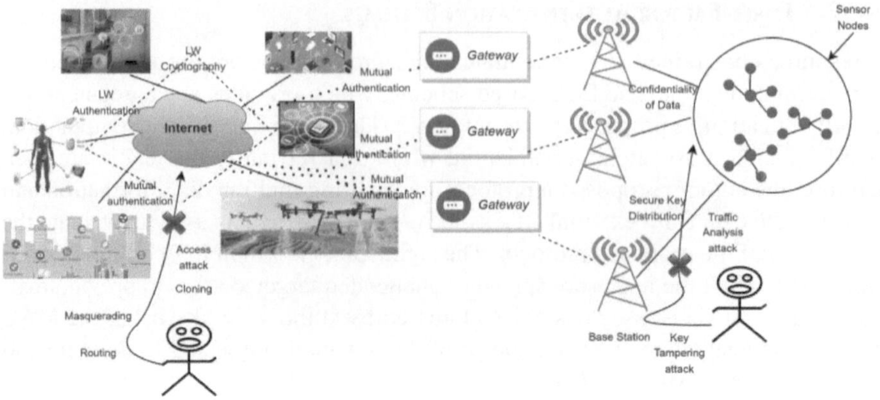

FIGURE 11.3 Authentication model.

11.9 LIMITATIONS OF EXISTING LIGHTWEIGHT CRYPTOGRAPHY

Considering the above-mentioned work, we analyzed lightweight cryptographically secured algorithms for IoT security has certain limitations:

1. **Secret key security**: By using lightweight cryptographic schemes, communicating entities share confidential information. However, the secret key is shared on the insecure classical channel which is susceptible to interception.
2. **Algorithms**: The lightweight cryptographic algorithms are based on mathematical primitives. Despite the advantages, there is no provision for how these algorithms will safeguard future IoT communications.
3. **Key size**: The asymmetric schemes are popular as they provide secure key generation and distribution. However, to increase the security for asymmetric algorithms, such as RSA, the key size should be increased. It makes encryption slower and increases the complexity.
4. **Memory constraints**: The AES algorithm ensures safe communication. However, considering the IoT resource-constrained nature, AES is not acceptable due to its high memory requirement to store larger blocks, key size, and s-boxes.

11.10 CONCLUSION AND FUTURE WORK

With the advancement of IoT devices around us, connected to the internet is increasing rapidly. The major issues concerning these applications are security, privacy, and authentication. Given the need for lightweight cryptography schemes for devices with limited computational power, efficient sensing abilities, and energy constraints. From the review, we have analyzed that RSA and AES schemes impress by their more hard mathematical structures. On the other side, hashing is considered the most suitable for ensuring authentication. In addition to that, block cipher techniques such as Feistel network-based SIMON and add rotate-XOR-based SPEK are most suitable for compact implementations. In terms of attacks and countermeasures, the security evaluation of major IoT applications is discussed. We have also proposed a novel architecture that provides data confidentiality and authentication to secure the IoT communication framework from cyberattacks. However, many recent cyberattacks have also appeared with the growth of LWC techniques. In addition, these LWC techniques are based on mathematical structures that are not secure with powerful futuristic computers. Therefore, advancements in cyber security algorithms and correspondingly increasing cyberattacks have opened new doors for researchers in the area of cybersecurity and in the field of lightweight cryptography and authentication techniques.

REFERENCES

1. Z. Shouran, A. Ashari, and T. Kuntoro, "Internet of Things (IoT) of Smart Home: Privacy and Security," *International Journal of Computer Applications*, vol. 182, no. 39, pp. 3–8, 2019, doi: 10.5120/ijca2019918450.
2. H. Nagdewani and P. S. Mehra, "A Complete Internet of Things based Home Security System," in *2022 Second International Conference on Computer Science, Engineering and Applications (ICCSEA)*, IEEE, 2022, pp. 1–5. doi: 10.1109/ICCSEA54677.2022.9936168.
3. V. Hassija, V. Chamola, V. Saxena, D. Jain, P. Goyal, and B. Sikdar, "A survey on IoT security: Application Areas, Security Threats, and Solution Architectures," *IEEE Access*, vol. 7, pp. 82721–82743, 2019, doi: 10.1109/ACCESS.2019.2924045.
4. P. S. Mehra, Y. B. Mehra, A. Dagur, A. K. Dwivedi, M. N. Doja, and A. Jamshed, "COVID-19 suspected person detection and identification using thermal imaging-based closed circuit television camera and tracking using drone in Internet of Things," *International Journal of Computer Applications in Technology*, vol. 66, no. 3–4, pp. 340–349, 2021.
5. M. binti Mohamad Noor and W. H. Hassan, "Current research on Internet of Things (IoT) security: A survey," *Computer Networks*, vol. 148, pp. 283–294, 2019, doi: 10.1016/j.comnet.2018.11.025.
6. M. Turkanović, B. Brumen, and M. Hölbl, "A novel user authentication and key agreement scheme for heterogeneous ad hoc wireless sensor networks, based on the Internet of Things notion," *Ad Hoc Networks*, vol. 20, pp. 96–112, 2014, doi: 10.1016/j.adhoc.2014.03.009.
7. S. Wang, J. Wang, and Z. Yu, "Privacy-Preserving Authentication in Wireless IoT: Applications, Approaches, and Challenges," *IEEE Wireless Communications*, vol. 25, no. 6, pp. 60–67, 2018, doi: 10.1109/MWC.2017.1800109.
8. A. Braeken, P. Porambage, M. Stojmenovic, and L. Lambrinos, "EDAAAS: Efficient distributed anonymous authentication and access in smart homes," *International Journal of Distributed Sensor Networks*, vol. 12, no. 12, 2016, doi: 10.1177/1550147716682037.
9. A. Dorri, S. S. Kanhere, R. Jurdak, and P. Gauravaram, "Blockchain for IoT security and privacy: The case study of a smart home," *2017 IEEE International Conference on Pervasive Computing and Communications Workshops, PerCom Workshops 2017*, no. January, pp. 618–623, 2017, doi: 10.1109/PERCOMW.2017.7917634.
10. N. Li and D. Liu, "Lightweight Mutual authentication for IoT and Its Applications," vol. 14, no. 8, 2017, doi: 10.1109/TSUSC.2017.2716953.

11. H. Rahimi, A. Zibaeenejad, P. Rajabzadeh, and A. A. Safavi, "On the security of the 5G-IoT architecture," *ACM International Conference Proceeding Series*, 2018, doi: 10.1145/3269961.3269968.
12. A. Riahi Sfar, E. Natalizio, Y. Challal, and Z. Chtourou, "A roadmap for security challenges in the Internet of Things," *Digital Communications and Networks*, vol. 4, no. 2, pp. 118–137, Apr. 2018, doi: 10.1016/j.dcan.2017.04.003.
13. S. A. Butt, J. L. Diaz-Martinez, T. Jamal, A. Ali, E. De-La-Hoz-Franco, and M. Shoaib, "IoT Smart Health Security Threats," *Proceedings – 2019 19th International Conference on Computational Science and Its Applications, ICCSA 2019*, no. October, pp. 26–31, 2019, doi: 10.1109/ICCSA.2019.000-8.
14. J. Song, Q. Zhong, W. Wang, C. Su, Z. Tan, and Y. Liu, "FPDP:Flexible Privacy-preserving Data Publishing Scheme for Smart Agriculture," *IEEE Sensors Journal*, vol. XX, no. Xx, pp. 1–1, 2020, doi: 10.1109/jsen.2020.3017695.
15. K. Tsiknas, D. Taketzis, K. Demertzis, and C. Skianis, "Cyber Threats to Industrial IoT: A Survey on Attacks and Countermeasures," *IoT*, vol. 2, no. 1, pp. 163–186, 2021, doi: 10.3390/iot2010009.
16. P. S. Mehra, M. N. Doja, and B. Alam, "Codeword Authenticated Key Exchange (CAKE) light weight secure routing protocol for WSN," *International Journal of Communication Systems*, vol. 32, no. 3, pp. 1–27, 2019, doi: 10.1002/dac.3879.
17. J. Sharma and P. S. Mehra, "Secure communication in IOT-based UAV networks: A systematic survey," *Internet of Things*, vol. 23, p. 100883, Oct. 2023, doi: 10.1016/J.IOT.2023.100883.
18. J. Sharma and P. S. Mehra, "A survey of security challenges and existing prevention methods in FANET," in Auerbach Publications eBooks, 2023, pp. 252–262. doi: 10.1201/9781003371380-24.
19. J. I. Ahmad, R. Din, and M. Ahmad, "Analysis review on public key cryptography algorithms," *Indonesian Journal of Electrical Engineering and Computer Science*, vol. 12, no. 2, pp. 447–454, Nov. 2018, doi: 10.11591/ijeecs.v12.i2.pp447-454.
20. Z. Liu, X. Huang, Z. Hu, M. K. Khan, H. Seo, and L. Zhou, "On emerging family of elliptic curves to secure internet of things: ECC comes of age," *IEEE Transactions on Dependable and Secure Computing*, vol. 14, no. 3, pp. 237–248, 2017, doi: 10.1109/TDSC.2016.2577022.
21. J. Fan, L. Batina, and I. Verbauwhede, "HECC Goes Embedded: An Area-Efficient Implementation of HECC," in *Selected Areas in Cryptography*, R. M. Avanzi, L. Keliher, and F. Sica, Eds., in Lecture Notes in Computer Science. Berlin, Heidelberg: Springer, 2009, pp. 387–400. doi: 10.1007/978-3-642-04159-4_25.
22. J. Hoffstein, J. Pipher, and J. H. Silverman, "NTRU: A ring-based public key cryptosystem," in *Algorithmic Number Theory*, J. P. Buhler, Ed., in Lecture Notes in Computer Science. Berlin, Heidelberg: Springer, 1998, pp. 267–288. doi: 10.1007/BFb0054868.
23. M.-J. O. Saarinen, "The BlueJay Ultra-Lightweight Hybrid Cryptosystem," in *2012 IEEE Symposium on Security and Privacy Workshops*, San Francisco, CA, USA: IEEE, May 2012, pp. 27–32. doi: 10.1109/SPW.2012.11.
24. S. K. Hafizul and I. Prosanta, "Efficient authentication protocol for secure multimedia communications in IoT-enabled wireless sensor networks," in Multimedia Tools and Applications, August 2017,doi: 10.1007/s11042-017-5376-4.
25. D. Mishra, P. Vijayakumar, V. Sureshkumar, R. Amin, S. H. Islam, and P. Gope, "Efficient authentication protocol for secure multimedia communications in IoT-enabled wireless sensor networks," *Multimedia Tools and Applications*, vol. 77, no. 14, pp. 18295–18325, Jul. 2018, doi: 10.1007/s11042-017-5376-4.

26. Z. Musliyana, T. Arif, and R. Munadi, "Security enhancement of advanced Encryption Standard (AES) using time-based dynamic key generation,"ARPN Journal of Engineering and Applied Sciences, vol. 10, pp. 8347–8350, Jan. 2015.
27. I. Bhardwaj, A. Kumar, and M. Bansal, "A review on lightweight cryptography algorithms for data security and authentication in IoTs," in *2017 4th International Conference on Signal Processing, Computing and Control (ISPCC)*, Solan, India: IEEE, Sep. 2017, pp. 504–509. doi: 10.1109/ISPCC.2017.8269731.
28. V. A. Thakor, M. A. Razzaque, and M. R. A. Khandaker, "Lightweight Cryptography Algorithms for Resource-Constrained IoT Devices: A Review, Comparison and Research Opportunities," *IEEE Access*, vol. 9, pp. 28177–28193, 2021, doi: 10.1109/ACCESS.2021.3052867.
29. A. Goutham Reddy, E.-J. Yoon, A. K. Das, and K.-Y. Yoo, "Lightweight authentication with key-agreement protocol for mobile network environment using smart cards," *IET Information Security*, vol. 10, no. 5, pp. 272–282, 2016, doi: 10.1049/iet-ifs.2015.0390.
30. M. J. Dworkin, "SHA-3 Standard: Permutation-Based Hash and Extendable-Output Functions," *NIST*, Aug. 2015, Accessed: Mar. 16, 2023. [Online]. Available: www.nist.gov/publications/sha-3-standard-permutation-based-hash-and-extendable-output-functions
31. D. Chawla and P. S. Mehra, "A Survey on Quantum Computing for Internet of Things Security," *Procedia Computer Science*, vol. 218, pp. 2191–2200, Jan. 2023, doi: 10.1016/j.procs.2023.01.195.
32. S. Yu, N. Jho, and Y. Park, "Lightweight Three-Factor-Based Privacy- Preserving Authentication Scheme for IoT-Enabled Smart Homes," *IEEE Access*, vol. 9, pp. 126186–126197, 2021, doi: 10.1109/ACCESS.2021.3111443.
33. H. L. Yeh, T. H. Chen, P. C. Liu, T. H. Kim, and H. W. Wei, "A secured authentication protocol for wireless sensor networks using Elliptic Curves Cryptography," *Sensors*, vol. 11, no. 5, pp. 4767–4779, 2011, doi: 10.3390/s110504767.
34. D. Chawla and P. S. Mehra, "A roadmap from classical cryptography to post-quantum resistant cryptography for 5G-enabled IoT: Challenges, opportunities and solutions," *Internet of Things*, vol. 24, p. 100950, Dec. 2023, doi: 10.1016/J.IOT.2023.100950.
35. O. Althobaiti, M. Al-Rodhaan, and A. Al-Dhelaan, "An efficient biometric authentication protocol for wireless sensor networks," *International Journal of Distributed Sensor Networks*, vol. 2013, 2013, doi: 10.1155/2013/407971.
36. R. Watro, D. Kong, S. Cuti, C. Gardiner, C. Lynn, and P. Kruus, "TinyPK: securing sensor networks with public key technology," Proceedings of the 2nd ACM workshop on Security of ad hoc and sensor networks, p. 59, 2004, doi: 10.1145/1029102.1029113.
37. Y. Lu, G. Xu, L. Li, and Y. Yang, "Anonymous three-factor authenticated key agreement for wireless sensor networks," *Wireless Networks*, vol. 25, no. 4, pp. 1461–1475, 2019, doi: 10.1007/s11276-017-1604-0.
38. D. Singh, B. Kumar, S. Singh, and S. Chand, "SMAC-AS: MAC Based Secure Authentication Scheme for Wireless Sensor Network," *Wireless Personal Communications*, vol. 107, no. 2, pp. 1289–1308, 2019, doi: 10.1007/s11277-019-06336-8.
39. D. Chawla and P. S. Mehra, "QSMAH: A novel quantum-based secure cryptosystem using mutual authentication for healthcare in the internet of things," *Internet of Things*, vol. 24, p. 100949, Dec. 2023, doi: 10.1016/J.IOT.2023.100949.
40. M. S. Umar and M. Q. Rafiq, "Select-to-Spawn: A novel recognition-based graphical user authentication scheme," in *2012 IEEE International Conference on*

Signal Processing, Computing and Control, Mar. 2012, pp. 1–5. doi: 10.1109/ISPCC.2012.6224382.

41. S. Challa *et al.*, "An efficient ECC-based provably secure three-factor user authentication and key agreement protocol for wireless healthcare sensor networks," *Computers & Electrical Engineering*, vol. 69, pp. 534–554, 2018.
42. C. Jacomme and S. Kremer, "An Extensive Formal Analysis of Multi-factor Authentication Protocols," *ACM Transactions on Privacy and Security*, vol. 24, no. 2, p. 13:1-13:34, Jan. 2021, doi: 10.1145/3440712.

12 Deep Learning Techniques for IoT Security

A Comprehensive Review and Future Directions

Aatif Jamshed, Diksha Chawla, Jatin Sharma, Deepakshi Dabas, and Pawan Singh Mehra

12.1 INTRODUCTION

The Internet of Things, also known as IoT [1], has swiftly matured into a technology that is changing paradigms by allowing the interconnection of many sensors, actuators, and other physical objects. This integrated ecosystem has had a transformative effect on many different industries, including healthcare, transportation, manufacturing, and agriculture. As a result, these businesses now have access to previously unimaginable prospects for automation, decision-making driven by data, and increased efficiency. On the other hand, the extensive deployment of IoT devices has resulted in substantial security concerns [2] that pose a risk to this technology's smooth integration and widespread adoption. Traditional security procedures are insufficient to protect the large and varied terrain of the IoT, necessitating novel and adaptable security solutions.

Deep learning [3–11] has taken the world by storm in artificial intelligence (AI). It's been knocking it out of the park, showing remarkable triumph in tricky endeavors like spotting images, understanding language, and even talking like a human. This sudden surge in AI has happened in the blink of an eye [12, 13]. Because of its ability to learn hierarchical representations [4] from enormous data sets and recognize detailed patterns, it is a great choice for addressing the complex and ever-evolving security threats [5] that the IoT faces.

The illustration shown in Figure 12.1 demonstrates how monitoring the devices connected to IoT may intelligently avoid zero-day threats [7]. The "normal" and "abnormal" interactions of IoT components and devices may be uncovered via data analysis that uses machine learning and deep learning [14], [15]. It is possible to

DOI: 10.1201/9781003405535-17

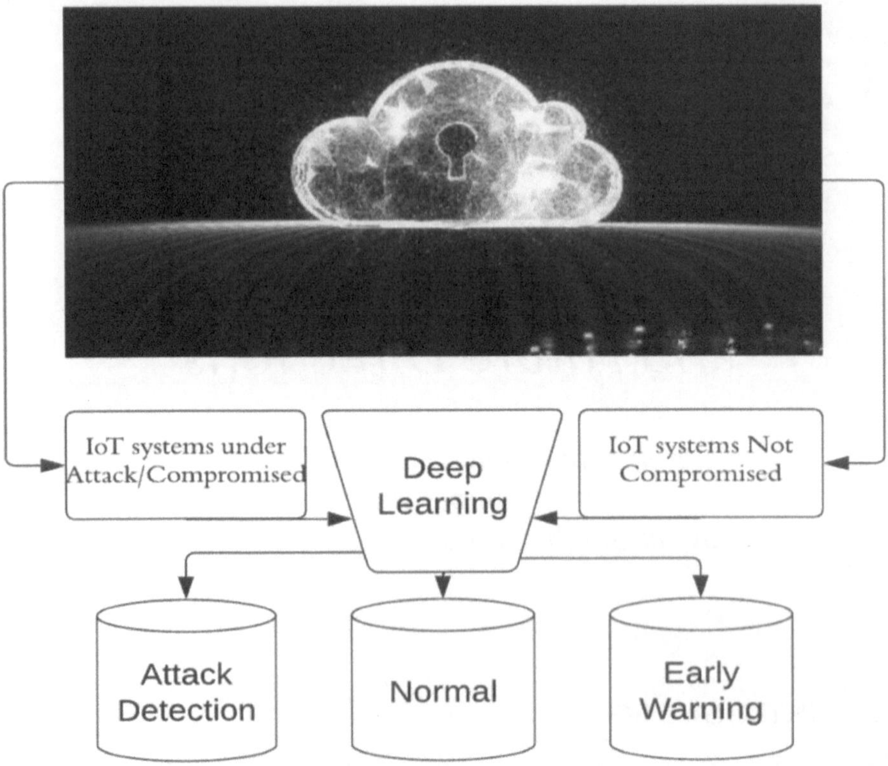

FIGURE 12.1 Deep leaning flow in attacks.

collect and analyze the input data from each component of the IoT system to uncover regular interaction patterns and early signs of potentially dangerous behavior. Deep learning (DL) approaches have the potential to accurately forecast incoming assaults that have not yet been recognized by capitalizing on the information gained from previous cases. This capacity to predict future assaults is especially significant since new attacks often take the form of modifications of existing ones. It is vital to replace secure device communication with security-based intelligence made possible by DL methods to guarantee the efficacy and safety of IoT systems. This is one of the most important steps that must be taken. This research aimed to conduct an in-depth analysis of the most recent and cutting-edge applications of DL that have been put into place to strengthen the safety of the IoT. Additionally, it lays the groundwork for future research endeavors.

12.1.1 The Development of IoT and Its Security Consequences

IoT has changed how electronic gadgets connect and share data, creating a dynamic, networked environment. From smart homes to industrial control systems, IoT devices are now everywhere, prioritizing device safety [16]. Conventional security methods cannot fulfill the different demands of each IoT device, making it challenging to secure them. A compromise in IoT security might lead to data theft, privacy concerns, and infrastructure failure [17].

12.1.2 Limitations of Conventional Security Mechanisms

The gadgets that operate in the IoT world is as slippery as an eel when it comes to being protected by regular security measures like firewalls and encryption. Since many devices connected to IoT have low computational capabilities, it might be difficult to deploy resource-intensive security mechanisms. The security landscape is made even more complicated by the absence of defined security frameworks and the regular occurrence of firmware vulnerabilities.

12.1.3 The Potential of Deep Learning in Internet of Things Security

The use of DL to address the shortcomings of more traditional security systems has shown incredibly positive potential, as proven by recent research. Even in extremely dynamic IoT contexts, it can detect complex and previously undiscovered threats, thanks to its capacity to learn and adapt to enormous amounts of data. In addition, DL techniques can be modified to function on resource-constrained IoT devices, ensuring that real-time security analysis will be performed effectively.

Figure 12.2's Intrusion Detection System (IDS) taxonomy helps industry, government, and investigators build an intelligent intrusion-detection system in the IoT ecosystem. Figure 12.2 [18] shows IDS DL taxonomy. The taxonomy covers key IoT security concerns and solutions. IoT security attacks, architecture layers, intrusion-detection systems, DL approaches, popular datasets, and classification algorithms are included in the taxonomy. The taxonomy's sections are interconnected as underlying causes of IoT security problems and/or solutions.

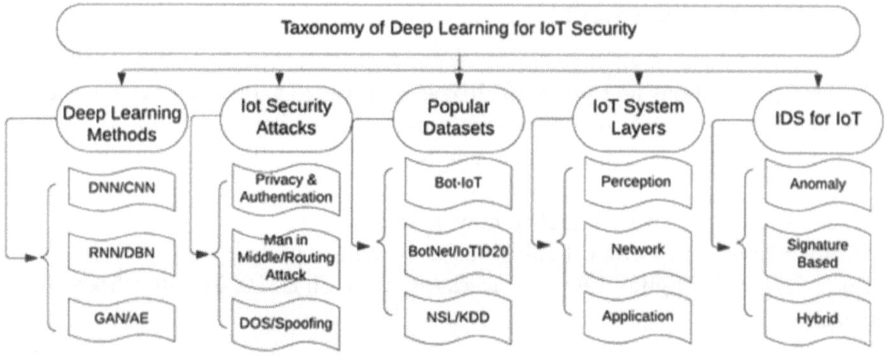

FIGURE 12.2 Deep learning taxonomy.

12.2 OBJECTIVES AND SCOPE

The primary objective of this research was to provide an extensive evaluation of the most current and state-of-the-art DL approaches to the issue of securing IoT. We intend to examine and categorize the various deep learning-based strategies currently used for IoT security. Find out what works and what doesn't about these methods and whether they may be used in tandem with other security measures. Learn how DL performs in intrusion detection, authentication, access control, anomaly detection, and threat intelligence related to IoT security. Let's spill the beans on the hurdles and drawbacks of using DL for IoT security, including the scarcity of data, the rock-solidness of the models, and the heavy processing load. Let's dive into the nitty-gritty of the privacy concerns and ethical dilemmas that come hand in hand with employing DL algorithms in IoT security. Cook up some fresh ideas to push the envelope in DL and make it a piece of cake to safeguard the ever-growing realm of IoT gadgets.

12.2.1 The Chapter's Framework

The chapter is organized as follows:

This study provides an in-depth analysis of DL approaches for IoT security and examines their uses and potential future developments. Setting the scenario, the Chapter 1 Introduction emphasizes the growth of IoT and the requirement for cutting-edge security solutions. The goals and scope of Chapter 2 are described, focusing on surveying and classifying DL approaches for IoT security, evaluating their efficacy, and addressing issues. IoT security-related topics like intrusion detection, authentication, access management, and anomaly detection are covered in depth. The difficulties and limitations of various DL for IoT security and methodology are discussed in Chapters 4 and 5. In this context, DL's ethical implications and privacy significances are covered in Chapter 6. Future research possibilities are examined in Chapter 7, including lightweight models, federated learning, and multimodal data fusion. The Summary and Conclusion highlights the importance of DL in protecting the IoT ecosystem, laying the groundwork for a more secure and resilient IoT future.

12.3 LITERATURE SURVEY

The authors looked through three databases and two different data sources for relevant studies and publications up to July 2023. The goal was to gather and extract relevant studies from the existing body of published research. The databases include Springer Link, the IEEE Xplore® digital library, and Science Direct. The data sources include Scopus and Google Scholar. The electronic connection to all of the databases and other data sources that were searched may be found here:

Springer Link (accessed on 1 July 2023) (http://link.springer.com).

IEEE Xplore® Digital Library (accessed on 1 June 2023) (http://ieeexplore.ieee.org).

Science Direct (accessed on 1 July 2023 at www.sciencedirect.com).

Scopus Database (accessed on 27 May 2023; available at: www.scopus.com).

Google Scholar (accessed on: 1 Feb 2023) (http://scholar.google.com).

Many investigators have conducted surveys on IoT security to give a clear picture of its weaknesses and lay the groundwork for further inquiry. Most IoT security assessments have left DL applications out in the cold. The studies examined network security, application security, access control, encryption, authentication, and other security-related issues in IoT systems.

Tariq et al. [1] thoroughly discuss IoT anomalies and security. We categorize and analyse IoT's multilayer security vulnerabilities, including connection, communication, and management protocols. The author investigated IoT security and set security objectives to evaluate IoT solutions.

Kumar [2] examined sustainable industrial IoT system challenges and opportunities. They discussed energy efficiency, resource optimization, waste reduction, and environmental mitigation. Bhachawat et al. [3] studied machine learning and DL for Alzheimer's and Parkinson's diagnosis. Model updates improved degenerative nerve disease classification. Mohsan et al. [4] identified risks and mitigation measures from the literature.

Ray [5] analyze Web3's innovations, impacts, and DApps. Das et al. [6] created the radio propagation environment without pilot signals in RIS. Ghiasi et al. [7] explained energy system cybersecurity. IoT intrusion-detection systems were divided into four categories by Heidari et al. [8]: signature-based, anomaly-based, and hybrid. Sarker et al. [19] advised cybersecurity and IoT researchers.

IoT communication security was emphasized by Görmüş et al. [20] after examining problems and solutions. Zarpelão et al. [21] investigated IoT infiltration.

Weber [18] examined the privacy and security of the IoT framework using legal and regulatory issues. Roman et al. [22] explored distributed IoT security and privacy. Researchers explored IoT security and privacy. Ransomware and IoT security [23].

Xiao et al. [24] reviewed IoT data privacy and security DL methods. Computing and communication overhead, backup security, and state monitoring hampered IoT ML adoption. Data mining and machine learning have also been studied [25, 26] for their potential use in detecting cyber-attacks. Data mining and machine learning handled cybersecurity, abuse, and anomalies [25, 26]. Unlike previous surveys, ours delves further into machine learning and DL to secure IoT. In this study, we look at the state of DL techniques for protecting IoT. In this chapter, we look at the potential uses of DL for IoT security and discuss the challenges and opportunities that lie ahead. This study provides a crucial road map for researchers looking to shift their focus from enabling secure communication between IoT components to adopting intelligent ways that assure end-to-end IoT security. Table 12.1 compares and contrasts various DL strategies for protecting and securing IoT devices, including their current and potential applications.

TABLE 12.1
Prospective Deep Learning Approaches for Enhancing the Security of Internet of Things (IoT) Systems

Technique	The Underlying Operational Concept	Pros	Cons	Application Potential in IoT Security
CNN	IoT security uses visual data for object recognition, intrusion detection, and tampering detection. Layers extract features. Feature-to-output layers are fully connected.	Automatic feature learning, generalization to new and unseen security threats.	Potential overfitting without proper regularization. Computational complexity during training.	Smart surveillance systems, traffic monitoring and anomaly detection.
RNN	RNNs handle sequence data, making them ideal for IoT security time-series analysis. Feedback links preserve information over time steps. RNN steps analyze input elements and update their hidden states, capturing data temporal dependencies.	Capture long-term dependencies, handle input sequences of varying lengths.	Suffer from vanishing or exploding gradients, training RNNs can be computationally expensive	Anomaly detection in sensor data, process speech data, enabling voice authentication.
LSTM	With the use of a memory cell, an input gate, a forget gate, and an output gate, it regulates the flow of information through the network. Long-term dependencies in sequential data may be learned using LSTMs for IoT security time-series analysis.	Address the vanishing gradient problem, reducing the risk of convergence issues.	LSTMs can lack interpretability, making it difficult to understand the specific factors that influence their decisions.	Based on prior sensor data, LSTMs may forecast IoT device maintenance needs, decreasing security risks from device faults.

Autoencoders	Autoencoders are unsupervised deep learning models. Encoders and decoders make them. The encoder compresses incoming data into a lower-dimensional latent space, while the decoder reconstructs it. Autoencoders reduce reconstruction error by teaching the model to compress and interpret input data.	Autoencoders can be trained to reconstruct clean data from noisy input.	Autoencoders can suffer from overfitting	Firmware integrity verification: autoencoders can be used to reconstruct and verify the integrity of firmware on IoT devices, detecting any modifications or tampering attempts.
GAN	GANs are competitively trained generator and discriminator neural networks. The generator creates synthetic data that resembles the dataset's genuine data, while the discriminator distinguishes real from false data. The generator seeks to trick the discriminator, while the discriminator seeks to better identify real from phony data. Adversarial training produces high-quality synthetic data.	GANs may add synthetic instances to training data and create realistic synthetic data that matches the original data distribution.	GANs may ignore certain modes in the data distribution, leading to incomplete or biased generation	Privacy-preserving data sharing and allowing secure sharing for collaborative security study without disclosing private information.
Capsule networks	Capsule networks are built from capsules, groupings of neurons that encode an object's posture, orientation, and appearance. Hierarchical capsules help the network grasp spatial connections between object pieces.	Capsule networks can capture spatial hierarchies and relationships, capsule networks employ dynamic routing algorithms.	Training capsule networks can be computationally expensive, especially for large-scale datasets	Stance estimation: capsule networks can estimate the stance and orientation of people or objects for security applications like person detection or gesture recognition.
Federated learning	Federated learning trains the model locally using its own data and delivers model changes to a central server. The server gathers updates to enhance the global model, iterating until convergence. Federated learning allows collaborative model training while protecting data.	Decentralized training: lets IoT devices locally train models, eliminating the need for a central server and enabling real-time model changes on the edge.	To prevent biases, training a model on varied and heterogeneous data from multiple devices or contexts is difficult.	Distributed intrusion detection, edge-based threat analysis.

12.4 METHODOLOGY

The study process includes a comprehensive assessment of the relevant scholarly articles, conference papers, and technical reports relating to DL applications in IoT security [27]. This review is carried out to find important information. When compiling a collection of articles from various sources, we will use academic resources such as Google Scholar, ACM Digital Library, and IEEE Xplore.

Securing IoT is essential owing to the widespread deployment of networked devices, which has the potential to result in major cybersecurity threats [28]. An attack might cause these hazards. DL techniques have shown that they have the potential to improve the security of IoT by identifying abnormalities, minimizing risks, and protecting data privacy. The following is a strategy, as shown in Figure 12.3, for using DL techniques [22] to IoT security.

Clearly describe the precise security goals you want to accomplish using DL techniques in the context of IoT. To understand the kinds of anomalies and assaults, you need to detect, prevent, and identify probable threat scenarios and attack routes.

Data Gathering and Preprocessing: Gather data from diverse IoT devices, sensors, and networks, ensuring the collected information encompasses lawful and illicit

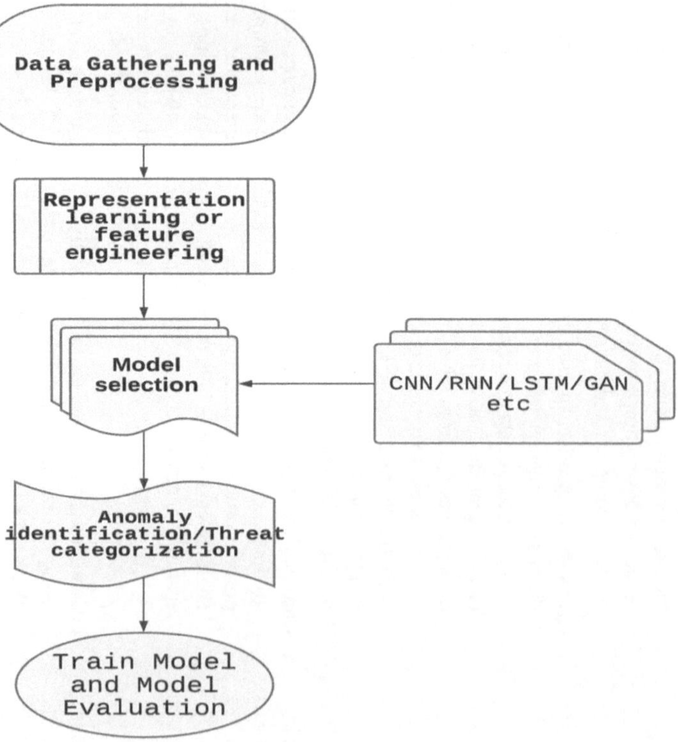

FIGURE 12.3 A strategy using deep learning.

activities. Several key steps must be undertaken to prepare the data for DL models adequately. These steps include reducing noise within the data, effectively managing any missing values and standardizing the data.

Representation learning or feature engineering: To successfully extract pertinent information from the raw IoT data, you may need to use representation learning methods like autoencoders or word embeddings, depending on the nature of the data.

Model selection: Choose the DL architectures most suited to the current IoT protection tasks. Models like convolutional neural networks (CNNs) for image data, long short-term memory (LSTM) networks for sequential data, and recurrent neural networks (RNNs) for time-series data are typical examples.

Split the preprocessed data into training and validation sets to train the DL model. To fine-tune hyperparameters and avoid overfitting, train the DL model using training data and utilize the validation set.

Anomaly identification: Autoencoders and generative models like variational autoencoders (VAEs) are examples of unsupervised learning techniques that may be used for anomaly identification. These models can recognize anomalies when they deviate from the learned representation of regular data.

Threat categorization: Apply supervised learning techniques for threat classification. Label your data with various attack or threat kinds and then train a classification model like CNN or LSTM to divide new occurrences into several threat categories.

Implement the trained model in a real-time monitoring system for IoT devices. Real-time Monitoring and Deployment. This will enable you to continually monitor for abnormalities or dangers and identify them as they emerge.

Model Evaluation: Use measures like accuracy, recall, F1-score, and area under the Receiver Operating Characteristic (ROC) curve to assess how well your DL model is doing. Adjust the model as needed.

Adaptation and Continuous Improvement: As the IoT ecosystem changes, new dangers might emerge. Update your DL models often to account for evolving risks and improve security controls.

Collaboration and Research: Exchange information and keep up with the most recent advancements in DL techniques for IoT security by working with researchers and the larger security community.

Different Types of Deep Learning: Discriminative, generative, and hybrid deep learning frequently use neural networks. Discriminative DL techniques include CNNs and RNNs. Different types of DL methods, such as autoencoders (AEs), deep

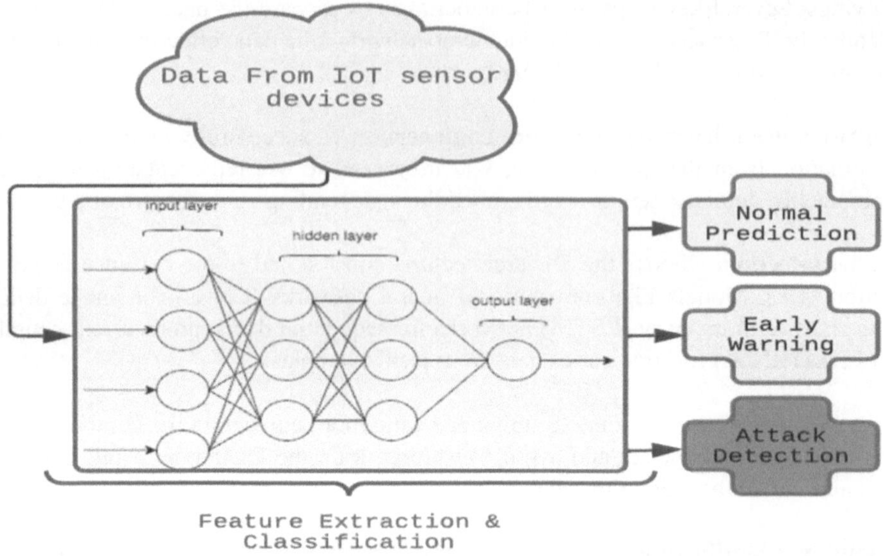

FIGURE 12.4 Deep learning working principle.

belief networks (DBNs), restricted Boltzmann machines (RBMs), generative adversarial networks (GANs), and ensemble DL networks (EDLNs) are all included in hybrid approaches [23–25]. CNNs eliminate the need for hand-crafted feature creation by automatically learning the most important qualities related to malware detection from raw data. The capacity to train a CNN to learn relevant features and perform classification in parallel is crucial to its use. This eliminates the need for a separate feature extraction phase, typically required in traditional Machine Learning (ML) approaches. Consequently, a comprehensive model is obtained, as depicted in Figure 12.4.

12.5 SIGNIFICANCE OF THE STUDY

The importance of this work rests in the fact that it provides an exhaustive and up-to-date summary of the cutting-edge DL techniques used to improve the security of IoT devices. The findings and conclusions of this study can help researchers, policymakers, and industry practitioners better grasp the potential of DL for IoT security and shape the future of secure IoT installations. This work has the potential to stimulate new lines of research by identifying research gaps and problems [26]. These new avenues of research could lead to the development of efficient and robust deep learning-based security solutions suited to IoT devices' specific characteristics. Ultimately, this research contributes to creating a trustworthy and safe ecosystem for IoT, which is necessary to realize the full potential of this game-changing technology [29].

12.6 FUTURE DIRECTIONS AND RESEARCH OPPORTUNITIES

IoT security needs to be improved by utilizing DL techniques due to IoT quickly changing environment and constantly increasing cybersecurity threats. As we look toward the future, some important areas and research possibilities become apparent:

DL researchers must concentrate on creating lightweight and effective models to implement deep learning-based security solutions on IoT devices with limited resources. Methods like model quantization, knowledge distillation, and model compression can be investigated to lessen model complexity and memory footprint while retaining a reasonable level of accuracy.

Federated learning holds great promise for safeguarding IoT devices that collect sensitive data. Federated learning is a privacy-preserving method. By allowing models to be trained locally on IoT devices without sharing raw data centrally, this decentralized learning paradigm reduces privacy issues while enhancing overall security.

Deep learning adversarial defense in IoT: As DL models for IoT security proliferate, adversarial attacks on these models could also rise. A significant topic for research is investigating and creating strong defenses against adversarial assaults on DL models tailored for IoT. IoT devices feature the fusion of many kinds of data, including audio, video, and sensor readings. Deep learning-based security solutions can be made more accurate and dependable by experimenting with methods to combine multimodal data and use complimentary information from other sources [30].

Deep learning for IoT with privacy protection: User privacy protection is paramount in IoT environments. Secure data sharing and model training can be made possible by research into privacy-preserving DL approaches, such as differential privacy and secure multi-party computation.

Explainable AI for IoT Security: DL models are sometimes viewed as "black boxes," which makes it difficult to comprehend how they make decisions. Creating explainable AI methods for IoT security models helps boost confidence and openness by enabling stakeholders to comprehend the thinking behind model outputs. IoT security needs to be collaborative and adaptive due to the dynamic nature of IoT ecosystems and the need to respond to new threats. It is crucial to research cooperative and adaptable DL methods that can constantly learn and update security policies in real time.

Benchmark Datasets for IoT Security: Developing uniform benchmark datasets specifically geared toward IoT security can promote comparison assessments of various DL models and techniques. This will facilitate reproducibility and allow researchers to evaluate the solutions they have offered objectively.

Secure Model Deployment and Updates: It's crucial to address the problems associated with secure model deployment and updates for IoT devices. The danger of malicious code injection should be reduced through research on methods to assure the integrity and validity of models during deployment and update operations.

IoT Security That Is Human-Centered: Researching how human psychology and behavior relate to IoT security can result in better user-friendly and safe IoT devices. Designing efficient authentication and access control systems that users are more likely to accept and adhere to requires understanding user behaviors and preferences.

Real-World Testbeds and Validation: Experiments and validations must be carried out in real-world IoT environments to evaluate the applicability and efficacy of deep learning-based security solutions. Empirical evaluations can be facilitated by creating real-world testbeds that replicate various IoT implementations. Collaborations between professionals from many fields, such as ML, cybersecurity, IoT engineering, and human-computer interface, can result in holistic and all-encompassing solutions that address the myriad problems associated with IoT security.

12.7 CONCLUSION

In conclusion, a promising approach to enhancing the resilience of IoT devices and networks against new attacks is incorporating DL techniques into IoT security. The promise of DL for IoT security may be unlocked by embracing these future directions and research opportunities, opening the door for a safer and more dependable IoT environment. Researchers and practitioners may work together to secure the revolutionary power of IoT while maintaining privacy, security, and dependability for consumers everywhere by tackling the issues and pushing the boundaries of knowledge. The chapter identifies gaps in existing research and suggests possible future avenues for developing deep learning-based IoT security. This emphasizes the significance of developing DL models that are lightweight and resource-efficient to accommodate the resource limitations of IoT devices. Furthermore, the study emphasizes the imperative for collaborative endeavors among academic institutions, private enterprises, and government entities to facilitate the effective implementation of DL in the realm of IoT security.

REFERENCES

1. Tariq, U., Ahmed, I., Bashir, A. K., & Shaukat, K. (2023). A Critical Cybersecurity Analysis and Future Research Directions for the Internet-of-Things: A Comprehensive Review. *Sensors*, 23(8), 4117.
2. Kumar, P. (2023). Enabling Sustainable Industrial IoT: A Comprehensive Review and Future Directions. *Transaction on Recent Devlopment in Industrial IoT*, 15(15), 1–16.
3. Bhachawat, S., Shriram, E., Srinivasan, K., & Hu, Y. C. (2023). Leveraging Computational Intelligence Techniques for Diagnosing Degenerative Nerve Diseases: A Comprehensive Review, Open Challenges, and Future Research Directions. *Diagnostics*, 13(2), 288.
4. Mohsan, S. A. H., Li, Y., Sadiq, M., Liang, J., & Khan, M. A. (2023). Recent advances, future trends, applications and challenges of internet of underwater things (iout): a comprehensive review. *Journal of Marine Science and Engineering*, 11(1), 124.
5. Ray, P. P. (2023). Web3: A comprehensive review on background, technologies, applications, zero-trust architectures, challenges and future directions. Internet-of-Things and Cyber-Physical Systems.

6. Das, S. K., Benkhelifa, F., Sun, Y., Abumarshoud, H., Abbasi, Q. H., Imran, M. A., & Mohjazi, L. (2023). Comprehensive review on ML-based RIS-enhanced IoT systems: basics, research progress and future challenges. *Computer Networks*, 224, 109581.
7. Ghiasi, M., Niknam, T., Wang, Z., Mehrandezh, M., Dehghani, M., & Ghadimi, N. (2023). A comprehensive review of cyber-attacks and defense mechanisms for improving security in smart grid energy systems: Past, present and future. *Electric Power Systems Research*, 215, 108975.
8. Heidari, A., & Jabraeil Jamali, M. A. (2022). Internet-of-Things intrusion detection systems: A comprehensive review and future directions. *Cluster Computing*, 26, 1–28.
9. Jamshed, A., Mallick, B., & Kumar, P. (2020). Deep learning-based sequential pattern mining for progressive database. *Soft Computing*, 24, 17233–17246.
10. Jamshed, A., Mallick, B., & Kumar Bharti, R. (2022). Grey wolf optimization (GWO) with the convolution neural network (CNN)-based pattern recognition system. *The Imaging Science Journal*, 70(4), 238–252.
11. Jamshed, A., Mallick, B., & Bharti, R. K. (2022, May). An Analysis of Sequential Pattern Mining Approach for Progressive Database by Deep Learning Technique." In *2022 6th International Conference on Intelligent Computing and Control Systems (ICICCS)* (pp. 1409–1415). IEEE.
12. Al-Garadi, M. A., Mohamed, A., Al-Ali, A. K., Du, X., Ali, I., & Guizani, M. (2020). A survey of machine and deep learning methods for Internet-of-Things (IoT) security. *IEEE Communications Surveys & Tutorials*, 22(3), 1646–1685.
13. Lin, T. (2020, November). "Deep learning for IoT." In *2020 IEEE 39th International Performance Computing and Communications Conference (IPCCC)* (pp. 1–4). IEEE.
14. Jindal, M., Gupta, J., & Bhushan, B. (2019, October). "Machine learning methods for IoT and their Future Applications. In *2019 International Conference on Computing, Communication, and Intelligent Systems (ICCCIS)* (pp. 430–434). IEEE.
15. Bagaa, M., Taleb, T., Bernabe, J. B., & Skarmeta, A. (2020). A machine learning security framework for iot systems. *IEEE Access*, 8, 114066–114077.
16. J. Sharma and P. S. Mehra, "Secure communication in IOT-based UAV networks: A systematic survey. *Internet of Things*, 23, 100883, doi: 10.1016/J.IOT.2023.100883.
17. D. Chawla, & P. S. Mehra, A roadmap from classical cryptography to post-quantum resistant cryptography for 5G-enabled IoT: Challenges, opportunities and solutions. *Internet of Things*, 24, 100950, doi: 10.1016/J.IOT.2023.100950.
18. Weber, R. H. (2010). Internet-of-Things–New security and privacy challenges. *Computer law & security review*, 26(1), 23–30.
19. Sarker, I. H., Furhad, M. H., and Nowrozy, R. (2021). AI-driven cybersecurity: An overview, security intelligence modeling and research directions. *SN Computer Science*, 2(3). doi: 10.1007/s42979-021-00557-0
20. Görmüş, S., Aydın, H., & Ulutaş, G. (2018). Security for the Internet-of-Things: a survey of existing mechanisms, protocols and open research issues. *Journal of the Faculty of Engineering and Architecture of Gazi University*, 33(4), 1247–1272.
21. Zarpelão, B. B., Miani, R. S., Kawakani, C. T., & de Alvarenga, S. C. (2017). A survey of intrusion detection in Internet-of-Things. *Journal of Network and Computer Applications*, 84, 25–37.
22. Roman, R., Zhou, J., & Lopez, J. (2013). On the features and challenges of security and privacy in distributed Internet-of-Things. *Computer networks*, 57(10), 2266–2279.

23. Yaqoob, I., Ahmed, E., ur Rehman, M. H., Ahmed, A. I. A., Al-garadi, M. A., Imran, M., & Guizani, M. (2017). The rise of ransomware and emerging security challenges in the Internet-of-Things. *Computer Networks*, 129, 444–458.
24. Xiao, L., Wan, X., Lu, X., Zhang, Y., & Wu, D. (2018). IoT security techniques based on machine learning: How do IoT devices use AI to enhance security? *IEEE Signal Processing Magazine*, 35(5), 41–49.
25. Buczak, A. L., & Guven, E. (2015). A survey of data mining and machine learning methods for cyber security intrusion detection. *IEEE Communications surveys & tutorials*, 18(2), 1153–1176.
26. Mishra, P., Varadharajan, V., Tupakula, U., & Pilli, E. S. (2018). A detailed investigation and analysis of using machine learning techniques for intrusion detection. *IEEE communications surveys & tutorials*, 21(1), 686–728.
27. C. Garg, D. Kumar Mishra, D. Raj, and P. Singh Mehra, P. (2022). "A Survey on Integration of Blockchain and IoT (BIoT): Open Issues, Challenges & Solution," In *Proceedings of the 2022 3rd International Conference on Intelligent Computing, Instrumentation and Control Technologies: Computational Intelligence for Smart Systems, ICICICT 2022* (pp. 880–886), doi: 10.1109/ICICICT54557.2022.9917632.
28. Roopak, M., Tian, G. Y., & Chambers, J. (2019, January). "Deep learning models for cyber security in IoT networks. In *2019 IEEE 9th annual computing and communication workshop and conference (CCWC)* (pp. 0452–0457). IEEE.
29. Manju, S. Singh, A. K. Dwivedi, A. K. Sharma, and P. S. Mehra, "Learning Automata Based Heuristics for Target Q-Coverage," in *ICRITO 2020 – IEEE 8th International Conference on Reliability, Infocom Technologies and Optimization (Trends and Future Directions)*, Institute of Electrical and Electronics Engineers Inc. (pp. 170–173). doi: 10.1109/ICRITO48877.2020.9197923.
30. Hathaliya, J. J., Tanwar, S., & Sharma, P. (2022). Adversarial learning techniques for security and privacy preservation: A comprehensive review. *Security and Privacy*, 5(3), e209.

13 An Empirical Study of the Concept Drift Driven Machine Learning for IoT Network Anomaly Detection

Meenal Jain, Anita Sahoo, and Pratibha Singh

13.1 INTRODUCTION

The Internet in these days is essential in all fields of work in education, business, healthcare, agriculture, traffic control systems, entertainment, security, and many other smart systems [1–5]. Internet of Things (IoT) has the outstanding capability of collaborating with other emerging technologies as artificial intelligence, machine learning (ML), ubiquitous computing, etc. The fast development of any country depends profoundly on fast-growing Internet services. These services are controlled by IoT devices connected through various technologies to communicate. Its exposure to the world makes it vulnerable towards attacks. It raises the importance of detecting the abnormal behaviour of the network due to irregular patterns that may degrade the performance of the IoT networks [1, 2]. An anomaly indicates a traffic/data pattern different from the network's normal pattern. There can be point anomaly or contextual anomaly. The point anomaly represents data points that fall outside the clusters of regular patterns. The anomaly that denotes a pattern irrelevant to the context of the present data may be relevant to the context of some other type of data [1]. The main challenges in detecting anomalies in IoT networks are heterogeneous systems, distributed environments, ignorance, and lack of awareness of using standard secure hardware and software [1]. The researchers have identified that ML algorithms and their enhanced versions have unlimited potential to detect anomalies. It has opened a new dimension to work on the massive data generated in IoT networks and protect it from various attacks. However, the machine learning (ML) algorithms have its challenges to solve. In a few of these areas, extracting the features from continuously changing data is difficult, requires fast computing resources, and requires a large amount of data for training. Thus, it requires enough sets of patterns representing irregular behaviour of different types [2].

DOI: 10.1201/9781003405535-18

Good algorithms for intrusion detection should handle the evolving nature of the samples' characteristics and extract important data features [3]. The concept drift-driven ML algorithms are more efficient in handling the challenges in anomaly detection than traditional ML algorithms. The concept drift-driven ML algorithms implement incremental learning and can handle the various drifts in anomalies. Thus, we decided to work on the concept drift solutions for anomaly detection that learn from experience in identifying the evolving patterns of anomalies.
The proposed work in this chapter focused on the following points:

- Studying the state-of-the-art ML-based methods evolved to detect the anomalies.
- Using the sampling technique k-means to create a highly representative subset.
- Handling drift in network streaming data, the Optimal Sliding Adaptive Windowing (OSAW) Developing Concept Drift Detection and Adaptation technique.
- Evaluating the performance of the proposed model using performance metrics.
- Experimenting on NSL-KDD and CIDDS (2017), two widely used publicly accessible datasets.

The remaining sections of this chapter are ordered as follows: Section 13.2 gives a detailed discussion of the contribution of machine learning algorithms in detecting the anomalies in IoT networks. The challenges that emerge in front of ML-based algorithms are due to unpredictable and evolving patterns of anomalies in IoT. Then, we reviewed how the concept drift-driven ML algorithms deal with this enormous problem. Section 13.3 proposes an effective Optimal Sliding and Adaptive Windowing (OSAW) algorithm with a k-means sampling technique. Section 13.4 shows the proposed technique's performance before and after drift handling on NSL-KDD and CIDDS (2017) datasets. Section 13.5 finally concludes this chapter.

13.2 LITERATURE REVIEW

IoT devices are heterogeneous devices connected via various technologies, connected in a distributed environment, and have resource constraints. This article [1] discusses the challenges of anomaly detection in IoT and the application of ML solutions. ML algorithms must extract useful features that discriminate normal network behaviour from irregular network behaviour. Other challenges are that ML models require extensive resources to compute and enormous data for training and are prone to adversarial attacks. This work has suggested using collaborative learning by combining blockchain with ML techniques for effective point or contextual anomaly detection. This article [2] has proposed a hybrid learning method for anomaly detection. It states that signature-based detection methods are insufficient and suggests using anomaly-based detection methods. It uses an enhanced support vector machine (SVM) algorithm to identify abnormal behaviour. It has also used temporal relationships among raw Internet packets for SVM learning. The enhanced SVM method is compared with the soft-margin SVM and one-class SVM on the MIT Lincoln Lab dataset. The enhanced SVM method is compared with the signature-based Network Intrusion Detection System (NIDS) method with real data. The enhanced SVM approach performs better than other systems, generates a lower false-positive rate, and can be better deployed in a real working environment. The authors in this article [3] discuss the state-of-the-art methods targeting core research challenges in IoT. They reviewed ML and deep learning methods for anomaly detection. These disputes are unexplained trends in data and the behaviour of devices that may occur due to unexpected noises, attacks, and anomalies. The algorithms must be able to compute large high-dimensional data streams flowing continuously in available limited memory space. It defines the taxonomy of anomaly detection, covering the techniques, nature of data, types of anomalies, unsupervised anomaly detection methods, supervised anomaly detection methods, evaluation criteria, etc. Most techniques are used in the literature on publicly available KDD99 datasets. In this article [4], the authors have addressed the challenges caused by the exponential growth of IoT networks by using mutual information to select the best features and deep neural networks. It presents the comparative analysis of deep learning models and their variants on the IoT-Botnet 2020 dataset. These different models include Recurrent Neural Network (RNN), Long Short-Term Memory (LSTM), Gated Recurrent Unit (GRU), and Convolutional Neural Network (CNN). The proposed model has improved the accuracy by 0.57%–2.6% and reduced the false alarm rate by 0.23%–7.98%.

This article [5] uses supervised and unsupervised ML algorithms to detect the sensor tempering in IoT systems remotely or physically. It has used an isolation forest-based method for identifying patterns in anomalies, validated through manual labelling and a decision tree (DT) is used to classify anomalies. The isolation forest with a DT has achieved 91.62% accuracy. This article [6] focused on ML-based algorithms for IoT security. This article has provided an exhaustive literature review and a comparative results analysis. The random forest algorithm has provided the best results with 99.3–99.9 accuracy. The combination of DT, K-Nearest Neighbors (KNN), and RF provides good attack detection results. After reviewing the literature, the random forest algorithm emerged as the best algorithm for anomaly detection. This article [7] has developed an ML-based system that can automatically detect attacks. It has used KNN, Logistic Regression (LR), SVM, multilayer perceptron, DT, and RF classifiers. The results on the BoT-IoT dataset are measured in terms of accuracy, precision, recall, Receiver Operating Characteristic (ROC), and F-score. The random forest classifier has shown comparatively better performance in identifying attacks and anomalies with 99.9% accuracy. Here, the authors worked with ML algorithms such as SVM, LR, artificial neural network (ANN), DT, and random forest (RF) [8]. The results obtained on the virtually created IoT environment identified that RF, DT, and ANN algorithms outperform other algorithms with 99.4% accuracy.

RF and DT perform better than ANN with other performance metrics. This article [9] stated that it is difficult to select the best algorithms because connecting devices behave differently when working in different domains and may face different types of anomalies. The ML algorithms are applied on the UNSW-NB15 dataset, which shows that the Averaged One Dependence Estimator (AODE) algorithm is effective and efficient in a distributed environment. Although the distributed estimator not perform better than its centralized form in terms of accuracy and execution time, it can be accepted and improved further with online learning. They created large IoT traffic datasets representing maximum IoT-type attacks for different real IoT environments [10]. The principal component analysis (PCA) with autoencoder is used for dimensionality reduction, and the deep learning architectures are optimized to attain better accuracy for identifying the types of attacks. The trained model achieves 99.89% accuracy for binary classification and 98.9% for multiclass classification. The model's performance is more stable when features are reduced in the range between 35 and 60.

In this paper [11], the authors focused on the appropriate choice of anomaly detection method in an IoT network. They applied variations of SVM as sequential minimal optimization (SMO), Lib SVM algorithm, and tree-based classifiers as reduces error pruning (REP) Tree, Logistic Alternating Decision (LAD) Tree, Random Forest (RF), and extended version of Adaptive boosting (AdaBoost), i.e. MultiBoost algorithm. The results are obtained on the UNSWNB15 dataset. It is an unbalanced dataset. MultiBoost algorithm shows the lowest performance, RF shows the best performance, REPTree gives smart solutions in less time, and LADTree performs well but with little delay. SMO and library for support vector machines (LibSVM) can also be recommended when there is no time constraint. This paper [12] has provided a comprehensive survey on the challenges of IoT networks and the roles of ML-based algorithms to solve those challenges. They compare supervised, unsupervised, semi-supervised, and reinforcement learning methods. These algorithms are applied for anomaly detection in traditional computer networks, cellular/wireless networks, Software-Defined Networking (SDN), IoT networks, cloud/edge/fog networks, and experimental datasets. Experimental datasets include KDD, UNSW-NB 15, and CICIDS2017, among others.

In his book, Bhattacharyya has discussed networks and anomalies, ML methods, approaches to detect anomalies in the network data, and methods for evaluation in detail [13]. It has defined requirements, types, and approaches for network anomaly detection. The various approaches for anomaly detection use supervised learning-parametric and non-parametric methods; unsupervised learning – clustering, outlier mining and association mining; probabilistic learning – hidden Markov model, Bayesian networks, naïve Bayes, Gaussian mixture model, EM algorithm; soft computing – genetic algorithm, ANN, ant colony methods, etc. Here [14], experiments use five ML algorithms: logistic regression, naïve Bayes, DT, RF, and ANN to predict the anomalies on the Kaggle dataset, categorizing it into eight categories. Firstly, the algorithms are applied to a complete dataset, achieving 99.4% average accuracy. Secondly, the data with 0 and 1 feature values are removed, achieving 99.99% accuracy. This paper [15] has discussed the potential attacks on IoT networks, showing that IoT botnets are a strong challenge for IoT networks, affecting the availability of IoT devices. This paper has reviewed the advantages and disadvantages of ML algorithms and suggested focusing more on using autoencoders for anomaly detection in future.

One of the main issues with ML algorithms for detecting anomalies is that the classifier learns to new/current traffic/data for predictions on recently trained data and that the classes become unbalanced as traffic changes over time, decreasing the accuracy of the classifier predictions. This issue was addressed by [16] and [17]. As the classifier's output suffers from concept drift, misclassification will rise. In order to achieve precise and effective results, it is crucial to identify certain concept drifts. Concept drift techniques in artificial intelligence and data science generally, specifically in pattern detection and data stream mining, have significantly improved traditional ML research. In the research, comprehensive concept drift surveys [18–21] have offered helpful information on how to deal with concept drift; however, particular study emphasis is solely on learning data stream, rather than examining concept drift adapting techniques.

To improve the efficiency of classifiers in a real-world application, [22] proposed some strategies to deal with concept drift. [23] provided a general structure that incorporates concept drift adaptability and active learning. [24] proposed the unsupervised clustering technique OLINDDA (OnLIne_Novelty and Drift_Detection_Algorithm) to identify abnormalities in the network. The analysis was carried out using the KDD99 dataset, and the outcomes illustrate the technique's ability to handle stream data. [25] proposed an ensemble incremental learning technique to prevent concept drift when identifying network anomalies. [26] classified abnormal traffic using an ANN-based classification algorithm in the concept drift network environment. Recent research [27] introduced the ARF (Adaptive_Random_Forest) technique, which enhances the RF (random forest) method with the concept drift detection approach, ADWIN [28]. Similar research may be found in [29], which employs the Hoeffding bound for identifying drift.

After reading these publications (which are described in Table 13.1), it became clear that while numerous strategies and procedures were employed to address concept drift, very little research had been done on the drift in IoT datasets or had been limited to a specific portion of the outdated dataset (KDD'99). These investigations lacked comprehensive depth, and it has been necessary to conduct a full analysis using both publicly accessible datasets (NSL-KDD and CIDDS (2017)) and to offer a concept drift detection-based model. The discussion of a concept drift-based detection proposed model follows in the next section.

13.3 METHODOLOGY

The entire architecture of the suggested IoT anomaly detection system is shown in Figure 13.1. In the first stage, the incoming IoT data streams are sampled using the k-means cluster sampling technique to create a highly representative subset. The ML model processes the data streams in the following stage. Suppose a concept drift is detected in the new data streams. In that case, the ML model will undergo retraining using the new concept data samples acquired through the recommended adaptive window mechanism (OSAW). This process aims to align the ML model with the updated concept inherent in the new IoT traffic data. As a result, the proposed

TABLE 13.1
A Summary of Algorithms and Dataset Used in the Literature

Problem Addressed	Method	Dataset
Surveyed ML models on mentioned datasets to detect anomalies and to reduce any kind of model corruption [1].	Suggested collaborative learning models combining ML models with blockchain technique.	Mostly used datasets are: N-BaIoT, CICIDS 2017, AWID, UNSW-NB15, NLS-KDD, Kyoto, KDD CUP 1999
Temporal relationships are used with enhanced SVM combining supervised and unsupervised learning with the benefit of low false alarms [2].	Used self-organized feature map (SOFM), passive TCP/IP fingerprinting (PTF), Genetic Algorithm (GA).	Live real-time dataset and MIT Lincoln Lab dataset.
To increase the performance of intrusion detection, the problem at hand involved the identification of the best numerical and categorical-numerical features [4].	To address this problem, mutual information was employed to determine the most suitable numerical and categorical-numerical features. Additionally, deep neural network-based intrusion detection models were implemented and compared.	IoT-Botnet 2020
The problem at hand involved identifying sensor tempering, specifically, detecting instances where the sensor data has been tampered with [5].	To address this problem, two different approaches were employed. Firstly, a supervised learning method called decision tree was utilized for sensor tampering identification. Secondly, an unsupervised learning approach known as isolation forest was applied to detect instances of sensor tempering.	Dataset collected from sensors installed in lab.
Analysing the performance of different classifiers [7].	Several classifiers were evaluated, including SVM, KNN, DT, ANN, logistic regression, random forest (RF), and multilayer perceptron. The performance of each classifier was assessed, and it was found that random forest (RF) outperformed.	The dataset from UNSW-Canberra-cyber with simulated IoT services and contains DDoS, DoS, Reconnaissance, information theft, and normal samples.

TABLE 13.1 (Continued)
A Summary of Algorithms and Dataset Used in the Literature

Problem Addressed	Method	Dataset
Identifying the classifier for anomaly detection [8].	It has identified that RF performs better than ANN, LR, DT, and SVM.	The synthetic dataset collected from Kaggle and created in virtual IoT environment.
Determining the most effective classifier for anomaly detection [9].	It found that distributed version of Averaged One Dependence Estimator (AODE) outperforms other 4 ML models.	UNSW-NB15 dataset.
Detecting different types of anomalies along with normal and malicious anomalies [10].	It proposed deep learning algorithm with hyperparameter optimization and feature reduction step using autoencoder.	Integrated dataset of five different types created from IoT public traffic.
bagging or boosting technique for anomaly detection [11].	SVM algorithms show low performance because of imbalanced dataset. Among ensemble classifiers REPTree is a smart solution, LADTree produces results with delay, and MultiBoost shows the lowest performance.	UNSW-NB15 dataset.
Predicting the anomalies in IoT traffic [14].	Decision tree and RF outperform other ML algorithms as logistic regression, ANN, and naïve Bayes.	Kaggle dataset of seven types of anomalies and one normal.
Identifying a novel cluster-based novelty detection technique for intrusion detection [24].	To find abnormalities in streaming data, the OLINDDA unsupervised clustering approach has been proposed.	Research was conducted using the KDD'99 dataset [30].
Identifying the statistical properties of the change in the status of the network data in real time [25].	To find abnormalities in streaming data, the ensemble approach has been proposed.	Research was conducted using the KDD'99 dataset [30].
Detecting the abnormalities in network traffic timely and effectively in concept drift-driven environment [26].	To identify anomalous network traffic in a concept-drifting context, ANN approach has been proposed [8, 26].	The generated testbed dataset was used for the analysis.

(continued)

TABLE 13.1 (Continued)
A Summary of Algorithms and Dataset Used in the Literature

Problem Addressed	Method	Dataset
Classifying the evolving real-time data streams in network [27].	To identify anomalous network traffic in a concept-drifting context, ARF_ADWIN approach has been proposed.	Experiments were done on airlines [31], covertype [30], electricity [32], spam assassin corpus [33] datasets.
Differentiating concept drift-driven data from noisy data in synthetic and real database [29].	To separate concept drift from noise in decision trees, a Hoeffding bound was presented.	Experiments ware done on KDDCup'99 [30].
Building machine learning models for good and speedy detection of attacks in network.	An effective C-ELM approach was suggested for quick learning [34].	The UNSW-NB15 [35] dataset and KDD [30] datasets were used in experiments.
Using sliding window for mining outliers from the categorical data [36].	A WCFPOF is being proposed to find network anomalies in categorical data streams. The sliding window handling drift has been applied.	Network anomaly detection was carried out using the KDD'99 [30] dataset.
Classifying the data stream considering the various types of concept drifts [37].	AUE technique being proposed to handle both gradual and sudden drifts.	The performance of the proposed model was done on covertype [30], Poker-Hand [30], and covertype [30] datasets.
Determining the optimum weight values for classifiers classifying the loose and rigorous concept-drifting data streams [38].	OWA technique suggested to address various types of notion drifts.	Experiments was done on KDDCup'99 [30].

model can maintain accurate cyber-attack detection while adapting to the constantly evolving new IoT traffic data patterns.

13.3.1 Proposed Algorithm: Optimal Sliding and Adaptive Windowing

To effectively identify concept drift and dynamically adjust to the ever-changing streams of IoT data for accurate analysis, this chapter introduces the OSAW model. A comprehensive insight into the OSAW methodology is provided by Algorithm 13.1,

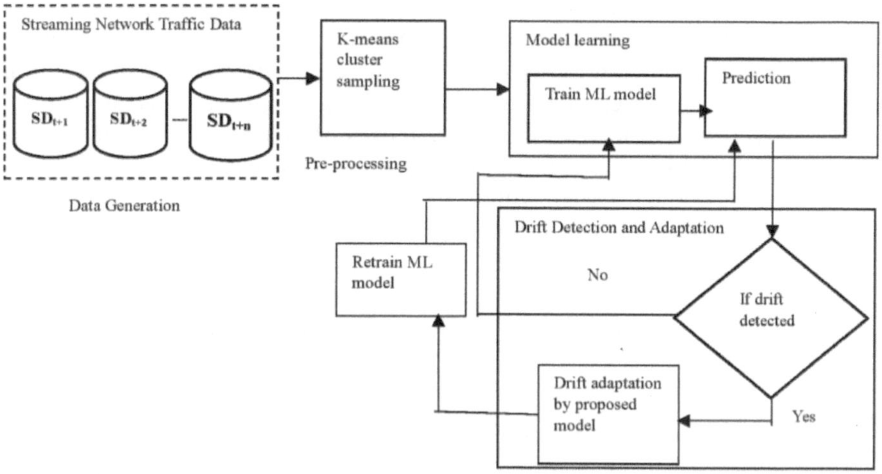

FIGURE 13.1 Drift Detection and Adaption Anomaly Detection Framework.

with its core function being "*Drift_Adapt*". Within this function, "*Drift_Adapt*", the primary focus is detecting concept drift in the streaming data. It employs input hyperparameter constants to update the ML model with new concept samples for drift adaptation.

The OSAW model employs two distinct types of windows: a sliding window (*SW*) utilized for concept drift detection and an Adaptive Window (*AW*) employed for storing fresh concept samples. The maximum size of the AW is represented as, while the size of the SW is denoted as p. Additionally, for concept drift detection, a pair of threshold values (thv), namely *i* and *j*, is employed to signify the warning and drift levels, respectively.

The core sequence of OSAW comprises the subsequent key stages. As each fresh data instance, denoted as k_{in}, arrives in the new data stream, its associated SW, SW_k, incorporates samples ranging from the $(k-p)_{th}$ to k_{th} instances. The accuracy metrics for both the ongoing and previous complete windows are indicated as SW_k and, respectively, are computed and juxtaposed. Should there be a decrease in accuracy by i per cent from the time of $k-p$ to k, the warning level is activated? Subsequently, the AW initiates the collection of incoming data instances as fresh concept samples (lines 6–12). Furthermore, if the accuracy of the SW continues to diminish by *j* per cent and reaches the drift level, the existing learner undergoes an update. This involves retraining the model using the newly accumulated concept samples from the AW (lines 13–18).

Additionally, the AW will persist in accumulating fresh data until one of the subsequent two criteria is met, thereby ensuring the creation of a dependable and consistent learner: (1) If, in comparison to the point where the drift initially emerged, the accuracy of the new concept descends to the warning threshold *i*, indicating the current learner's incapacity to comprehend the new concept, necessitating an update. (2) The AW expands until it reaches the maximum size of p'_{max}, a measure taken to satisfy both memory and real-time requirements. This triggers an additional learner update, leveraging the samples present in the AW to enhance the learner's reliability further. After this potential drift detection and learner update, the system transitions back to its normal state, preparing for any forthcoming drift detections (lines 26–35).

Conversely, if the accuracy of the SW stabilizes or even increases to the regular level during the warning phase, it will be treated as a false alarm. Subsequently, the AW will be reset, and the system will revert to its normal state, prepared to monitor any possible subsequent drift occurrences (lines 19–24).

Algorithm 13.1: Optimal Sliding and Adaptive Windowing (OSAW)

Input:

Str: streaming network traffic data,
$i; j$: th_v for warning and drift
p: the constant size of the sliding_window,
p'_{max}: upper limit on the adaptive window's size,
Model: A Machine Learning model,

Output:

Acc_{max}: Performance Metrics.

1 *Drift_Adapt(Str, Model, i, j, p,* $p'_{\max}$):
2 $AW' \leftarrow \varnothing$; // Adaptive window initialization
3 $St \leftarrow 0$; // States: normal, drift, warning
4 *for* each sample $n_k \in Str$do
5 SW_k // last p instances sliding window of n_k
6 $AccSW_k \leftarrow accuracy(SW_k)$; // Present window accuracy
7 $AccSW_{k-p} \leftarrow accuracy(SW_{k-p})$; // Previous window accuracy
8 $if\,(St == 0)\,\&\&\,(AccSW_k < i * AccSW_{k-p})$ then// Transition to warning from normal state
9 $AW' \leftarrow AW' \cup \{n_k\}$// Start collecting new instances in AW'
10 $St \leftarrow 1$; // In warning_state
11 *end*
12 $if\,(St == 1)$ then // warning_state
13 $p' \leftarrow Size(AW')$;
14 $ifAccSW_k < j * AccSW_{k-p}$ then//Current window accuracy falls to drif_level
15 $St \leftarrow 2$; // Enter drift state
16 $f \leftarrow k$; // Store first new concept window accuracy as baseline
17 $Model' \leftarrow Retrain\ Classifier\ on\ AW'$; // classifier retrained on new_concept_samples

$elseif\left(AccSW_k \geq i * AccSW_{k-p}\right) \| p' == p'_{max}\right)$ then // False alarm: return to normal state or remain constant
 $AW' \leftarrow \emptyset$// free the adaptive_window
 $St \leftarrow 0$// back to normal_state
$else$ // remain in the warning_state
 $AW' \leftarrow AW' \cup \{n_k\}$ // AW' Continue collecting new instances
end
end
$if\,(St == 2)$ then// In a drift state
 $p' \leftarrow Size(AW')$;
 $if(AccSW_k < i * AccSW_{f+p}) \| p' == p'_{max})$ then
  $Model' \leftarrow Retrain\ Classifier\ on\ AW'$; // Construct a robust classifier
  $AW' \leftarrow \emptyset$// free the new concept window
  $St \leftarrow 0$// Back to a normal state
 $else$
  $AW' \leftarrow AW' \cup \{n_k\}$// AW' keeps collecting new instances
 end
end
end
$return\ Acc_{max}$;

13.4 RESULTS AND DISCUSSION

13.4.1 Results and Analysis

The experiments were shown on a Windows 10 system equipped with the following specifications: An Intel (R) Core i3 processor running at 2.0 GHz and 4 GB of RAM. Python scripting was employed for processing purposes. To evaluate the efficiency of the suggested concept drift detection and adaptation using the OSAW approach in IoT data streams, the CIDDS (2017) [39] and NSL-KDD [40] datasets were employed.

13.4.2 The Used Datasets

The NSL-KDD dataset stands out as the predominant publicly available dataset for anomaly detection. The dataset comprises 148,517 records, each comprising 41 features, with the 42nd feature representing the label class.

The latest publicly accessible dataset utilized by researchers engaged in anomaly detection is the CIDDS (2017) dataset. Comprising 671,241 records, each entry comprises ten features, with the 11th indicating the label class. Among these ten features, eight are continuous, while two are categorical. We transform the continuous features into discrete values to enhance the classifier's classification performance, employing four categories to categorize each continuous attribute.

It was observed that within the CIDDS (2017) dataset, two features – "type of service" and "flows" – solely contain values of 0.0 and 1.0. Similarly, the "number of outbound commands" feature in the NSL-KDD dataset exclusively consists of 0.0 values. Consequently, it was decided to remove instances associated with these specific parameters.

13.4.3 OSAW Performance Over CIDDS (2017) and NSL- KDD Datasets

Table 13.2 compares diverse supervised ML techniques' performance using the separate CIDDS (2017) and NSL-KDD datasets. K-Means clustering was employed to label the unlabelled dataset. The resultant clusters were then utilized to train the classifier model. Across both datasets, there was an observable reduction in accuracy values.

Table 13.3 shows the accuracy of our proposed method for detecting network anomalies based on concept drift on two separate datasets. Throughout the two datasets, we

TABLE 13.2
Performance of Machine Learning Techniques before Drift Handling

	CIDDS (2017) [39]				NSL-KDD [40]			
Method	Acc (%)	Precision (%)	Recall (%)	F1-Score (%)	Accuracy (%)	Precision (%)	Recall (%)	F1-Score (%)
KNN	82	79	75	77	77	83	77	77
LR	81	80	77	78	70	76	70	70
RF	86	89	88	88	77	83	77	76
SVM	88	86	84	85	81	85	81	81

TABLE 13.3
Effectiveness of Machine Learning Approaches Following Drift Handling

	CIDDS (2017) [39]			NSL-KDD [40]				
Method	Acc (%)	Precision (%)	Recall (%)	F1-Score (%)	Accuracy (%)	Precision (%)	Recall (%)	F1-Score (%)
Proposed OSAW & KNN	91.2	90.8	92.3	92.7	81.6	85.7	81.7	81.5
Proposed OSAW & LR	94.4	92.4	93.6	93.1	85.4	76.4	70.7	70.8
Proposed OSAW & RF	96.2	93.5	92.1	91.7	88.5	85.2	88.4	87.2
Proposed OSAW & SVM	99.8	99.1	99.2	99.1	91.3	88.3	91.7	89.8

TABLE 13.4
Evaluating Different Approaches for Detecting Concept Drift Using the NSL-KDD Dataset

Reference Id	Drift Detection Technique	Model	Acc. (%)	Precision (%)	Recall (%)	F1-Score (%)
[41]	HDDM as the method of detecting drift.	Multilayer perceptron (MLP)	88.0	85.6	88.2	87.8
[42]	Margin_Density based detecting drift	Support vector Machine (SVM)	89.4	87.4	88.2	86.6
[43]	Active learning based anomaly detection	Naïve Bayes (NB)	89.2	88.5	89.1	89.4
Proposed approach	Optimal Sliding and Adaptive Windowing (OSAW)	Support Vector Machine (SVM)	91.3	88.3	91.7	89.8

have attained high accuracy scores. As shown in Table 13.3, the proposed method can obtain the best accuracy of 99.8% and 91.3% on CIDDS (2017) and NSL-KDD when the drift is handled, but the accuracy of the SVM model falls noticeably to only 88% and 81% when the drift is not handled. This puts importance on the development of drift adaptation techniques.

Table 13.4 displays a comparison between the proposed OASW & SVM model and the innovative adaptive techniques. Linear support Vector Classification has been employed due to its greater versatility in selecting penalties and loss functions, making it more adaptable for handling large sample sizes. To enhance model accuracy, there are several parameters that need to be tuned. Three key parameters, namely loss, penalty, and C (regularization), have been introduced. It shows that the proposed adaptive model performs better than all previous techniques in terms of accuracy. In contrast, the four alternative methods being compared, namely SAM-KNN, ARF, Stochastic Regularized Probability (SRP), and Hierarchical Attention Transformer (HATT), exhibit significantly lower accuracy levels compared to the proposed method, ranging from 94.11% to 96.58%.

13.5 CONCLUSION

The work presented in this chapter understands the challenges, emphasizes one of the major problems for IoT networks, explores the possible solutions, and implements and analyses the results of concept drift-driven ML algorithms to determine whether it has the potential to detect abnormal behaviour in the IoT network. It reviews the state-of-the-art ML methods that detect the concept drift in anomalies. K-means cluster sampling is used to sample the data streams to identify the representative subset of the dataset. The new data samples are collected using the OSAW algorithm. The K-nearest neighbour, logistic regression, RF, and SVM algorithms are applied to this CIDDS (2017) and NSL-KDD dataset sampled dataset. The experimental analysis has shown the difference in the performances (accuracy, precision, recall, and F1-score) of K-nearest neighbor, logistic regression, RF, and SVM with drift handling and without drift handling on CIDDS (2017) and NSL-KDD dataset. The accuracy of SVM with drift handling reaches 99.8% and 91.3% for CIDDS (2017) and NSL-KDD datasets, respectively, which is better than the results attained without handling drifts. We create a perspective here to accept the advancement in ML that better reacts to the dynamism in IoT network anomalies. We aim to expand on this research by exploring the following directions: creating controlled network environments to generate attacks or anomalies and presenting the obtained results. To handle conceptual drift, we will employ various mechanisms to detect and address such drift. Furthermore, we plan to conduct an analysis of distributed Denial of Service (DDoS) attacks, considering the vulnerabilities of services like Heartbleed and exploring brute force attacks on the web.

REFERENCES

1. A. Diro, N. Chilamkurti, V. D. Nguyen, and W. Heyne, "A comprehensive study of anomaly detection schemes in IoT networks using machine learning algorithms," *Sensors*, vol. 21, no. 24, pp. 8320, 2021.
2. T. Shon and J. Moon, "A hybrid machine learning approach to network anomaly detection," *Information Sciences*, vol. 177, no. 18, pp. 3799–3821, 2007.
3. R. Al-amri, R. K. Murugesan, M. Man, A. F. Abdulateef, M. A. Al-Sharafi, and A. A. Alkahtani, "A review of machine learning and deep learning techniques for anomaly detection in IoT data," *Applied Sciences*, vol. 11, no. 12, pp. 5320, 2021.
4. Z. Ahmad, A. Khan, K. Nisar, I. Haider, R. Hassan, M. R. Haque, and J. J. Rodrigues, "Anomaly detection using deep neural network for IoT architecture," *Applied Sciences*, vol. 11, no. 15, pp. 7050, 2021.
5. A. K. Pathak, S. Saguna, K. Mitra, and C. Åhlund, "Anomaly detection using machine learning to discover sensor tampering in IoT systems," in *ICC 2021-IEEE International Conference on Communications*, pp. 1–6, IEEE, June 2021.
6. S. H. Haji and S. Y. Ameen, "Attack and anomaly detection in IoT networks using machine learning techniques: A review," *Asian journal of research in computer science*, vol. 9, no. 2, pp. 30–46, 2021.
7. H. Tyagi and R. Kumar, "Attack and Anomaly Detection in IoT Networks Using Supervised Machine Learning Approaches," *Revue d'Intelligence Artificielle*, vol. 35, no. 1, pp. 11–21, 2021.
8. M. Hasan, M. M. Islam, M. I. I. Zarif, and M. M. A. Hashem, "Attack and anomaly detection in IoT sensors in IoT sites using machine learning approaches," *Internet of Things*, vol. 7, p. 100059, 2019.
9. M. Nawir, A. Amir, N. Yaakob, and O. B. Lynn, "Effective and efficient network anomaly detection system using machine learning algorithm," *Bulletin of Electrical Engineering and Informatics*, vol. 8, no. 1, pp. 46–51, 2019.

10. L. Aversano, M. L. Bernardi, M. Cimitile, R. Pecori, and L. Veltri, "Effective anomaly detection using deep learning in IoT systems," *Wireless Communications and Mobile Computing*, vol. 2021, p. e9054336, 2021.
11. V. Timčenko and S. Gajin, "Machine learning based network anomaly detection for IoT environments," in ICIST-2018 conference, 2018.
12. S. Wang, J. F. Balarezo, S. Kandeepan, A. Al-Hourani, K. G. Chavez, and B. Rubinstein, "Machine learning in network anomaly detection: A survey," *IEEE Access*, vol. 9, pp. 152379–152396, 2021.
13. D. K. Bhattacharyya and J. K. Kalita, *Network anomaly detection: A machine learning perspective*, Crc Press, 2013.
14. I. Mukherjee, N. K. Sahu, and S. K. Sahana, "Simulation and modeling for anomaly detection in IoT network using machine learning," *International Journal of Wireless Information Networks*, vol. 30, pp. 1–17, 2022.
15. R. Alhajri, R. Zagrouba, and F. Al-Haidari, "Survey for anomaly detection of IoT botnets using machine learning auto-encoders," *International Journal of Applied Engineering Research*, vol. 14, no. 10, pp. 2417–2421, 2019.
16. M. Choras, R. Kozik, R. Renk, and W. Hołubowicz, "The concept of applying life-long learning paradigm to cybersecurity," in *International Conference on Intelligent Computing*, Springer, Cham, pp. 663–671, Aug. 2017, doi: 10.1007/978-3-319-63315-2_58.
17. I. Zliobaite, M. Pechenizkiy, and J. Gama, "An overview of concept drift applications," in *Big data analysis: new algorithms for a new society*, Springer, Cham, pp. 91–114, 2016, doi: 10.1007/978-3-319-26989-4_4.
18. B. Krawczyk, L. L. Minku, J. Gama, J. Stefanowski, and M. Woniak, "Ensemble learning for data stream analysis: A survey," *Information Fusion*, vol. 37, pp. 132–156, 2017.
19. S. Ramírez-Gallego, B. Krawczyk, S. García, M. Wozniak, and F. Herrera, "A survey on data preprocessing for data stream mining: Current status and future directions," *Neurocomputing*, vol. 239, pp. 39–57, 2017.
20. J. Gama, "A survey on learning from data streams: Current and future trends," *Progress in Artificial Intelligence*, vol. 1, no. 1, pp. 45–55, 2012.
21. J. A. Silva, E. R. Faria, R. C. Barros, E. R. Hruschka, A. de Carvalho, and J. Gama, "Data stream clustering: A survey," *ACM Computing Surveys*, vol. 46, no. 1, pp. 1–31, 2013.
22. Nidhi, V. Mangat, V. Gupta, R. Vig, "Methods to investigate concept drift in big data streams," *Knowledge Computing and Its Applications*, vol. 1, no. 1, p. 51, 2018, doi: 10.1007/978-981-10-6680-1_3.
23. I. Zliobaite, A. Bifet, B. Pfahringer, G. Holmes, "Active learning with drifting streaming data," *IEEE Transactions on Neural Networks and Learning Systems*, vol. 25, no. 1, pp. 27–39, Jan. 2014.
24. E. J. Spinosa, F. de Leon, A. P. de Carvalho, and J. Gama, "Cluster-based novel concept detection in data streams applied to intrusion detection in computer networks," in *Proceedings of the 2008 ACM Symposium on Applied Computing*, pp. 976–980, Mar. 2008, doi: 10.1145/1363686.1363912.
25. X. Yuan, R. Wang, Y. Zhuang, K. Zhu, J. Hao, "A concept drift based ensemble incremental learning approach for intrusion detection," in *2018 IEEE International Conference on Internet of Things and IEEE Green Computing and Communications and IEEE Cyber, Physical and Social Computing and IEEE Smart Data*, pp. 350-357, Jul. 2018, doi: 10.1109/Cybermatics_2018.2018.00087.

26. R. H. Tsaih, S. Y. Huang, M. C. Lian, Y. Huang, "ANN mechanism for network traffic anomaly detection in the concept drifting environment," in *2018 IEEE Conference on Dependable and Secure Computing*, pp. 1–6, Dec. 2018, doi: 10.1109/DESEC.2018.8625108.
27. H. M. Gomes, A. Bifet, J. Read, J. P. Barddal, F. Enembreck, B. Pfahringer, G. Holmes, and T. Abdessalem, "Adaptive random forests for evolving data stream classification," *Machine Learning*, vol. 106, pp. 1469–1495, 2017.
28. A. Bifet and R. Gavalda, "Learning from time-changing data with adaptive windowing," in *Proceedings of the 2007 SIAM International Conference on Data Mining*, vol. 7, 2007, Art. no. 2007.
29. P. Li, X. Wu, X. Hu, and H. Wang, "Learning concept-drifting data streams with random ensemble decision trees," *Neurocomputing*, vol. 166, pp. 68–83, 2015.
30. M. Lichman, "UCI machine learning repository," 2013. [Online]. Available: http://archive.ics.uci.edu/ml
31. A. Bifet, G. Holmes, R. Kirkby, and B. Pfahringer, "MOA: Massive online analysis," *Journal of Machine Learning Research*, vol. 99, pp. 1601–1604, 2010.
32. M. Harries, "Splice-2 comparative evaluation: Electricity pricing," Tech. Rep., The University of South Wales, 1999.
33. I. Katakis, G. Tsoumakas, E. Banos, N. Bassiliades, and I. Vlahavas, "An adaptive personalized news dissemination system," *Journal of Intelligent Information Systems*, vol. 32, no. 2, pp. 191–212, 2008.
34. C. R. Wang, R. F. Xu, S. J. Lee, and C. H. Lee, "Network intrusion detection using equality constrained-optimization-based extreme learning machines," *Knowledge-Based Systems*, vol. 147, pp. 68–80, 2018, doi: 10.1016/j.knosys.2018.02.015.
35. "UNSW-NB Dataset," https://research.unsw.edu.au/projects/unsw-nb15-dataset.
36. Q. Wu and S. Ma, "Detecting outliers in sliding window over categorical data streams," in *2011 Eighth international conference on fuzzy systems and knowledge discovery (FSKD)*, vol. 3, pp. 1663–1667, IEEE, July 2011, doi: 10.1109/FSKD.2011.6019780.
37. D. Brzezinski and J. Stefanowski, "Reacting to different types of concept drift: The accuracy updated ensemble algorithm," *IEEE Transactions on Neural Networks and Learning Systems*, vol. 25, no. 1, pp. 81–94, Jan. 2014.
38. P. Zhang, X. Zhu, and Y. Shi, "Categorizing and mining concept drifting data streams," in *Proceedings of the 14th ACM SIGKDD International Conference on Knowledge Discovery and Data Mining*, 2008, pp. 812–820.
39. "CIDDS (2017)," 2017, accessed: 2020-05-25. www.hs-coburg.de/forschung/forschungsprojekte-oeffentlich/informationstechnologie/cidds-coburg-intrusion-detection-data-sets.html
40. "NSL-KDD," accessed: 2020-05-25. http://nsl.cs.unb.ca/NSL-KDD/.
41. D. Pradheep, R. Gokul, V. Naveen, and J. Vijayarani, "Anomaly intrusion detection based on concept drift," *Global Journal of Computer Science and Technology*, 2020.
42. T. S. Sethi and M. Kantardzic, "On the reliable detection of concept drift from streaming unlabeled data," *Expert Systems with Applications*, vol. 82, pp. 77–99, 2017, doi: 10.1016/j.eswa.2017.04.008.
43. P. Alaei and F. Noorbehbahani, "Incremental anomaly-based intrusion detection system using limited labeled data," in *2017 3th International Conference on Web Research (ICWR)*, 2017, Apr 19: IEEE, pp. 178–184, doi: 10.1109/ICWR.2017.7959324.

14 Internet of Things (IoT) Security Using Blockchain

Diksha Chawla, Pawan Singh Mehra, Jatin Sharma, Deepakshi Dabas, and Aatif Jamshed

14.1 INTRODUCTION

The utilization of IoT in diverse industrial fields has been undergoing rapid and exponential growth over recent years. The impact of IoT applications, such as smart urban systems, advanced healthcare solutions, intelligent transportation networks, environmental tracking, wearable technology, the Industrial Internet, and more, has a significant influence on the everyday lives of individuals [1]. Nonetheless, the widespread adoption of IoT introduces numerous security and trust-related hurdles. Frequently positioned in distant locales, IoT devices face susceptibility to data tampering, resulting in the dissemination of deceptive data, incongruities, and potential hazards to privacy and security. Even if vouched for by other layers of the IoT ecosystem, data sourced from various devices might harbor malicious intent and remain untrustworthy. To maintain security and trust criteria such as authentication, access control, diversity, integrity, and confidentiality, it is necessary to transform information into a secure and reliable format [2]. Relying solely on conventional security measures like cryptographic methods such as RSA, AES, and ECC [3] falls short of ensuring data integrity on this vast scale, thereby significantly impeding the potential future integration of IoT. It's important to highlight that within this context, primary access verification is typically carried out by a centralized entity, which is susceptible to a single point of failure.

The concept of blockchain technology, initially introduced by Satoshi Nakamoto [4], offers a shared, digital, and decentralized ledger. It has already found extensive application in cryptocurrency dealings and has emerged as a pivotal technology for fostering further advancements across diverse Internet of things (IoT) contexts [5]. As a distributed, immutable, and tamper-proof ledger database, blockchain can tackle the pressing security concerns within the realm of IoT, primarily focusing on ensuring data integrity and dependability [6]. Blockchain enables software applications to transmit and log transactions or events, forming tamper-resistant ledgers accessible solely to

DOI: 10.1201/9781003405535-19

designated entities. It additionally guarantees that data transmission and reception occur in a reliable, open, and distributed manner, following a peer-to-peer approach. Interest in implementing blockchain within IoT networks has already materialized within academic and industrial circles, aiming to furnish enhanced security measures. In the IoT framework, many heterogeneous devices share information with sufficient network connectivity [7]. In such networks, integrating blockchain with IoT creates a distributed peer-to-peer network, eliminating the Man-In-Middle attack [8].

Motivated by the earlier discussion, in this chapter, we discussed IoT security issues, existing solutions, and a detailed overview of blockchain technology. Then, we discuss the integration of blockchain with IoT. The blockchain-based security solutions concerning IoT framework are also discussed in our work.

14.2 LITERATURE SURVEY

Ghorbani et al. [9] proposed a lightweight encryption technique that mutually authenticates sensor nodes in IoT communication. Their scheme provides resistance from perfect forward secrecy and mutual authentication. Gope et al. [10] analyze IoT applications. As the different devices are involved in IoT communication, each has a unique location and can be identified using Radio-Frequency Identification (RFID) tags. The IoT has a huge potential to generate data. Such data are considered big data in the IoT environment. Therefore, Wazid et al. [11] analyzed security requirements in IoT-based cloud big data environments. As big data allows the user to query for real-time data, it should be accessed by only authorized users. Li and Liu [3] analyze the smart city [3] applications. In smart city automation, sensor nodes are connected to provide information such as air quality, public services, and urban planning. The author proposed a public key encryption scheme to establish a secure channel. Their scheme is based on secure key generation. The lightweight encryption techniques are based on public and private key encryption techniques to secure IoT data from attackers.

Jahid et al. [12] in his work discussed the role of 6G networks in enhancing the quality of experience (QoS) for IoT applications. The authors discussed various aspects of IoT, such as decentralization, immutability, limitations in spectrum resources, innate privacy and security, compatibility, secrecy, and the evolving landscape of intelligent application domains such as industrial IoT (IIoT) and Industry 5.0 applications. Mistry et al. [13] in his work discussed the advancement in IoT for industrial automation. The authors discussed the role of blockchain in secure information exchange for industrial IoT-based applications. Dorri et al. [8] discussed the role of blockchain in smart home security. The authors discussed the role of home miners and the secret information miners maintain in the form of transactions. In detail, discussion on IoT-based smart applications and their security issues are discussed by Kumar and Sharma [6]. The author also discussed the trust issues concerning IoT frameworks and blockchain-based solutions. Different attacks on IoT-layered architecture, such as phishing attacks, Denial of Service (DoS), and node capture attacks, are also discussed.

14.3 IOT SECURITY CONCERNS AND SOLUTIONS

14.3.1 Security Concerns

1. **Confidentiality:** The user data must remain confidential while sharing data over a communication channel. However, In IoT communication, hackers can explore various vulnerabilities, such as accessing user data by analyzing the network traffic. It may launch various attacks to compromise the confidentiality of data, including tracing attacks, key attacks, brute force attacks, and Man-In-Middle attacks.
2. **Integrity:** As devices exchange vast amounts of data, if one device's security is compromised, it can lead to compromise, leading to integrity breaches down the line.
3. **Authentication:** Mutual authentication among IoT devices is a critical aspect. In an IoT communication scenario, if the authentication is not strong, the adversary may launch impersonation attacks, replay attacks, and masquerade attacks.

14.3.2 Solutions for Securing IoT

Existing solutions for securing IoT communication are as follows:

1. **Authentication solutions:** To protect IoT communication from unauthorized access, existing researchers considered the following best practices:
   a. **Secure key management:** An asymmetric encryption scheme known as private key provides public and private key security. If the data is encrypted using the private key, it can only be decrypted using the public key. Such key is distributed using classical communication channels and hence vulnerable to eavesdropping attacks.
   b. **Two-factor authentication:** This scheme requires multiple factors (e.g., password and OTP) to authenticate the communicating entities. Such schemes are often vulnerable to password-guessing, impersonation, and reply attacks.
   c. **Three-factor authentication:** Such schemes strengthen the security provided by adding extra features and overcoming the deficiencies of two-factor authentication schemes. They ensure that even if an attacker gains partial information about the user's smart card, biometric data, or password, they cannot compromise the entire system. Such schemes are often vulnerable to Man-In-Middle attacks, Phishing Attacks, and credential stuffing [14].
2. **Confidentiality Solutions**: To ensure that only legitimate users can decrypt the information, existing researchers have adopted the following schemes:
   a. **Cryptographic primitives**: Many schemes based on symmetric and asymmetric cryptographic primitives such as RSA, ECC, and AES provide security to IoT data (stored or in transient) [3]. These schemes are susceptible to known-plaintext attacks (KPA), chosen-plaintext attacks

(CPA), differential cryptanalysis, linear cryptanalysis, and Man-In-Middle attack.

b. **Secret key**: To maintain the confidentiality of data, secure key distribution is important in IoT communication. Such keys must be transferred using the secure channel, and a key protection mechanism must be implemented. The existing methods, such as The Diffie–Hellman key exchange method, are based on the discrete logarithm problem (DLP) and are vulnerable to index calculus as well as the Pollard rho attack on DLP.

3. **Integrity solutions**: In IoT communications, integrity must be maintained to prevent any unauthorised entity to modify the message content. To secure the integrity of data, existing schemes relied on [15]:
   a. **Digital signatures**: IoT devices can sign their data using the private key, and recipients can access that data using their public key to ensure the data hasn't been tampered with.
   b. **Hash function:** The function produces a unique output by taking the input data. Any changes in the data change the hash function, providing evidence of tempering. However, such hash codes can easily be cracked down with rainbow tables.

14.4 BLOCKCHAIN

14.4.1 Overview

Blockchain technology (BCT) was first proposed by Satoshi Nakamoto [4]. In recent years, blockchain technology has gained momentum as a rising innovation in academic and industrial spheres. Blockchain provides common distributed immutable public ledger technology [12]. Blockchain technology, without the involvement of the central trusted party, can record and verify transactions between participants. To overcome the disadvantage of centralized databases, decentralized architectures can be used for peer-to-peer (P2P) communications among the nodes. Within blockchain technology, a ledger utilizes public key cryptography to uphold an expanding series of information blocks. Every individual block within a blockchain is composed of a sequence of blocks, each holding a hash, timestamp, random number, and certain transactions from the preceding block. The block is appended to the blockchain by the mining process. The miners performing the mining process add the new block to the blockchain. To create a block, certain designated nodes called miners engage in solving a computationally intensive cryptographic challenge called proof of work (PoW). The miner that successfully solves this challenge first is responsible for adding the new block to the blockchain.

The new block is added by applying secure SHA 256 [12]. Based on the data from its parent block, the hash of the parent block is connected and retained within the child block. The hash value is generated according to the content of the block; any changes in the block will change its hash value. In blockchain, private key security is provided for each node through digital signatures. As shown in the architecture of

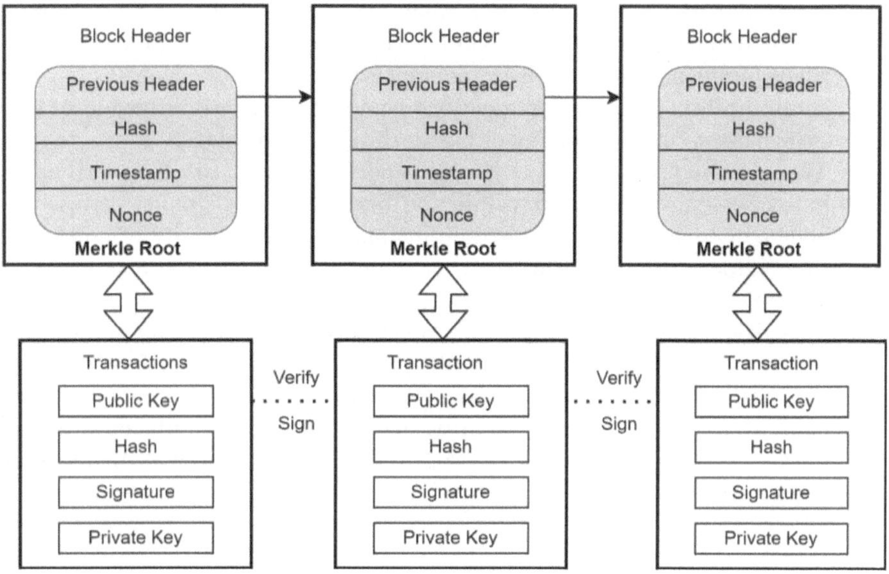

FIGURE 14.1 Blockchain architecture.

blockchain in Figure 14.1, every block possesses a sole parent and child, containing header details and transaction information. The header includes essential elements like the transaction count, timestamp, Nonce, block version, difficulty proof, the parent block's hash, and a combined hash of all transactions in the sequence. Conversely, a series of linked blocks form a chain that progressively expands as blockchain users conduct transactions.

When a miner discovers a nonce, selected randomly, that results in the hash of the complete block falling below the specified nBits threshold, the miner shares the block with the entire network. Once other miners authenticate and approve the block, it becomes the most recent addition to the blockchain. Subsequently, the mining process proceeds for the subsequent block.

14.4.2 Characteristics

The fundamental characteristics of blockchain that make it attractive for IoT systems are:

1. **Ledger:** It functions as a decentralized and communal database that holds records of all executed network transactions. The ledger guarantees multiple entities' simultaneous authentication and confirmation of each transaction. Once transactions are completed, the information becomes immutable and cannot be altered or removed.
2. **Decentralization:** Blockchain employs a variety of consensus mechanisms to verify transactions and uphold the distributed ledger without the need for

a central authority, such as a government or a bank. Blocks can be duplicated and disseminated across the network to mitigate the risk of failures, ensuring adaptability and enhancing data endurance.

3. **Immutability:** Confirmed transactions cannot be erased within the blockchain system due to the interlinking of blocks through hash values.
4. **Transparency:** Every participating node possesses identical privileges to access transaction data within the blockchain. In simpler terms, any node can engage with others while maintaining the integrity of confirmed transactions. However, there are no constraints imposed on the overall system's operation.
5. **Security**: Within blockchain's framework of asymmetric cryptography, each transaction block is associated with a distinct cryptographic key, ensuring the preservation of data integrity. The core characteristics encompassing consensus algorithms, immutability, and anonymous identification ensure the entire chain's dependability, privacy, and security. These attributes work cohesively to maintain the unaltered integrity of previous transactions.

14.4.3 Protocols and Technologies

14.4.3.1 Consensus Protocol

Blockchain ensures that all participants agree on the block to be added to blockchain. Blockchain networks utilize consensus protocols to ensure the uniformity of state data among their members. These protocols define the criteria for validating a block or transaction and specify which participant has the authority to add transactions to the record through a process called 'mining'. The consensus blockchain algorithms are proof of work (PoW) and proof of stake (PoS). In PoW, cyber-attacks can be avoided by circulating a puzzle to every participant who may try to add a block to the blockchain. The participant has to show expensed efforts in the process. In PoS, the miner should have more cryptocurrency to have more processing. It decreased the computational expenses, unlike PoW, which demands greater processing power for puzzle resolution.

14.4.3.2 Practical Byzantine Fault Tolerance

The scenario entails peers attempting to achieve consensus, yet within the peer group, there could be untrustworthy individuals who might sabotage the consensus-building process. The potential tactics of these betrayers encompass ignoring messages, disseminating false information, counterfeiting others' messages, and engaging in 'two-face' behavior, where a node communicates contradictory viewpoints to different nodes. These approaches can potentially trigger Byzantine failures in networks that depend on consensus.

14.4.3.3 Smart Contract

A smart contract represents executable code capable of automated execution. This functionality enables the automatic processing of unique transactions using predefined algorithms and self-triggered computations. Nevertheless, the code itself could be exposed to vulnerabilities and potential attacks. When utilizing smart contracts for safeguarding IoT data integrity, exploring methods that guarantee the security of the contract's code and logic is valuable. Within a peer-to-peer blockchain network, all nodes can cryptographically examine the activities of smart contracts through an automated transaction protocol. Blockchain can be categorized into four types:

1. **Permissionless**: Individuals possessing computational capacity engage in open, public blockchain networks to conduct transactions and contribute to the mining nodes and consensus mechanism, facilitating the incorporation of new transaction blocks into the chain.
2. **Permissioned**: Participation in the scheme is limited to authorized users who are eligible based on specific business requirements. Approved entities are expected to possess connections to pre-existing applications.
3. **Public**: The permissionless proof-based security system of the public access chain ensures network dependability for all users, irrespective of their initial entry-level identity.
4. **Private**: In a private blockchain network, there is enhanced flexibility through increased control over transaction rules, which can be altered through consensus agreement.

14.5 REQUIREMENT OF BLOCKCHAIN IN IOT

The incorporation of blockchain technology into the IoT realm emerges from crucial needs that aim to tackle the constraints and difficulties presented by conventional centralized systems. Below are fundamental rationales elucidating the necessity of blockchain within the IoT context.

14.5.1 Decentralization

Conventional IoT frameworks frequently depend on central entities for validation and verification, a situation that can result in vulnerabilities and susceptibility to failures due to centralization. The distributed consensus mechanisms inherent in blockchain eradicate the necessity for intermediaries, nurturing trust and dependability among peers within a decentralized network.

14.5.2 Smart Contracts

The smart contract feature of blockchain allows for automated and self-triggered agreements founded on predetermined conditions. This proves especially advantageous in IoT contexts, where devices can independently initiate actions in response to real-time data inputs, obviating the requirement for intermediaries.

14.5.3 Interoperability

IoT environments encompass a range of devices, protocols, and platforms that might encounter challenges in seamless communication. Blockchain offers a standardized and interoperable framework that facilitates diverse IoT devices interacting and exchanging data across various networks.

14.6 BLOCKCHAIN DEPLOYMENT IN IOT

In this section, we discussed IoT-based applications and the existing research in securing IoT-based communication.

14.6.1 Agriculture

Intelligent farming employs contemporary advancements like IoT, Global Positioning System (GPS), and big data to enhance both the volume and excellence of agricultural yields. Data such as temperature, light, soil moisture, and humidity can be stored within a centralized management system and assessed using specific Artificial Intelligence (AI) algorithms. Distributed ledger technologies (DLTs) are seen as having the highest capability to enhance efficiency, transparency, and traceability for crops within these agricultural supply networks. Leng et al.'s [16] work considers extensive research on blockchain-integrated systems. It offers a clear and secure method for recording transactions within the agricultural supply chain. The primary benefit offered by DLTs is improved traceability. Food safety remains a matter of public concern; agriculture systems based on blockchain and IoT can ensure trustworthy food safety by collecting and revealing details about food production, processing, storage, and distribution. Caro et al. [17] allow for the tracking of agricultural products within the supply chain.

14.6.2 Industry 4.0

In the current age, achieving full automation of industrial and commercial procedures has become a tangible achievement. Significant technological advancements, combined with their integration into various industries, have led to the rise of a novel production approach termed Industry 4.0. Within the context of Industry 4.0, the IoT is anticipated to provide potentially transformative solutions for managing business process systems. Kapitonov et al. [18] proposed utilizing decentralized systems (blockchain technology) for effective and secure communication among autonomous agents within a multi-agent system. Thorough studies of blockchain within industrial IoT are conducted [19].

14.6.3 Smart V2V Communications

Smart transportation systems are among the upcoming applications bolstered by the technical capabilities offered by 6G network services. Nevertheless, a notable security concern regarding Intelligent Transportation Systems (ITS) revolves around their inclination towards centralization, potentially rendering centralized authorities vulnerable to temporary disruptions caused by external malicious attacks. Mistry et al. [13] for intelligent transportation system proposed ecosystem characterized by autonomy, security, trust, and decentralization, facilitated by blockchain technology.

14.6.4 UAV Communications

In recent times, unmanned aerial vehicles (UAVs)-based communications have gained significant traction in civilian applications, branching out from their traditional military uses. Jahid et al. [12] discussed that the blockchain technology within the drone industry enhances identity management and air traffic control with increased precision and enhanced security. Mistry et al. [13] in their work discussed blockchain-based security for autonomous UAVs.

14.7 SECURING IOT USING BLOCKCHAIN

To ensure the security and trust in the IoT communication framework, blockchain technology provides benefits such as:

1. **Interoperability**: This interoperability in IoT communication can have many different forms. The present solution depends upon many gateways and third parties. However, an automated solution is required to resolve the interoperability issues which can have from heterogeneous devices, networks, and protocols. Blockchain provides the perfect solution. Blockchain operates in a decentralized and automated fashion, thus making all interoperability dependencies reliant upon it.
2. **Privacy**: Preserving data confidentiality presents a significant hurdle for all information systems, including IoT. Whether applied in contexts like smart residences, healthcare, or vehicular systems, the privacy of data transmitted within IoT must be upheld. For example, a patient integrated into an IoT-driven healthcare system might wish to keep her medical details confidential from others. Blockchain transactions are safeguarded through cryptographic measures, offering an extra level of security for IoT data during both its transmission and storage. This encryption guarantees that solely authorized entities can access and comprehend the data.
3. **Authentication**: In IoT communication, mutual authentication is the foremost requirement. As symmetric and asymmetric solutions are available for secure communication, most of the solutions are based on public key methods. Despite the entire methods available, ensuring authentication in the IoT communication framework is a crucial task. A private blockchain method can be implemented for ensuring mutual authentication among communicating entities. Smart contracts can be used to establish trust among communicating entities.

4. **Integrity**: Weaknesses in the integrity of IoT data not only hinder its proper functioning but also render it susceptible to a compromised platform for potential attackers. To maintain the integrity of hash functions, RSA and AES schemes are used which are based on hard mathematical primitives [20]. However, these cryptographic schemes are susceptible to attacks such as secret key attacks, man-in-the-middle attacks, and attackers can manipulate data. To maintain the data integrity, blockchain features provide solutions such as decentralization, distributed ledger, and immutability.

14.8 A COMPARATIVE ANALYSIS: AUTHENTICATION TECHNIQUES OF IOT

In our work, we analyzed IoT bringing advancements in future communication technologies. IoT can be developed with inter- and intra-device connectivity like a machine-to-machine (M2M), a device-to-anything (D2X), and a vehicle-to-anything (V2A). These concepts provide huge benefits by interconnecting devices, sensors, and base stations. Besides technological challenges, security and privacy made IoT implementation more complicated. Many researchers are working on authentication techniques such as (a) public key encryption, which is used to secure IoT communication by sharing secret keys with the legitimate user. Recently, [21] analyzed that due to the complexity of integer factorization, the popularity of public key encryption techniques such as RSA is increasing continuously [22]. (b) Private key encryption: The same key is used at both ends to send and receive messages. The symmetric key cryptography algorithm, such as AES, is popular for encrypting data without sacrificing keys [23]. Symmetric algorithms such as Height, Tiny Encryption Algorithm(TEA), and RC5 provide solutions for possible attacks such as saturation-related and differential attacks [1]. (c) Hashing: Hash-based authentication is widely used for ensuring message integrity verification, digital signature, and biometric authentication. The hash-based [24] digital signature schemes are analyzed based on RSA, ECC, and ECDSA. The security of these schemes relies on factoring large composite integers and computing discrete logarithms [23]. In the IoT framework, three-factor authentication provides resistance to impersonation attacks [25]. The sensor network is a critical element of IoT communication. Therefore, Huang et al. [26] proposed wireless authentication schemes for wireless sensor networks based on two-level security. In the realm of blockchain, its inherent emerging traits such as decentralization, public and private encryption, and smart contracts are pivotal. Blockchain of Things (BoT) is envisioned as the pivotal facilitator for the widespread adoption of IoT. This convergence aligns with the transformative potential of next-generation wireless communication technologies. The self-regulating collaboration of IoT devices achieved through smart contracts (when contractual conditions are met) within a blockchain environment facilitates their interaction without the need for intermediaries. Table 14.1 shows a comparison of various IoT security mechanisms based on a centralized and distributed architecture.

TABLE 14.1
Comparative analysis of authentication schemes

Ref.	Objective	Centralized/ Distributed	Security analysis	Results	Merits	Demerits
[13]	Industrial automation for 5G-enabled IoT.	Distributed	Not Defined	Analysis of IoT applications and requirement of Blockchain.	Industrial automation.	Security was not explored.
[27]	Smart home privacy issues.	Centralized	Eavesdropping, replay, User impersonation attack.	Classical attacks on smart homes are thoroughly investigated.	Smart home privacy is discussed.	Implementation details are not described.
[12]	The potential of 6G and Blockchain.	Distributed	Authentication, Privacy and security	A detailed roadmap is provided	Blockchain integration with IoT is discussed.	Implementation challenges are not discussed.
[13]	IoT application's security concerns.	Distributed	Authentication, privacy, security, and autonomy.	The industrial aspect of IoT security is discussed.	Blockchain for IoT-based applications.	Limitations and challenges are not discussed.
[5]	IoT security concerns.	Distributed	Trust management issues	Analyzed attacks on the IoT communication framework.	Blockchain trust-based solutions provided.	Challenges to implementing blockchain not described.
[28]	Privacy-preserving issues are discussed.	Distributed	Eavesdropping, collusion attacks	JAVA platform is utilized.	Implementation details are discussed.	IoT and blockchain integration challenges are not identified.
[29]	The land records problem is discussed.	Distributed	Data loss, integrity, and security.	The blockchain platform is proposed.	Distributed immutable ledgers.	Blockchain techniques are not discussed in detail.

14.9 CONCLUSION

With the advancement of IoT, devices around us connected to the internet is increasing rapidly. The major issues concerning these applications are security, privacy, and authentication. We provided the Blockchain features, potentials, integration of blockchain with IoT to resolve the above-mentioned issues. This survey paper offers a valuable source for gaining a clear comprehension of the current research contributions of blockchain technology in preparation for the upcoming implementation of IoT communication.

REFERENCES

1. M. binti Mohamad Noor and W. H. Hassan, "Current research on Internet of Things (IoT) security: A survey," *Computer Networks*, vol. 148, pp. 283–294, 2019, doi: 10.1016/j.comnet.2018.11.025.
2. D. Chawla and P. S. Mehra, "QSMAH: A novel quantum-based secure cryptosystem using mutual authentication for healthcare in the internet of things," *Internet of Things*, vol. 24, p. 100949, Dec. 2023, doi: 10.1016/J.IOT.2023.100949.
3. N. Li and D. Liu, "Lightweight Mutual authentication for IoT and its Applications," *IEEE Transactions on Sustainable Computing*, vol. 14, no. 8, 2017, doi: 10.1109/TSUSC.2017.2716953.
4. R. Chanana, A. K. Singh, R. Killa, S. Agarwal, and P. S. Mehra, "Blockchain Based Secure Model for Sensor Data in Wireless Sensor Network," in *2020 6th International Conference on Signal Processing and Communication (ICSC)*, IEEE, Mar. 2020, pp. 288–293. doi: 10.1109/ICSC48311.2020.9182776.
5. D. Chawla and P. S. Mehra, "A Survey on Quantum Computing for Internet of Things Security," *Procedia Computer Science*, vol. 218, pp. 2191–2200, Jan. 2023, doi: 10.1016/J.PROCS.2023.01.195.
6. R. Kumar and R. Sharma, "Leveraging blockchain for ensuring trust in IoT: A survey," *Journal of King Saud University – Computer and Information Sciences*, vol. 34, no. 10, pp. 8599–8622, Nov. 2022, doi: 10.1016/j.jksuci.2021.09.004.
7. J. Sharma and P. S. Mehra, "Secure communication in IOT-based UAV networks: A systematic survey," *Internet of Things*, vol. 23, p. 100883, Oct. 2023, doi: 10.1016/J.IOT.2023.100883.
8. A. Dorri, S. S. Kanhere, R. Jurdak, and P. Gauravaram, "Blockchain for IoT security and privacy: The case study of a smart home," *2017 IEEE International Conference on Pervasive Computing and Communications Workshops, PerCom Workshops 2017* , no. January, pp. 618–623, 2017, doi: 10.1109/PERCOMW.2017.7917634.
9. R. Lu, K. Heung, A. H. Lashkari, and A. A. Ghorbani, "A lightweight privacy-preserving data aggregation scheme for fog computing-enhanced IoT," *IEEE Access*, vol. 5, pp. 3302–3312, 2017, doi: 10.1109/ACCESS.2017.2677520.
10. P. Gope, R. Amin, S. K. Hafizul Islam, N. Kumar, and V. K. Bhalla, "Lightweight and privacy-preserving RFID authentication scheme for distributed IoT infrastructure with secure localization services for smart city environment," *Future Generation Computer Systems*, vol. 83, pp. 629–637, 2018, doi: 10.1016/j.future.2017.06.023.
11. M. Wazid, A. K. Das, R. Hussain, G. Succi, and J. J. P. C. Rodrigues, "Authentication in cloud-driven IoT-based big data environment: Survey and outlook," *Journal of Systems Architecture*, vol. 97, pp. 185–196, 2019, doi: 10.1016/j.sysarc.2018.12.005.

12. A. Jahid, M. H. Alsharif, and T. J. Hall, "The convergence of blockchain, IoT and 6G: Potential, opportunities, challenges and research roadmap," *Journal of Network and Computer Applications*, vol. 217, p. 103677, Aug. 2023, doi: 10.1016/j.jnca.2023.103677.
13. I. Mistry, S. Tanwar, S. Tyagi, and N. Kumar, "Blockchain for 5G-enabled IoT for industrial automation: A systematic review, solutions, and challenges," *Mechanical Systems and Signal Processing*, vol. 135, p. 106382, 2020, doi: 10.1016/j.ymssp.2019.106382.
14. D. Chawla and P. S. Mehra, "A roadmap from classical cryptography to post-quantum resistant cryptography for 5G-enabled IoT: Challenges, opportunities and solutions," *Internet of Things*, vol. 24, p. 100950, Dec. 2023, doi: 10.1016/J.IOT.2023.100950.
15. F. A. Alaba, M. Othman, I. A. T. Hashem, and F. Alotaibi, "Internet of Things security: A survey," *Journal of Network and Computer Applications*, vol. 88, pp. 10–28, 2017, doi: 10.1016/j.jnca.2017.04.002.
16. "Research on agricultural supply chain system with double chain architecture based on blockchain technology | Future Generation Computer Systems." Accessed: Aug. 11, 2023. [Online]. Available: https://dl.acm.org/doi/abs/10.1016/j.future.2018.04.061
17. M. R. Pincheira Caro, M. S. Ali, M. Vecchio, and R. Giaffreda, "Blockchain-based traceability in agri-Food supply chain management: A practical implementation." *Conference: 2018 IoT Vertical and Topical Summit on Agriculture - Tuscany (IOT Tuscany)*, p. 4, 2018. doi: 10.1109/IOT-TUSCANY.2018.8373021.
18. A. Kapitonov, I. Berman, S. Lonshakov, and A. Krupenkin, "Blockchain Based Protocol for Economical Communication in Industry 4.0," in *2018 Crypto Valley Conference on Blockchain Technology (CVCBT)*, Jun. 2018, pp. 41–44. doi: 10.1109/CVCBT.2018.00010.
19. D. Miller, "Blockchain and the Internet of Things in the Industrial Sector," *IT Professional*, vol. 20, no. 3, pp. 15–18, May 2018, doi: 10.1109/MITP.2018.032501742.
20. W. J. Buchanan, S. Li, and R. Asif, "Lightweight cryptography methods," *Journal of Cyber Security Technology*, vol. 1, no. 3–4, pp. 187–201, Oct. 2017, doi: 10.1080/23742917.2017.1384917.
21. J. I. Ahmad, R. Din, and M. Ahmad, "Analysis review on public key cryptography algorithms," *Indonesian Journal of Electrical Engineering and Computer Science*, vol. 12, no. 2, pp. 447–454, Nov. 2018, doi: 10.11591/ijeecs.v12.i2.pp447-454.
22. G. Yang, Q. Huang, D. S. Wong, and X. Deng, "Universal authentication protocols for anonymous wireless communications," *IEEE Transactions on Wireless Communications*, vol. 9, no. 1, pp. 168–174, Jan. 2010, doi: 10.1109/TWC.2010.01.081219.
23. V. Hassija, V. Chamola, V. Saxena, D. Jain, P. Goyal, and B. Sikdar, "A Survey on IoT Security: Application Areas, Security Threats, and Solution Architectures," *IEEE Access*, vol. 7, pp. 82721–82743, 2019, doi: 10.1109/ACCESS.2019.2924045.
24. J. Buchmann, E. Dahmen, and M. Szydlo, "Hash-based Digital Signature Schemes," In: Bernstein, D. J., Buchmann, J., Dahmen, E. (eds) *Post-Quantum Cryptography*. Springer, Berlin, Heidelberg, 2009, pp. 35–93. doi: 10.1007/978-3-540-88702-7_3.
25. Y. Lu, G. Xu, L. Li, and Y. Yang, "Anonymous three-factor authenticated key agreement for wireless sensor networks," *Wireless Networks*, vol. 25, no. 4, pp. 1461–1475, 2019, doi: 10.1007/s11276-017-1604-0.
26. H. F. Huang, Y. F. Chang, and C. H. Liu, "Enhancement of two-factor user authentication in wireless sensor networks," *Proceedings – 2010 6th International Conference on Intelligent Information Hiding and Multimedia Signal Processing, IIHMSP 2010*, pp. 27–30, 2010, doi: 10.1109/IIHMSP.2010.14.

27. Z. Shouran, A. Ashari, and T. Kuntoro, "Internet of Things (IoT) of Smart Home: Privacy and Security," *International Journal of Computer Applications*, vol. 182, no. 39, pp. 3–8, 2019, doi: 10.5120/ijca2019918450.
28. P. C. Sharma, M. R. Mahmood, H. Raja, N. S. Yadav, B. B. Gupta, and V. Arya, "Secure authentication and privacy-preserving blockchain for industrial internet of things," *Computers and Electrical Engineering*, vol. 108, p. 108703, May 2023, doi: 10.1016/j.compeleceng.2023.108703.
29. V. Thakur, M. N. Doja, Y. K. Dwivedi, T. Ahmad, and G. Khadanga, "Land records on Blockchain for implementation of Land Titling in India," *International Journal of Information Management*, vol. 52, p. 101940, Jun. 2020, doi: 10.1016/j.ijinfomgt.2019.04.013.

15 Secure Communication and Authentication in IoT-Based UAV Networks

Jatin Sharma, Pawan Singh Mehra, Diksha Chawla, Deepakshi Dabas, and Aatif Jamshed

15.1 INTRODUCTION

Unmanned Aerial Vehicles (UAVs) are aircraft that can fly autonomously and with self-guidance under remote control. These dynamic and fast-moving flying machines need consistent communication dependability, which may be attained by utilising the Mavlink protocol, which bridges the communication gap between UAVs and the Ground Control Station (GCS). The concept of the Internet of Things (IoT) describes a network of interconnected entities outfitted or acquired with flight controllers, detectors, and software, facilitating the exchange of information among drones via wireless communication. Consequently, drone swarms share similarities with the IoT framework. These networks comprise a collection of interlinked drones empowered by Internet connectivity while being gathered with integrated detectors and flight controllers. A total of 863,728 drones are enrolled in the USA as per FAA (Federal Aviation Administration) [1]. The drone market is projected to witness an annual growth of around 6.5%, potentially doubling by the year 2025. This technology has garnered significant attention across domains such as rescue operations, payload delivery logistics, photography, and other tasks that are challenging for a single UAV to accomplish alone. Consequently, coordination among UAVs necessitates the exchange of commands to fulfil these multifaceted objectives. These networks of UAVs establish substantial volumes of data encompassing visual, telemetry, and detector measurements. Prevailing studies on secure UAV design predominantly centred on computationally intensive methods. A primary shortcoming of prior research lies in the absence of an in-depth exploration into the entire cycle of attack durations and security weaknesses concerning UAV communication. To address these limitations, it is crucial to examine existing countermeasures and propose innovative approaches thoroughly. This chapter delves into both physical and logical attacks, presenting authentication strategies that offer insights into future directions.

DOI: 10.1201/9781003405535-20

15.1.1 Motivation

In spite of the evident benefits offered by UAV networks, their implementation is hampered by security considerations. The issue of UAV security was brought to the forefront in 2011 when the RQ-170 drone was gathered by Iran force [2]. The potential for adversaries to launch network-based assaults looms whenever UAV networks are engaged in tasks such as combat communication and reconnaissance in perilous situations. An illustrative case emerged during the Russia-Ukraine conflict of 2022 [3]; Ukrainian military units formed compact squads with the objective of directly engaging Russian UAVs by employing laser-based weaponry. An expert [4] elucidates how attackers can exploit vulnerabilities by interfering with the connections linking UAVs and base stations. A recent examination underscores the lack of security in drone networks. UAVs containing confidential data related to military operations are susceptible to compromise. Particularly in security-sensitive targets, encompassing surveillance, recovery, and heightened security measures is imperative. However, existing studies often present only concise evaluations or surveys, focusing on a limited array of security threats or countermeasures. The absence of a comprehensive exploration in the drone network domain has prompted the need for this contemporary study. The work endeavours to bridge the existing discrepancies and address the security requirements. Therefore, there is a necessity for a preliminary investigation into authentication mechanisms.

15.1.2 Contribution

- Presents a survey on secure communication and authentication in IoT-based UAV networks.
- Provides a summary of related work.
- This work compares with an existing survey using security parameters like mutual authentication, un-traceability, etc. in a tabular layout.
- This work discusses the state-of-art authentication techniques.

15.1.3 Outline of the Chapter

The arrangement of the segments is structured as follows: Section 15.2 provides an overview of UAVs. Section 15.3 delves into the pertinent work. Security requirements for secure communication and authentication in IoT-based UAV networks is discussed in Section 15.4. Section 15.6 describes the secure communication and authentication mechanisms. Lastly, Section 15.5 encapsulates the concluding remarks and future directions.

15.2 OVERVIEW OF UNMANNED AERIAL VEHICLES

This section covers the complete details of UAVs.

15.2.1 UAV

An UAV refers to an aircraft or vehicle that operates without human intervention, capable of functioning autonomously and independently. UAVs have gained prominence

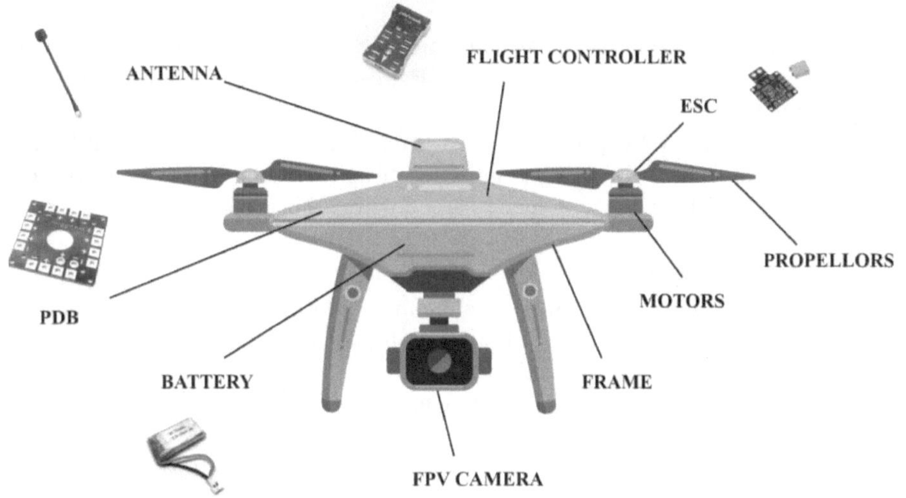

FIGURE 15.1 UAV.

for their contributions in areas such as surveillance and search and rescue operations on the battlefield. The network of UAVs extends its reach to diverse fields, including wireless hotspot provision, smart urban environments, remote sensing applications, agricultural endeavours, and more. Figure 15.1 displays the constituents of the UAV elements [5].

The primary constituents of UAVs encompass:

- Motors
- Flight controller
- Power distribution board
- Electronic speed controllers
- Antenna
- Radio receiver
- Battery
- FPV camera
- Propellers
- Video transmitter

15.2.2 Category of UAVs

15.2.2.1 Weight-based [6]

a. **Nano UAV:** aircraft with a weight of less than 250 g
b. **Micro UAV:** aircraft with a weight greater than 250 g and less than 2 kg
c. **Small UAV:** aircraft with a weight greater than 2 kg and less than 25 kg
d. **Medium UAV:** aircraft with a weight greater than 25 kg and less than or equal to 150 kg
e. **Large UAV:** aircraft with a weight greater than 150 kg

15.2.2.2 Altitude-based [7]

a. **High altitude UAV(HAUAV):** This category of aircraft functions at extremely elevated altitudes surpassing 10 km. Balloons, airplanes, and dirigibles fall within this classification
b. **Low altitude UAV(LAUAV):** This aircraft classification is versatile and adaptable, capable of flying at altitudes of up to 10 km. Vertical Takeoff and Landing (VTOL) craft are encompassed within this group.

15.2.2.3 Wings-based [7]

a. **Fixed-wing UAV:** This refers to a compact pilotless aircraft with stationary wings, requiring a runway for both takeoff and landing. These aircraft exhibit substantial range and resilience.
b. **Rotatory-wing UAV:** The rotary-wing vehicle may exist in singular or multirotor configurations, including VTOL UAVs. This wing design offers exceptional manoeuvring capabilities, facilitating precise takeoffs and controlled flight.
c. **Hybrid-wing UAV:** It involves the integration of both stationary and rotary-wing elements.

15.2.2.4 Range-based [7]

a. **Short-range UAV:** UAV capable of covering distances of up to 10 km.
b. **NATO UAV**: UAV capable of flying distances of up to 50 km.
c. **Tactical UAV**: UAV capable of flying distances by 50 km.
d. **Moderate altitude and extended endurance:** UAV capable of travelling distances by 200 km.
e. **Elevated altitude and prolonged endurance UAV:** UAV with boundless endurance.
f **Superspeed UAV:** aircraft capable of covering distances exceeding 200 km.

15.2.3 Connections in IoT-Based UAV Networks

UAVs possess the capability to establish communication through data exchange amongst each other independently, without relying on any existing infrastructure. Additionally, they can be seamlessly coordinated through a base station.

There exist four categories of linkage, encompassing the subsequent [8]:

a. **Air-2-Air communication (A2A):** Within A2A communication, a single UAV has the capacity to share data with another UAV and assume the role of a relay node, thereby extending the range of connectivity.
b. **Air-2-Ground communication (A2G)**: In this connection, a drone possesses the capability to transmit crucial data to the base station. Subsequently, the base station holds the capacity to issue specific task-related control commands to the UAVs.

c. **Space/Satellite 2 Air communication (S2A):** This connection involves the transfer of information between drones and satellite apparatus. This type of connection finds utility across diverse UAV operations, encompassing tasks such as remote sensing and information transfer.
d. **Mixed communication:** This form of communication involves combining the previously mentioned three types of communication, namely S2A, A2A, and A2G, in order to achieve expanded coverage and enhanced data rates.

15.2.4 Features of IoT-Based UAV Networks

The IoT-based UAV networks' features are the following [8]:

a. **Expense:** The affordability of small UAVs within FANET (Flying Adhoc Network) stems from their reduced operational costs.
b. **Resilience:** The presence of multiple UAVs ensures resilience in the event of a single UAV node experiencing a mission failure.
c. **Acceleration:** The increased number of UAVs results in expedited mission completion.
d. **Scalability:** The network of multiple UAVs exhibits scalability, allowing for the inclusion of additional UAVs within FANET. Consequently, this expansion facilitates the broadening of the mission's coverage area [9].
e. **Reliability**: Unfavourable atmospheric conditions can negatively impact A2G communication. In such scenarios, A2A communication can effectively address connectivity challenges, ensuring the dependability of the network.

15.3 PERTINENT WORK

Within this segment, a compilation of pertinent research endeavours concerning secure communication and authentication within IoT-based UAV networks is presented. While the number of published works in this domain may be limited, this chapter diligently strives to encompass the most significant contributions that are shown in Table 15.1, accompanied by a comparative analysis in Table 15.2.

Lin et al. [10] examined the security and privacy issues relevant to the Internet of drones (IoD) settings. They highlighted matters like verifying the authenticity of drones and ground stations, safeguarding the identity and whereabouts privacy of drones, maintaining the confidentiality of data transferred from drones to the cloud, and ensuring secure data exchange among cooperating drones. The authors also put forth remedies to protect the identity and location privacy of drones as well as the confidentiality of outsourced data.

Teng et al. [11] introduced an authentication procedure aimed at facilitating communication between drones. This approach employs Elliptic Curve Cryptography (ECC) and Public Key Infrastructure (PKI). While their method offers certain security benefits, its implementation could be burdensome for devices with limited resources, such as drones.

TABLE 15.1
Summary of Pertinent Work

Reference	Year	Mechanism	Limitations
Lin et al. [10]	2018	Privacy-preserving authentication and identity-based encryption	• The inherent limitations of resources within the IoD architecture hinder the efficient implementation of Identity-based Encryption (IBE). • No mutual authentication
Teng et al. [11]	2019	Elliptic Curve Cryptography (ECC) and Public Key Infrastructure (PKI)	• Implementation could be burdensome for devices with limited resources, such as drones.
He et al. [12]	2019	Proxy blind signature scheme	• Length of the signature is large. • Signature processing requires more time.
Tian et al. [13]	2019	Efficient privacy-preserving authentication	• Security features such as man-in-the-middle and mutual authentication are missing. • No formal security assessment method.
Huang et al. [14]	2020	Triangular-oriented strategy	• Exorbitant expenses for communication and storage.
Gope et al. [15]	2020	One-way hash and PUF	• It falls short in terms of preserving the privacy of location.
Zhang et al. [16]	2020	One-way hash and 2-factor authentication	• Un-traceability is not part of its package.
Alladi et al. [17]	2021	Drone-MAP	• No security assessment method like the random oracle model was provided. • Does not meet all security requirements like Identity verification and physical attack resistance.
Rajasekaran et al. [18]	2022	Anonymous mutual and batch authentication	• Bilinear groups are expensive and hard to implement due to complex in nature. • Trusted servers can be vulnerable to adversary attacks.
Salam et al. [19]	2023	SELTHA (Secure, Efficient, and Lightweight TOTP-based Hashed Authentication)	• No formal security analysis approach like ProVerif and Scyther etc.

TABLE 15.2
Comparative Analysis of Pertinent Work

Security Requirements	[10]	[11]	[12]	[13]	[14]	[15]	[16]	[17]	[18]	[19]	This Work
Mutual authentication	✖	✓	✖	✖	✖	✓	✓	✓	✓	✓	✓
Man-in-the-middle	✖	✓	✖	✖	✖	✖	✓	✖	✓	✓	✓
Replay attack	✖	✓	✖	✓	✖	✓	✓	✓	✖	✓	✓
Brute-force attack	✖	✖	✖	✖	✖	✖	✖	✖	✖	✓	✓
Physical capture	✓	✖	✓	✖	✖	✓	✓	✖	✖	✓	✓
Session key attack	✖	✓	✖	✖	✖	✓	✓	✓	✓	✓	✓
Un-traceability	✖	✖	✖	✖	✖	✖	✓	✓	✖	✖	✓
Anonymity	✖	✓	✖	✖	✖	✓	✓	✖	✓	✖	✓

✖, not discussed; ✓, discussed.

He et al. [12] put forth an ECC-driven proxy blind signature system for drone networks. This proxy signature, based on blind identification, intentionally omits any details about the drone's identity when acting as a delegated signer to preserve anonymity. However, this particular approach doesn't align with our authentication objectives and may expose vulnerabilities to harmful drone incursions. Furthermore, an effective proxy signature mechanism should possess comparable information about the proxy signer as it does for the original signer.

Tian et al. [13] introduced an authentication framework designed to establish secure communication among drones. This system incorporates an online/offline signature method and leverages mobile edge computing technology. In addition to ensuring robust authentication, the suggested strategy takes into account the imperative of safeguarding the privacy of drones' identities and whereabouts. The authors fail to describe security features such as man-in-the-middle, mutual authentication, etc.

Huang and Savkin [14] proposed a novel technique for establishing charging stations for drones. This method employed a triangular-oriented strategy. Furthermore, charging stations that experienced lower footfall or remained unoccupied were systematically eliminated through recursive processes. It's important to note that this research does not primarily address the aspects of communication and storage expenses.

Gope et al. [15] put forth a proposal aimed at guaranteeing the tangible safety of drones. This scheme relies on the employment of physically unclonable functions and hash operations. While the scheme effectively safeguards the physical integrity of drones, it falls short in terms of preserving the privacy of their location.

Zhang et al. [16] unleashed an efficient authentication and key agreement (AKA) blueprint, hinging entirely on a unidirectional secure hash function. Through this strategy, drones and users engage in reciprocal authentication. Despite its robustness against diverse security menaces, this strategy neglects the inclusion of measures for concealing location privacy or fortifying against physical vulnerabilities.

Alladi et al. [17] presented a novel authentication protocol reliant on Physical Unclonable Functions (PUFs) for 5G networks with drone assistance. This protocol enables a reciprocal authentication process between a drone and a base station, leading to the establishment of a confidential key. Nevertheless, their approach employs PUF solely on the drone's end, which falls short of constituting a fully reciprocal authentication mechanism. The drone's challenge-response pairs (CRPs) are retained openly in the base station's database, devoid of encryption. This presents a potential risk, such as the exposure of CRPs, given the possibility of the base station being compromised or exploited by an unscrupulous insider.

Rajasekaran et al. [18] present an effective mutual authentication scheme with anonymity, aiming to mitigate privacy and security concerns while enhancing communication efficiency between the UAV and the user. The study introduces various strategies, including anonymous batch authentication within FANET, which streamlines the authentication process for a substantial cluster of drones concurrently, thereby minimising computational burdens.

Salam et al. [19] introduced the SELTHA (Secure, Efficient, and Lightweight TOTP-based Hashed Authentication) methodology with the aim of enhancing the security of drone communication, encompassing interactions with both the ground control server and drone-to-drone connections. The methodology comprises a trio of phases, collectively forming a comprehensive procedure. The significance of ensuring communication security is paramount and carries critical importance, primarily due to the wireless communication inherent to drones within the architecture of IoD. This wireless nature renders the drone network a highly appealing target for cybercriminals seeking to exploit vulnerabilities, particularly when security protocols are absent or when weak protocols are in play.

15.4 SECURITY REQUIREMENTS FOR SECURE COMMUNICATION AND AUTHENTICATION

This section covers the security properties which are required for authentication in IoT-based UAV networks.

15.4.1 Mutual Authentication

In the context of UAV and GCS, mutual authentication denotes a security procedure where both the UAV and GCS validate each other's identities prior to initiating communication or sharing information. This mechanism guarantees the legitimacy and authorisation of both entities, thereby thwarting any unauthorised access and potential security vulnerabilities [19].

15.4.2 Man-in-the-Middle Attack

In the scenario of a man-in-the-middle attack targeting an UAV and a GCS, a malevolent third party intervenes in and manipulates the communication link connecting the

UAV and the GCS. This aggressor strategically places themselves between the two entities, masquerading as each side to potentially pilfer data, introduce fraudulent information, or cause disturbances in the communication flow. This deceitful manoeuvre undermines the reliability and security of the communication pathway [19].

15.4.3 Replay Attack

In the context of an UAV network, a replay attack involves the unauthorised seizure and subsequent playback of previously recorded data exchanges or communications. In this situation, a malicious actor captures genuine data transmitted between UAVs and later replays it to trick the network or cause operational disruption. This could lead to unauthorised entry, propagation of inaccurate directives, or data manipulation, consequently eroding the network's dependability and security [19].

15.4.4 Brute-Force Attack

In IoT-linked UAV networks, a brute-force attack involves a systematic and exhaustive approach by a malicious individual to test all conceivable permutations of passwords or encryption keys. This endeavour is undertaken with the objective of unauthorised network entry. In this scenario, the attacker sets their sights on the security protocols within the UAV network integrated into the Internet of Things (IoT). Through repetitive trials of various combinations, the attacker strives to uncover the accurate password or key, compromising the network's security posture and potentially obtaining command over the UAVs or gaining access to sensitive data. This attack capitalises on feeble or easily deducible passwords, underscoring the vital need for robust security measures to counteract such impending dangers effectively.[19]

15.4.5 Physical Capture Attack

In UAV networks, a physical capture attack entails a malicious individual physically seizing control of an UAV to compromise its functioning, data, or payload. This attack commonly demands direct physical proximity to the UAV, granting the attacker the ability to manipulate its components, alter software configurations, or abscond with sensitive data. The attacker may employ diverse tactics, including intercepting the UAV during flight, taking control while grounded, or utilising deceptive techniques to establish authority. This attack presents a notable menace to the security and reliability of UAV operations, highlighting the critical significance of implementing rigorous physical security measures to avert unauthorised access and tampering [19].

15.4.6 Session Key Attack

In UAV networks, a session key attack involves a malevolent endeavour to undermine the security of session keys employed to encrypt and ensure the confidentiality of communication among elements within the UAV network. Within this attack category, an unapproved individual or entity strives to intercept, manipulate, or exploit these session keys with the aim of gaining entry to the network without authorisation or deciphering encrypted data. Such an assault holds the potential to result in unapproved data access, unsanctioned manipulation of UAVs, or the revelation of sensitive information. The prevention of session key attacks holds paramount importance to uphold the integrity and security of UAV networks, necessitating the implementation of resilient encryption protocols and stringent security measures [19].

15.4.7 Un-Traceability

Un-traceability within UAV networks pertains to the quality of guaranteeing that activities and transactions occurring in the network cannot be readily associated with particular UAVs or individuals. In this framework, un-traceability strives to obscure the source and destination of data or communication, delivering a degree of anonymity and confidentiality for both UAVs and their operators. This aspect thwarts the straightforward identification of which UAVs partake in specific actions by external observers, amplifying security measures and safeguarding confidential information [17].

15.4.8 Anonymity

Anonymity within UAV networks signifies a condition wherein the identities of entities, encompassing UAVs and operators, engaged in network communications and activities remain hidden or undisclosed. This trait strives to guarantee that actions and interactions occurring cannot be readily linked to particular individuals or devices. Through upholding anonymity, UAV networks elevate privacy and security measures, introducing complexity for unauthorised entities seeking to trace or pinpoint participants. As a result, sensitive information and operational integrity are safeguarded effectively [18].

15.5 SECURE COMMUNICATION AND AUTHENTICATION MECHANISMS

This section discusses various authentication mechanisms in IoT-based UAV networks. In the absence of authentication and the inability to confirm message authenticity, counterfeit signals can be maliciously introduced. The deployment of authentication systems that exclusively ensure the credibility of authorised entities within IoT-driven UAV networks stands indispensable for rectifying this issue.

15.5.1 Mutual Authentication Scheme

The devised scheme addresses the challenge of disassociation frames within compromised drones. This comprehensive arrangement encompasses the base station, interface, and drones themselves. The central concept revolves around harnessing encoded communication channels linking these pivotal components [20, 21].

15.5.2 Sentinel Mechanism

The core objective driving the SENTINEL mechanism, a robust and streamlined identification method tailored for drones, is the establishment of reciprocal credibility between drones and base stations. This entails the UAV initiating a takeoff request directed at the base station, which subsequently evaluates its endorsement. Upon approval, the base station logs the drone's identity and key within the drone log repository, thereby enabling universal access across all base stations [22].

15.5.3 Secure Identity Verification Mechanism

This groundbreaking framework was constructed upon the formidable underpinning of elliptic curve cryptosystems [23].

15.5.4 Discernible Yet Privacy-Enforcing Authentication

This method finds its prowess in drone applications, characterised by deploying hash functions and public key method [24].

15.5.5 Authentication-Secured Encryption

The innovators present a triadic authorised control strategy forged upon secured encryption designed for the dynamic realm of IoT. This ingenious technique capitalises on a trinity of authentication facets: password, mobile device, and biometrics [25].

15.5.6 Fortified Framework for Safeguarding Privacy During Authentication

This strategic blueprint hinged on the prowess of mobile edge computing. The pivotal constituents encompass esteemed authority, UAVs poised for action, and the arsenal of devices for edge computing on the move. The architectural marvel, a one-signature blueprint, is meticulously woven, boasting unparalleled efficiency in signature key generation and evolution. This masterstroke not only detects replay assaults but also quashes the spectre of repudiation perils [15].

15.5.7 Stealthy Pathways Empowered Lightweight Authentication

Within this strategy, the innovator harnesses covert channels nestled within the fabric of physical layers. Employing this tactical approach, many threats are vanquished, including message replay, impersonation, eavesdropping, and menacing Man-in-the-Middle (MITM) incursions [26].

15.5.8 Cipher Concordance and Streamlined Verification Paradigm

The unleashed blueprint deploys an exchange protocol based on Diffie Hellman and engrafts a symmetrical key cipher mechanism, forging an ironclad pact for message secrecy and unassailable integrity. The riveting ballet of authentication unfurls as recipients and drones meld their identities by exclusive-or or fortified by an unyielding one-way hash ritual. Within this audacious construct, a mutual authentication covenant thrives in full force [16].

15.5.9 Streamlined Key Accord in Remote User Authentication

The author conceived this scheme to empower ground users with direct dominion over UAV data operations [27].

15.5.10 Dual-Phase, Agile Mutual Authentication Paradigm

The envisaged system is tailored for interconnected drones operating within the software-defined network method. Identification unfolds in a binary sequence: initial authentication transpires among the lead UAV and the base station, succeeded by authentication between the subordinate drone and its lead counterpart. Baptised as PARTH, the protocol stands as a bastion, guaranteeing mutual authentication, unwavering integrity, and sanctuary for session keys [28].

15.5.11 Authentication Framework for Aggregated Data

The innovative strategy put forth by the researcher advocates the adoption of an identity cipher technique coupled with EC methodologies for the authentication of information aggregates. This tactical manoeuver not only bolsters data security but also serves to curtail the burden of computations and communication expenses [29].

15.5.12 Encryption Paradigm with Homomorphic Properties

The innovative method introduced by the author is termed as sophisticated linear homomorphic authenticated encryption (LinHAE). This technique is meticulously crafted to thwart adversarial endeavours like forgery and eavesdropping. Unlike conventional encryption approaches, LinHAE distinguishes itself by eschewing the storage of secret keys within the controller, reserving its utility exclusively for Ground Control Station (GCS) deployment [30].

15.5.13 Security Protocol Anchored in Mutual Authentication for UAV to UAV and UAV to GCS Communication

The author put forward a pair of meticulously designed secure protocols tailored to serve military applications [31].

a. **Protocol for Ensuring Security between UAVs:** This protocol ensures robust communication integrity among the UAVs, achieved through a combination of key exchange and reciprocal authentication.

b. **Protocol for Ensuring Security between UAVs and Ground Control Stations:** Embedded within this protocol's architecture is an imperative mandate for impervious communication channels between the UAV and its ground control citadel. The sanctity of this conduit guarantees the impervious exchange of vital telemetry data and status revelations, all transpiring within the fortified nexus between the UAV and GCS. A rigorous scrutiny of the secure method fabric was undertaken, employing both the formidable Burrows-Abadi-Needham (BAN) logic and Scyther tool.

15.5.14 Secure Hashing Scheme

The innovator has introduced a private framework centred around hash message authentication. This design fortifies against various threats, including forgery attacks, denial of service onslaughts, pilfered breaches, playback assaults, and fake gambits. Moreover, this architecture curtails storage and computational burdens, resulting in diminished overhead [32].

15.5.15 Protocol for Authentication and Key Accord with Three Factors

The author has introduced a protocol for AKA employing a triad of factors. This scheme leverages the Boyko-Peinado-Venkatesan (BPV) pre-calculation alongside the 4Q technique. The protocol's safety is fortified with a rigorous approach, guaranteeing the assurance of future security. Moreover, empirical investigations on hardware reveal that 4Q curve strategies manifest a striking efficiency gain, performing approximately 4–5 instances swifter than traditional elliptic curve methods [33].

15.5.16 Strong Key Management Protocol for Authentication

The SKMPA protocol relies on two cryptographic methods: authenticated encryption and EC cryptography. A random oracle model and a Scyther tool are used to verify. This technique prevents tampering and playback attacks. The testing confirms that the SKMPA protocol effectively counters hidden threats. This method is efficient, requiring less computation and storage [34].

15.5.17 Scheme for Lightweight Verification with Two Factors

The author proposed a plan based on an asymmetric cryptographic method. The prior approach faced the risk of losing access to UAV networks driven by IoT due to issues in key management. The introduced technique has demonstrated superior security compared to alternative approaches. It's robust against various attacks, including phishing and replay assaults [35].

15.5.18 Encryption Approach Utilising Symmetric and ECC Techniques

This technique guarantees the safety of both drones and recipients by incorporating symmetric encryption and ECC. It ensures anonymity and defense against offline attempts to guess passwords and a wide range of threats, including tampering [36].

15.5.19 MAC Protocol (Message Authentication Code Protocol)

This particular approach employs a dual-phase methodology [37, 38]

a. **Validation of data:** The cluster head node employs data validation as a filtering process.
b. **Cluster identification:** At the destination corner, the validation phase of the group is employed to guarantee the accuracy and authenticity of the data. However, this approach is susceptible to five distinct types of attacks, including propagation delay, packet falsification, manipulation of data, replay, and grey hole attacks.

15.5.20 Temporal Credentials-Based Authentication

The new security system for UAV networks connected to the Internet of Things (IoT) uses time-limited access and allows quick and private user verification. It ensures that both users and the network verify each other's identity. The system uses three ways to confirm a user's identity: their phone, unique physical traits, and a password [39].

15.5.21 PUF-Based Drone-2-Drone Mutual Authentication Protocol

The procedure enables a pair of drones to authenticate one another through PUFs and create a confidential key for encrypted communication. This technique employs the concept of extended CRPs (eCRPs). This eCRP concept effectively tackles the challenge of refraining from storing plaintext CRPs at the verifier, consequently thwarting any attempts at disclosing CRPs [40].

15.5.22 Access and Handover Authentication Schemes

This scheme guarantees the UAV's ironclad direct entry into satellite networks. At the same time, the ground terminals pull off a secure network service heist courtesy of the UAV. And hold on tight because we're introducing a tailor-made handshake authentication routine to guarantee those terminals a rock-solid transition to a fresh UAV. Brace yourselves as these protocols flex formidable security muscles and flaunt streamlined, turbocharged performance [41].

15.6 CONCLUSION AND FUTURE SCOPE

Fortifying communication security is paramount and carries a critical weight due to the wireless communication intrinsic to drones within IoT-based UAV network framework. This wireless mode of communication serves as a magnet for cyber malefactors aiming to infiltrate the drone network by exploiting vulnerabilities stemming from either the absence of robust security protocols or the presence of feeble ones. This paperwork starts introducing UAVs, its categories, features, and types of connections in IoT-based UAV networks. Pertinent work describes the current work done by the authors, with limitations for each, through the summary table. This work also presents a comparative analysis of the previous work with cutting-edge solutions for authentication. Future work can include formal security analysis like ProVerif and Random Oracle Model for the proposed protocol. Lightweight Blockchain-based authentication and Quantum Key distribution for key exchange can be utilised for mutual authentication in future work.

REFERENCES

1. "Drones by the Numbers," *Federal Aviation Administration*. www.faa.gov/node/54496
2. D. He, S. Chan, and M. Guizani, "Communication Security of Unmanned Aerial Vehicles," *IEEE Wirel. Commun.*, vol. 24, no. 4, pp. 134–139, Aug. 2017, doi: 10.1109/MWC.2016.1600073WC.
3. S. Tiwari, "Ukraine Shoots 'Laser Weapons' at Russian Suicide Drones," Dec. 09, 2022. [Online]. Available:https://eurasiantimes.com/ukraine-shooting-laser-weapons-at-russian-suicide-drones-report/
4. V. K. Ralegankar *et al.*, "Quantum Cryptography-as-a-Service for Secure UAV Communication: Applications, Challenges, and Case Study," *IEEE Access*, vol. 10, pp. 1475–1492, 2022, doi: 10.1109/ACCESS.2021.3138753.
5. J. Sharma and P. S. Mehra, "Secure Communication in IOT-based UAV networks: A systematic survey," *Internet Things*, vol. 23, p. 100883, Oct. 2023, doi: 10.1016/j.iot.2023.100883.
6. DGCA India, "Know about UAS types." https://digitalsky.dgca.gov.in/home
7. V. Chamola, P. Kotesh, A. Agarwal, Naren, N. Gupta, and M. Guizani, "A Comprehensive Review of Unmanned Aerial Vehicle Attacks and Neutralization Techniques," *Ad Hoc Netw.*, vol. 111, p. 102324, Feb. 2021, doi: 10.1016/j.adhoc.2020.102324.
8. J. Sharma and P. S. Mehra, "A survey of security challenges and existing prevention methods in FANET," in B. Alam and M. Alam (Eds.), *Intelligent Data Analytics, IoT, and Blockchain*, CRC Press, 2023, pp. 252–262. doi: 10.1201/9781003371380-24.
9. A. Dwivedi, A. Sharma, and P. Mehra, "Energy-Aware Routing Protocols for Wireless Sensor Network Based on Fuzzy Logic: A 10-Years Analytical Review," *EAI Endorsed Trans. Energy Web*, p. 166548, Jul. 2018, doi: 10.4108/eai.6-10-2020.166548.
10. C. Lin, D. He, N. Kumar, K.-K. R. Choo, A. Vinel, and X. Huang, "Security and Privacy for the Internet of Drones: Challenges and Solutions," *IEEE Commun. Mag.*, vol. 56, no. 1, pp. 64–69, Jan. 2018, doi: 10.1109/MCOM.2017.1700390.
11. L. Teng *et al.*, "Lightweight Security Authentication Mechanism Towards UAV Networks," in *2019 International Conference on Networking and Network Applications (NaNA)*, Daegu, Korea (South): IEEE, Oct. 2019, pp. 379–384. doi: 10.1109/NaNA.2019.00072.
12. L. He, J. Ma, R. Mo, and D. Wei, "Designated Verifier Proxy Blind Signature Scheme for Unmanned Aerial Vehicle Network Based on Mobile Edge Computing," *Secur. Commun. Netw.*, vol. 2019, pp. 1–12, Apr. 2019, doi: 10.1155/2019/8583130.

13. Y. Tian, J. Yuan, and H. Song, "Efficient Privacy-Preserving Authentication Framework For Edge-Assisted Internet of Drones," *J. Inf. Secur. Appl.*, vol. 48, p. 102354, Oct. 2019, doi: 10.1016/j.jisa.2019.06.010.
14. H. Huang and A. V. Savkin, "A Method of Optimized Deployment of Charging Stations for Drone Delivery," *IEEE Trans. Transp. Electrification*, vol. 6, no. 2, pp. 510–518, Jun. 2020, doi: 10.1109/TTE.2020.2988149.
15. P. Gope and B. Sikdar, "An Efficient Privacy-Preserving Authenticated Key Agreement Scheme for Edge-Assisted Internet of Drones," *IEEE Trans. Veh. Technol.*, vol. 69, no. 11, pp. 13621–13630, Nov. 2020, doi: 10.1109/TVT.2020.3018778.
16. Y. Zhang, D. He, L. Li, and B. Chen, "A lightweight authentication and key agreement scheme for Internet of Drones," *Comput. Commun.*, vol. 154, pp. 455–464, Mar. 2020, doi: 10.1016/j.comcom.2020.02.067.
17. T. Alladi, V. Venkatesh, V. Chamola, and N. Chaturvedi, "Drone-MAP: A Novel Authentication Scheme for Drone-Assisted 5G Networks," in *IEEE INFOCOM 2021 – IEEE Conference on Computer Communications Workshops (INFOCOM WKSHPS)*, Vancouver, BC, Canada: IEEE, May 2021, pp. 1–6. doi: 10.1109/INFOCOMWKSHPS51825.2021.9484594.
18. A. S. Rajasekaran, A. Maria, F. Al-Turjman, C. Altrjman, and L. Mostarda, "Anonymous Mutual and Batch Authentication with Location Privacy of UAV in FANET," *Drones*, vol. 6, no. 1, p. 14, Jan. 2022, doi: 10.3390/drones6010014.
19. W. Salam, S. K.-R. Raazi, and N. H. Ansari, "SELTHA: Secure, Efficient and Lightweight Authentication Mechanism for Unmanned Aerial Vehicle Network," in *2023 7th International Multi-Topic ICT Conference (IMTIC)*, Jamshoro, Pakistan: IEEE, May 2023, pp. 1–7. doi: 10.1109/IMTIC58887.2023.10178552.
20. D. Chawla and P. S. Mehra, "QSMAH: A novel quantum-based secure cryptosystem using mutual authentication for healthcare in the internet of things," *Internet of Things*, vol. 24, p. 100949, Dec. 2023, doi: 10.1016/J.IOT.2023.100949.
21. D. Chawla and P. S. Mehra, "A Survey on Quantum Computing for Internet of Things Security," *Procedia Comput. Sci.*, vol. 218, pp. 2191–2200, 2023, doi: 10.1016/j.procs.2023.01.195.
22. G. Cho, J. Cho, S. Hyun, and H. Kim, "SENTINEL: A Secure and Efficient Authentication Framework for Unmanned Aerial Vehicles," *Appl. Sci.*, vol. 10, no. 9, p. 3149, Apr. 2020, doi: 10.3390/app10093149.
23. Y. Kirsal Ever, "A Secure Authentication Scheme Framework for Mobile-Sinks Used in the Internet of Drones Applications," *Comput. Commun.*, vol. 155, pp. 143–149, Apr. 2020, doi: 10.1016/j.comcom.2020.03.009.
24. C.-L. Chen, Y.-Y. Deng, W. Weng, C.-H. Chen, Y.-J. Chiu, and C.-M. Wu, "A Traceable and Privacy-Preserving Authentication for UAV Communication Control System," *Electronics*, vol. 9, no. 1, p. 62, Jan. 2020, doi: 10.3390/electronics9010062.
25. S. Mandal, B. Bera, A. K. Sutrala, A. K. Das, K.-K. R. Choo, and Y. Park, "Certificateless-Signcryption-Based Three-Factor User Access Control Scheme for IoT Environment," *IEEE Internet Things J.*, vol. 7, no. 4, pp. 3184–3197, Apr. 2020, doi: 10.1109/JIOT.2020.2966242.
26. X. H. Cao, X. Du, and E. P. Ratazzi, "A Light-Weight Authentication Scheme for Air Force Internet of Things," in *ICC 2019 – 2019 IEEE International Conference on Communications (ICC)*, Shanghai, China: IEEE, May 2019, pp. 1–6. doi: 10.1109/ICC.2019.8761538.
27. M. Wazid, A. K. Das, N. Kumar, A. V. Vasilakos, and J. J. P. C. Rodrigues, "Design and Analysis of Secure Lightweight Remote User Authentication and Key Agreement

Scheme in Internet of Drones Deployment," *IEEE Internet Things J.*, vol. 6, no. 2, pp. 3572–3584, Apr. 2019, doi: 10.1109/JIOT.2018.2888821.

28. T. Alladi, V. Chamola, Naren, and N. Kumar, "PARTH: A Two-Stage Lightweight Mutual Authentication Protocol for UAV surveillance networks," *Comput. Commun.*, vol. 160, pp. 81–90, Jul. 2020, doi: 10.1016/j.comcom.2020.05.025.
29. W. Hong, L. Jianhua, L. Chengzhe, and W. Zhe, "A Provably Secure Aggregate Authentication Scheme for Unmanned Aerial Vehicle Cluster Networks," *Peer-Peer Netw. Appl.*, vol. 13, no. 1, pp. 53–63, Jan. 2020, doi: 10.1007/s12083-019-0718-9.
30. J. H. Cheon *et al.*, "Toward a Secure Drone System: Flying with Real-Time Homomorphic Authenticated Encryption," *IEEE Access*, vol. 6, pp. 24325–24339, 2018, doi: 10.1109/ACCESS.2018.2819189.
31. Y. Ko, J. Kim, D. G. Duguma, P. V. Astillo, I. You, and G. Pau, "Drone Secure Communication Protocol for Future Sensitive Applications in Military Zone," *Sensors*, vol. 21, no. 6, p. 2057, Mar. 2021, doi: 10.3390/s21062057.
32. S. U. Jan, F. Qayum, and H. U. Khan, "Design and Analysis of Lightweight Authentication Protocol for Securing IoD," *IEEE Access*, vol. 9, pp. 69287–69306, 2021, doi: 10.1109/ACCESS.2021.3076692.
33. N. Zhang, Q. Jiang, L. Li, X. Ma, and J. Ma, "An Efficient Three-Factor Remote User Authentication Protocol Based on BPV-FourQ for Internet of Drones," *Peer-Peer Netw. Appl.*, vol. 14, no. 5, pp. 3319–3332, Sep. 2021, doi: 10.1007/s12083-021-01130-5.
34. M. Tanveer, A. U. Khan, N. Kumar, and M. M. Hassan, "RAMP-IoD: A Robust Authenticated Key Management Protocol for the Internet of Drones," *IEEE Internet Things J.*, vol. 9, no. 2, pp. 1339–1353, Jan. 2022, doi: 10.1109/JIOT.2021.3084946.
35. D. Chawla and P. S. Mehra, "A roadmap from classical cryptography to post-quantum resistant cryptography for 5G-enabled IoT: Challenges, opportunities and solutions," *Internet of Things*, vol. 24, p. 100950, Dec. 2023, doi: 10.1016/J.IOT.2023.100950.
36. S. Hussain, S. A. Chaudhry, O. A. Alomari, M. H. Alsharif, M. K. Khan, and N. Kumar, "Amassing the Security: An ECC-Based Authentication Scheme for Internet of Drones," *IEEE Syst. J.*, vol. 15, no. 3, pp. 4431–4438, Sep. 2021, doi: 10.1109/JSYST.2021.3057047.
37. J. Sun *et al.*, "A data authentication scheme for UAV Ad hoc network communication," *J. Supercomput.*, vol. 76, no. 6, pp. 4041–4056, Jun. 2020, doi: 10.1007/s11227-017-2179-3.
38. P. S. Mehra, "LBECR: Load Balanced, Efficient Clustering and Routing Protocol for Sustainable Internet of Things in Smart Cities," *J. Ambient Intell. Humaniz. Comput.*, vol. 14, no. 8, pp. 10493–10515, Aug. 2023, doi: 10.1007/s12652-022-03706-2.
39. J. Srinivas, A. K. Das, N. Kumar, and J. J. P. C. Rodrigues, "TCALAS: Temporal Credential-Based Anonymous Lightweight Authentication Scheme for Internet of Drones Environment," *IEEE Trans. Veh. Technol.*, vol. 68, no. 7, pp. 6903–6916, Jul. 2019, doi: 10.1109/TVT.2019.2911672.
40. K. Lounis, S. H. H. Ding, and M. Zulkernine, "D2D-MAP: A Drone to Drone Authentication Protocol Using Physical Unclonable Functions," *IEEE Trans. Veh. Technol.*, vol. 72, no. 4, pp. 5079–5093, Apr. 2023, doi: 10.1109/TVT.2022.3224611.
41. X. Ren *et al.*, "A Novel Access and Handover Authentication Scheme in UAV-Aided Satellite-Terrestrial Integration Networks Enabling 5G," *IEEE Trans. Netw. Serv. Manag.*, pp. 1–1, 2023, doi: 10.1109/TNSM.2023.3246732.

16 A Comprehensive IoT Security Framework Using Edge Computing
Approaches and Challenges

Chintan Singh, Himanshu Khajuria, and Biswa Prakash Nayak

16.1 INTRODUCTION

Over time, Internet Protocol mainstay network and core of it have become increasingly interconnected; they have progressively become the destination of storage, processing, and network key functions. The bulk of information collected from Internet of Things (IoT) nodes are increasing exponentially, making careful processing of the data imperative to speed up response time and effectively distribute network congestion. The restrictions of the cloud's big-data processing have recently come to light. By 2020, it is expected that 50 billion devices are going to be interconnected to the internet [1]. By enabling the ability to monitor, interpret, control, and optimize physical systems, IoT applications offer unprecedented levels of convenience to our lives. Unfortunately, this also opens up a new realm of cyber threats that can put our systems at risk [2].

The lines between the physical and digital realms are blurring more and more every day. With advances in sensing, communication, and microcontroller tech, we can now seamlessly blend the two together. This integration allows us to access, control, and monitor physical objects and environments remotely, giving us greater insight and control of our surroundings. By combining the physical and the digital, we can develop more efficient and cost-effective solutions to our everyday problems [3]. IoT has been revolutionizing technology, making it faster and more powerful. A robust and resilient architecture must be implemented to take advantage of these innovations. Fog/edge computing (FEGC) is a good fit for the problem, as it promises to deliver real-time control and analytics with millisecond response time. FEGC is ideal for developing and configuring a scalable and adaptable IoT system capable of providing reliable sensing, actuation, and analysis. Additionally, FEC can manage many geographically distributed "things" and schemes with self-motivated agility. In other words, FEGC helps ensure an efficient and reliable IoT system [4].

DOI: 10.1201/9781003405535-21

In essence, omnipresent devices should be able to interoperate with one another via an array of networks for the sole purpose of providing services, sharing data, and supporting a wide range of applications. Issues with connectivity, latency, capacity, service, and privacy have inhibited the implementation of IoT. So, it is essential to have mechanisms that can handle big things and objects. IoT and cyber-physical systems (CPhS) have fragile architectural foundations. Service-based designs were described in previous studies [3].

Substantial research has been conducted on developing innovative edge-based IoT haven solutions. In addition to edge-based security architecture strategies, these initiatives also incorporate firewalls [5], intrusion detection systems [6], authentication and authorization protocols [7], and privacy-maintenance strategies [8]. In contrast, IoT edge security research is still in its initial stages. Further edge-based mechanisms for IoT require constant development. There is also a limited amount of descriptive assessment that can accurately depict the current state of the art in the field of study.

16.1.1 Motivation and Contribution

The chapter titled "A Comprehensive IoT Security Framework Using Edge Computing: Approaches and Challenges" is an essential contribution to the field of IoT security. This chapter offers innovative edge-based IoT security frameworks to address the growing need for robust security frameworks in the rise of IoT. By reviewing available alternatives and proposing new research avenues and cutting-edge research issues, this chapter offers a thorough understanding of the challenges and approaches to IoT security using edge computing.

This chapter's contribution is crucial for shaping the future of the internet and ensuring that IoT remains a safe and secure technology for everyone to use. By using edge computing-based services, we can overcome the limitations of cloud computing while ensuring that our data is secure and our systems are protected against cyber threats. The security frameworks proposed in this chapter can help organizations and individuals protect their data and devices from malware, ransomware, and other cyber threats.

Moreover, this chapter's contribution highly motivates researchers and professionals working there. It highlights the most pressing challenges that need to be addressed. By following the guidelines proposed in this chapter, researchers and professionals can design and implement secure IoT systems resilient to cyber-attacks. In summary, the chapter titled "A Comprehensive IoT Security Framework Using Edge Computing: Approaches and Challenges" provides a comprehensive overview of IoT security using edge computing, making it a must-read for anyone interested in this field.

16.1.2 Fog/Edge Computing (FEGC)

FEGC hovers just on the fringes of the cloud, promising improved data storage and access. By utilizing distributed computing and storage resources, these technologies can help optimize data storage and increase security while reducing latency and improving scalability. FEGC is one of the two technologies used to overcome the challenges of processing and transmitting data quickly and efficiently. Edge computing (EGC) is a subdivision of fog computing (FGC), and the power consumption of nodes is low when compared to FGC [9]. EGC best suits environments requiring data processing, while FGC adds a decentralized enterprise layer between the data source and a centralized cloud platform. EGC centres where data is collected, while FGC focuses on the space of facts amid the source and the cloud [10].

Globally recognized computer enterprises like Microsoft Azure, OneDrive, IBM, Mega, Google Cloud system, and Amazon web portals are propagating the cloud computing (CLCG) approach. The disadvantages of this centralized approach have been highlighted in various literature [11, 12]. The increase in machine-to-machine communication exacerbates the issue, making the use of the CC model problematic for managing network congestion.Valuable data and applications may experience significant dormancy because of the comparatively considerable distance between both data hubs. Business monopolization is another clear uptake, as the wealthiest multinational corporations can afford to build and maintain their cloud infrastructure and develop proprietary protocols [13]. Disdaining the introduction of FEGC, which bids small- to medium-sized organizations greater economic prospects, CLCG still maintains its crucial role in the conceptual model. Such activities expand but might be affected by the changes occurring inside the interconnection. Such activities can be preconceived; however, they could be affected by the dynamic alterations occurring inside the network, like processing and storage abilities, bandwidths, security risks, network failures, and financial restrictions. The demonstration of the CLCG model with services that are included in it are illustrated in Figure 16.1, which are software

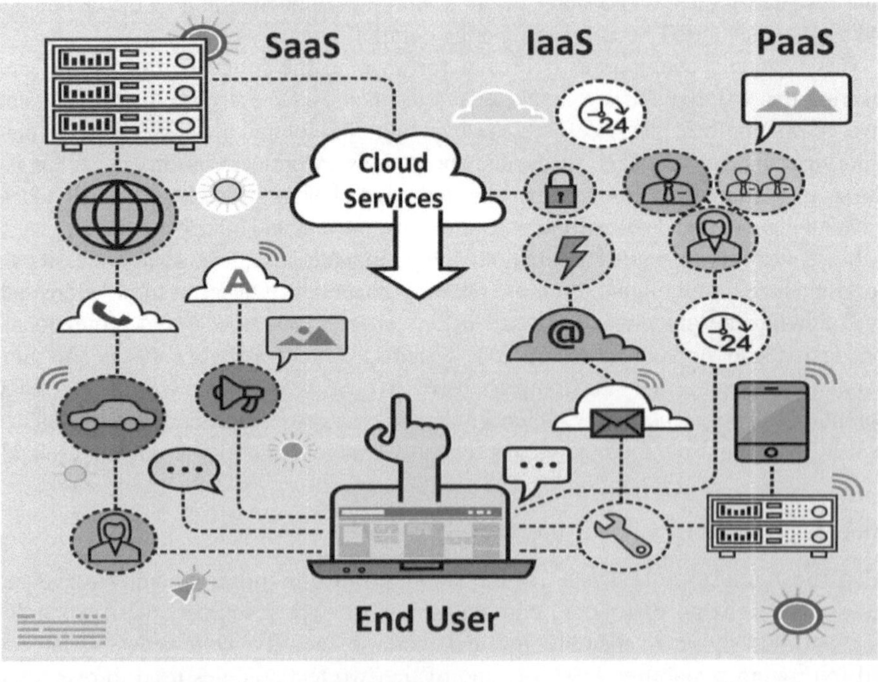

FIGURE 16.1 Cloud Computing CLCG Model.

as a service (SaaS), infrastructure-as-service (IaaS), and platform as a service (PaaS). SaaS is a popular way of distributing applications, where customers can access and use programs maintained and operated by the service provider and are accessed through a web browser. One of the most common examples of SaaS is a web-based email service, such as Gmail or Outlook. Other examples include customer relationship management systems, such as Salesforce, and enterprise resource planning systems, such as Microsoft Dynamics. Infrastructure use as a service is a CLCG model. Provider sets up, monitors, and maintains the network, servers, storage, and other components needed for the organization's applications and services. For example, Amazon Web Services (AWS) provides a suite of IaaS products, including virtual machines, object storage, load-balancing tools, and databases.

FGC devices might not always be at the network edge but close by, which is something to consider. Inversely, edge devices typically remain in the network layer and operate as the pre-initial contact for IoT devices. In short, EGC devices, although being close to both, are typically closer to IoT end devices than FGC devices. There has been a widespread misunderstanding of the concepts of FGC and EGC. Most researchers see FGC as a part of the EGC of IoT and microdata centre conceptions [14]. FGC and EGC have services close towards end-users; although FGC is located in edge devices on the network, often one or two network trips away from the edge, whereas EGC is a member of edge devices. The EGC platform falls under the category of restricted devices since it has restricted energy and storage. Expanding IoT applications may increase resource competition and inefficiencies [15]. EGC has higher resource conflict than FGC due to its proximity to IoT endpoints. Also, EGC focuses more on the underlying domain than FGC does on the infrastructure domain. The rationale for integrating FEGC is that each has unique characteristics that can benefit a system. FEGCs are secure, scalable, open, autonomous, reliable, agile, hierarchically organized, and programmable. These characteristics enhance the system's scalability, flexibility, and efficiency and are inherent to EGC. As such, the decision to bring these architectures together is based on the unique ways they can complement each other. Such peculiarities about FEGC are:

1. They offer low latency, low distortion, and intelligence inside the system and are mostly integrated wirelessly.
2. They are positioned between end clients and data centres, complementing the cloud's features.
3. Both of them allow for the virtualization of applications at the network's edge and employ a virtualized IaaS platform.
4. They both provide CLCG across numerous geographic regions to alleviate the workload on the cloud.

Relying on performance, FGC and EGC can intertwine and coexist within the cyber-physical system (CYPS).

16.1.3 Cyber-Physical System (CPhS/CPS)

CPhS emerges for upcoming era engineering structure. The term "cyber-physical system" was attributed by Helen Gill in about 2006 [16]. Although having no relation to physical processes, the terms "cyber-physical system" and "cyber security" are commonly used interchangeably. CPhS is the term used to describe the intimate integration of computations, algorithms, and physical devices. It can be defined as the confluence associated with computational processes. Embedded computers and network controls help to evaluate physical processes, typically using feedback loops, where the physical processes influence computations and vice versa.The CPhS cyberspace focuses on computation and administrators. The CPhS upholds the traits listed below [6]:

1. Adaptivity: The system can adapt to changing conditions or environments in order to achieve its goals or objectives.
2. Integration: The physical components, sensors, actuators, and embedded processors are all integrated into a single and unified system that performs a specific task or tasks.
3. Robustness: The system can operate reliably even in the presence of faults or noise.
4. Real-time Response: The system is able to respond to time series data and provide accurate information that rely upon analysed datasets.
5. Scalability: The system can scale up or down when more or fewer components are needed.
6. Security: The system is designed to protect data and is resistant to malicious cyber-attacks.

Physical systems are connected to cyber systems through unifaces. Such unifaces are made up of elements such as communication networks, transducers, effectors, digitizers, and synthesizers. Figure 16.2 demonstrates architecture based on the IoT devices in cyber world. According to Figure 16.2, CPhS span multiple physical

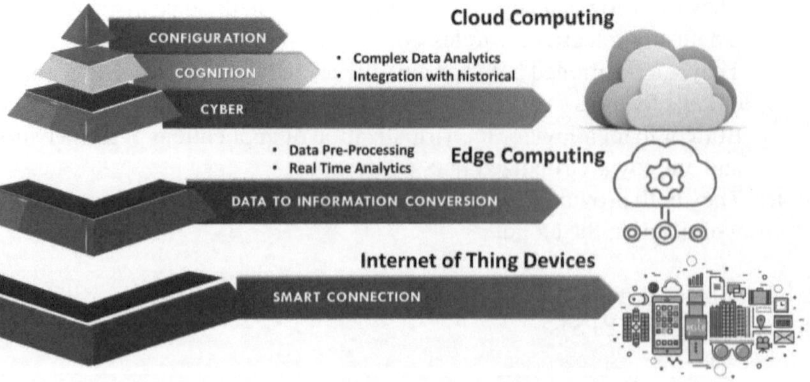

FIGURE 16.2 IoT-based cyber-physical system (CYPS) paradigm.

devices, communiqué and system technologies, and applications within every CPhS domain. The integration of CPhS horizontally through the IoT emphasizes the need for protocol, middleware, and interface homogenization within the control plane across heterogeneous networks. It is crucial to note that physical connections alone are insufficient, and uniformity across networks is also essential for successful interconnection.

16.2 EDGE COMPUTING-BASED SECURED IOT FRAMEWORK (EGCIOT)

Since EGC emerged, many researchers have investigated EGC-based techniques to solve IoT security problems. The term EGC-based secured IoT framework (EGCIoT) was first adopted by Lin et al. (2017) to emphasise the enormous potential that may come from effectively integrating the FEGC paradigm into the IoT architecture [3]. Here, we provide a layperson's explanation of the recommended architecture and discuss its advantages and disadvantages. The EGCIoT expands on the underlying IoT architecture and more efficiently accommodates most IoT requirements by implementing the distributed FEGC paradigm. There is no commonly accepted architecture for the IoT, as has already been noted. The frameworks that the authors have offered in various research papers fully demonstrate their perspectives on the IoT. This section focuses on various architecture and models derived for EGCIoT.

16.2.1 Previous Literature

This section tabulates the previous findings incorporated in different domains of IoT applications and provides them with a secure framework via EGC. Table 16.1 shows

TABLE 16.1
Previous Literature for Various Frameworks Securing IoT Using EGC and Their Limitations

Author	Methodological Framework	Result or Advantages	Limitations	Reference
Hu et al., 2017	Identification and resolution in IoT that is implemented on edge devices.	Propose a comprehensive solution consisting of a security framework for IoT that ensures the confidentiality, availability of data, and integrity.	Unbaling capture of other secret features of biometrics other than face recognition	[17]

(continued)

TABLE 16.1 (Continued)
Previous Literature for Various Frameworks Securing IoT Using EGC and Their Limitations

Author	Methodological Framework	Result or Advantages	Limitations	Reference
Yigitoglu et al., 2017	Foggy framework for continuous automated IoT in EGC	Foggy, the framework, enables the dynamic provisioning of resources and automated deployment of applications within fog computing architectures.	The resource requirements of Foggy, such as computational power, memory, or network bandwidth, might not be effectively addressed. The research might not sufficiently address the challenges related to integrating foggy with existing fog computing infrastructures, platforms, or protocols.	[18]
Porambage et al., 2018	Multi-access EGC framework	Secure and efficient authorization framework can enable real-time, high-bandwidth, and low-latency access to radio network resources.	Long end-to-end latency in network. The research might not address the challenges that arise when scaling up the scheduling approach to handle a hefty number of microservices or edge devices in a real-world IoT deployment	[19]
Hsu et al., 2018	Reconfigurable edge-based security framework (RESIOT)	Low computational cost, high felicity in key management, and higher compatibility in deploying latest security algorithm.	Higher processing time as compared to other architecture for IoT	[20]

TABLE 16.1 (Continued)
Previous Literature for Various Frameworks Securing IoT Using EGC and Their Limitations

Author	Methodological Framework	Result or Advantages	Limitations	Reference
Casado-Vara et al., 2018	Distributed blockchain-based architecture in edge layer for IoT	Improved data superiority, detection of falsified data in network, open thread technology used via Rasberry pi.	Complex mathematical formulations, vague result for analysis of functions to find cooperatives temperature	[21]
Chen et al., 2018	Edge-specific device supporting network architecture for IoT-based manufacturing	Results in technical deployment of architecture in smart industries.	Implementing real-time control loop feedback is complex, and time consumption is high to analyse the data streams to generate decisions	[22]
Guardo et al., 2018	Exploited fog-based framework for IoT in agriculture	The proposed framework utilizes a two-tier fog architecture and its available resources to minimize data transmission to the offsite location. This approach can effectively improve computational load balancing and reduce waiting times.	No discussion on other environmental conditions and factors that can affect proposed framework, also need prototype that can implant the framework and can surveillance requirement of farmlands	[23]

(continued)

TABLE 16.1 (Continued)
Previous Literature for Various Frameworks Securing IoT Using EGC and Their Limitations

Author	Methodological Framework	Result or Advantages	Limitations	Reference
Khafajiy et al., 2018	Three-layered frameworks for securing IoT in healthcare system using EGC	The framework facilitates seamless collaboration among fog nodes, ensuring efficient resource management and job allocation. This results in a healthcare system scenario with high quality of service, specifically low latency.	Dearth of data validations and integrity of proposed framework. The research may not provide sufficient insights into the practical implementation aspects	[24]
Tuli et al., 2019	FogBus	FogBus utilizes blockchain, authentication, and sensitive data can be protected from unauthorized access using encryption operations. With its streamlined and platform-agnostic software systems, FogBus offers effortless deployment, scalability, and cost-effectiveness.	Research may not provide a comprehensive comparative analysis of FogBus with other existing frameworks or Efficient Fog Computing (EFGC) are two emerging technologies that offer a variety of solutions for businesses and organizations. It may not address the potential challenges, such as deployment considerations, integration with	[25]

TABLE 16.1 (Continued)
Previous Literature for Various Frameworks Securing IoT Using EGC and Their Limitations

Author	Methodological Framework	Result or Advantages	Limitations	Reference
			existing systems, or interoperability issues that could arise during the implementation process.	
Sodhro et al. 2019	AI-based framework for EGC on industries	A framework is proposed that introduces adaptive timing for sensing and transmitting processes in portable IoT devices. Additionally, the framework presents a data reliability model for edge AI-based IoT devices operating on hybrid TPC.	Lack of cases for proper monitoring of framework in industrial application	[26]
Wang et al., 2019	Satellite EGC for IoT in aerospace	A novel framework for aerospace is proposed that enables dynamic software loading in orbit. The satellite's corresponding hardware structure and software architecture are presented, and the results show that the	The scalability of the satellite edge computing system might not be thoroughly addressed in the research. It may not consider the challenges that arise while increasing readability to IoT devices or accommodate increasing data volumes in real-world aerospace applications	[27]

(continued)

TABLE 16.1 (Continued)
Previous Literature for Various Frameworks Securing IoT Using EGC and Their Limitations

Author	Methodological Framework	Result or Advantages	Limitations	Reference
		space edge computing system outperforms traditional satellite constellations to reduce time and consumption of energy.		
Junxia li et al., 2020	Software-defined network edge framework	Authenticity of IoT devices on edge servers in healthcare system.	Simulation study has been performed using Software-Defined Networking (SDN) but research may not provide sufficient insights into the practical implementation aspects	[28]
Lawal et al., 2020	Hybrid anomaly mitigation framework	The network traffic flow can be accurately classified using an extreme gradient boosting algorithm, while blacklisted IP addresses can be detected more swiftly.	Need exploration of botnets attacks to develop signatures for the framework. Does not include a comprehensive comparative analysis with existing microservice scheduling techniques or frameworks	[29]

TABLE 16.1 (Continued)
Previous Literature for Various Frameworks Securing IoT Using EGC and Their Limitations

Author	Methodological Framework	Result or Advantages	Limitations	Reference
Jha et al., 2020	IoT sim edge framework	This framework holds the diverse working of architecture of edge IoT, making it easy for users to test their frameworks and infrastructure with flexibility and ease.	Simulation framework should be implanted on physical level and more test cases are required to promote effectiveness	[30]

the literature findings available with the methodological framework used by various researchers, along with their findings and limitations.

16.2.2 Architecture at Edge Layer

EGC brings forth a platform to produce and devise comprehensive and innovative sanctuary strategies for IoT applications. Such strategies are premeditated to meet the security requirements of endpoint devices to minimize the security burden placed on these devices by shifting the protection to this layer. By keeping safety apparatuses EGC, it is possible to reduce the security risks associated with the limited resources available on IoT devices. Edge-based architecture consists of mainly three security architectures device-centric, user and terminal security.

Device-centric: A device-oriented EG security strategy for IoT devices is essential. Billions of IoT devices are tangible and physical objects that are embedded in the world and provide information, as well as deciding crucially to govern the physical realm. While user-centric IoT security focuses on protecting users, device-centric IoT security tailors security measures to each specific device, considering factors such as resources, the sensitivity of the data it collects, and the potential impact of any actions. Moreover, device security can also be taken into account. Two notable examples of strategies that use the edge layer to implant specified EGC security strategies are EdgeSec and ReSIoT [20].

User-centric: User satisfaction is essential for the success of IoT applications. Thanks to thousands of IoT devices interconnection, users can access large resources across multiple assets, like PCs, androids, advanced TVs, and smartwatches. However, security is a major concern when using IoT applications [31]. On the one hand, users may not be able to access a trusted, secure device. As a result, it is unsafe to leave security to the user.

End-to-end security: Security between IoTs and the cloud is often desired for many IoT applications. However, achieving this type of security in IoT is difficult due to the complexity of devices. EGC is used to connect disparate nodes and the cloud. Mukherjee et al. (2017) outlined the middleware's management of security functions, including message authentication codes (MACs), encryption algorithms, authenticators, and secure sessions for androids [32].

16.2.3 Firewall at EGC

Many nodes have limited resources and cannot support resource-intensive security applications like firewalls. Additionally, due to the vast number of IoT devices, implementing and managing firewalls for each device would be excessively costly. The most economical else effective solution is provided by firewalls situated at the edge. An illustration of an edge-based firewall architecture may be seen in Figure 16.3. Firewall policies generated by IoT apps are converted into a collection of flow policies. The collection of flow policies that result from the identification and resolution of any conflicts is then transformed into a set of distributed firewall

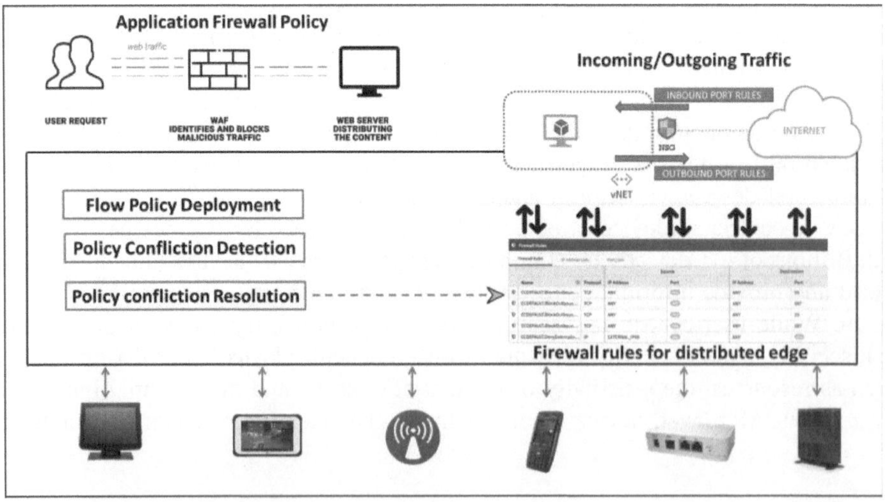

FIGURE 16.3 An edge distributed firewall for edge layer.

rules that are applied at the edge. Every traffic is then at risk of inspection based on these rules. Deploying the firewall at the edge layer offers several advantages. Firstly, updating the firewall is easier and more manageable as it is centralized conceptually.

Secondly, for some IoT applications, a particular IoT subsystem might be managed by an edge device and the setting up of a firewall to meet the security requirements of that subsystem. The edge layer can record user and end device movement and credentials.

16.2.4 Edge Computing-Based Intrusion Detection System

A multitude of IoT devices were breached by attackers in 2016 and utilized to execute a Distributed Denial of Service(DDoS) assault on multiple Domain Name System (DNS) servers belonging to Dye Inc. This resulted in a considerable loss due to the disruption of internet connectivity across a large region. The advanced system could have detected the DDoS attack early and minimized the resulting damage. Edge-based intrusion detection systems demonstrate various benefits with the additional information available at the edge [33]. Figure 16.4 illustrates a conceptual design for such a system. This design incorporates a distributed traffic monitoring service that gathers live network traffic data. Intrusion detection algorithms are executed on each edge device to analyse the traffic. Additionally, traffic information from various edge devices is examined to detect collaboration intrusion. The outcomes of the finding procedure are then enforced by the network managers installed at the edge endpoints.

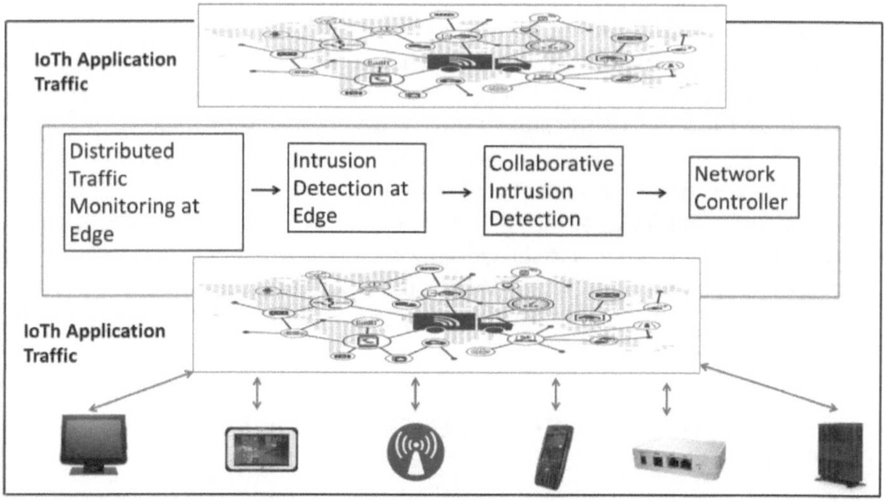

FIGURE 16.4 Intrusion detection systems for distributed edge computing.

Similarly, an EGC-based anomaly analysis was reported by Yaseen et al. (2016) [34]. The proposed system is designed to identify a particular yet key attack, selectively forwarding in the system of mobile IoTs. IoT devices are used as watchers to measure their surrounding devices' packet decreeing rates [34]. Edge servers collect, consolidate, and distribute relevant data from watchdogs to other edge servers. Then, a vote mechanism is used for identification of any suspected activities.

16.2.5 Edge-Based Authentication and Authorization Framework

Authentication and authorization are critical security measures that can help prevent various attacks, such as distributed denial of service attacks [35]. IoT system architectures are anticipated to rely on authentication and authorization mechanisms to ensure end-to-end security, but this is extremely difficult to achieve for various reasons. As an illustration, mutual authentication is complicated to set up due to the difficulty of establishing a direct transmission. Moreover, numerous conventional authentication methods, such as those utilizing digital signatures, are unsuitable for end devices in IoT. For example, Elsayed et al. (2023) proposed an illustration focusing on EGC serves as a proxy for authentication and authorization purposes [36]. In this concept, multi-factor authentication and authorization are implemented via control system pathways and an on-demand reputation monitoring service is used to confirm the authenticity of identities [37]. To verify the information you know (such as a login and password), what you have (such as a token or key), and/or what you are (such as biometrics), two-factor authentication often employs more than two separate authentication techniques. An IoT user sends an inquiry comprising physiological and identification data to the gateway during the authentication and authorisation procedure. The gateway proceeds to authenticate the user and determine their level of authorization.

Elsayed et al. (2023) proposed a novel control system gateway that acts as a proxy for authentication and authorization in IoT. The access point offers immediate form identity surveillance services, multiple authentication methods and authorization, or multi-factor certification. Verification forms various aspects of identity, including knowledge (e.g. username/password and security questions), possession (such as a token, a key, a certificate, or a smart fob), and/or biometric traits. In the authentication and authorization process, the IoT user submits a request along with their biometric and identity information to the gateway, which then authenticates the user and determines their authorization level [36]. To mutually authenticate an IoT user and an IoT device, Sha et al.'s (2020) research suggests a solution that leverages an edge device as an intermediate [38].

16.3 THREE-LAYER SECURITY FRAMEWORK USING EDGE COMPUTING

The architecture of this design is traditional and takes into account three primary layers: the sensing or data acquisition layer, the network layer, and the user layer.

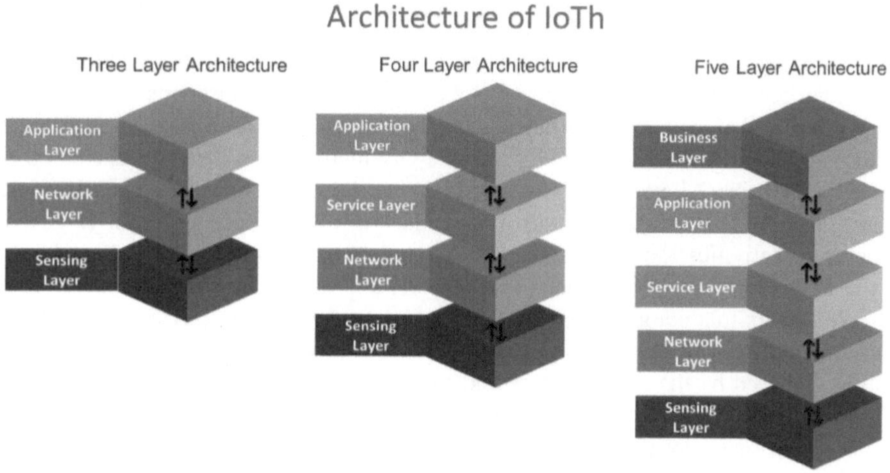

FIGURE 16.5 Three, Four and Five-layer framework/architecture of IoT.

These three layers combined make up the whole system as shown in Figure 16.5. The sensing layer is tasked with acquiring data, the network layer enables communication and data transfer, and the application layer is used to make sense of the data gathered by the two layers above.

The sensing layer collects data using various devices like Radio-Frequency Identification (RFID) readers, sensors, wireless identification and sensing platforms (WISP), and actuators. Nodes will work together within a certain radio range to ensure secure data communication in the IoT network. Data can be collected through the sensing layer and sent to the network layer. The network layer is in charge of transferring data between multiple networks. It supports internet access and computing devices like CC platforms and utilizes modern technologies such as 5G/LTE, Bluetooth, and Wi-Fi. This layer is essential for ensuring the data is routed and sent to the right destinations quickly and securely. The application layer serves different needs depending on token requests sent from the network layer. It is home to a variety of IoT applications that are deployed with middleware functionalities. As fog/edge deployments become more widespread, applications that are made by different vendors will need to ensure compatibility and interoperability to function seamlessly within diverse and distributed network environments. [39].

The three-layer architecture may seem straightforward, but when analysing the second layer, work can see the complexity of implementing services like data mining and analytics. This requires careful planning and consideration to ensure that the functionality of these services is retained while they are integrated into the architecture

16.4 FOUR-LAYER ARCHITECTURE USING EDGE COMPUTING

To enhance information services in the IoT, the service layer is introduced between the network and app layers. Irrespective of the technological platforms they use, this layer offers a framework which allows businesses and institutions to build, implement, and integrate these services. A Service-Oriented Architecture (SoA) is another name for the service layer, which organizes the workflow of coordinated services and makes it possible to reuse both software and hardware. Designing, deploying, and integrating activities independent of the technological platform is possible, thanks to the SoA [40]. The sensor, network, service, and application layers are the four different layers that comprise the IoT architecture, which is depicted in Figure 16.5. With an emphasis on each service layer, we will briefly traverse through each layer. Data is gathered by the sensing layer and sent to the network layer. The information is then transmitted from the network layer to the provider layer, which is processed before being made available to the application layer. The application layer subsequently uses the data to make choices.

1. Sensing layer: This serves as the framework for the architecture and is in charge of gathering, analysing, and extrapolating data from physical devices. The higher layers receive the data after that.
2. Network layer: Course decisions are made in this area, which supports transmitting data across various networks and topologies.
3. Service layer: This layer, referred to as the interface or middleware layer, provides a wide range of services. It can be broken down into four components. These components are responsible for identifying, locating, and executing services; managing service execution; managing service invocations; and managing data security. The service composition sublayer, a part of SoA, also provides features for composing specific services from networked objects to create tailored applications [37].
    a. Service management: Each object is managed and provided with primary functional requirements through this process. The service layer provides a variety of functionalities, such as QoS, management, semantics, and the deployment of other areas at runtime to meet application requirements. It also maintains a service repository to identify service-object pairs within the IoT network.
    b. Service interface: A bridge connecting all offered services is the service's gateway. Links are required to make corporate processes less difficult [41].
4. Application layer: The app layer, which is at the top of the physical layer and provides end-user assistance according to system operations, is situated there. The middleware layer is separate from the application layer from the traditional three-tier architecture. Rather, the program's layer gives instructions. The establishment of inconsistently distributed infrastructures and applications, such as smart dwellings, intelligent public transit, smart businesses, and smart

medical care, is made possible by the service/middleware level, which offers an interactive interface through standard web service protocols and service composition technologies.

16.5 FIVE-LAYER FRAMEWORK

The aim of this model is to offer more advanced services with a business-oriented approach that extends beyond the traditional application layer. This model comprises the sensing, network, service, application, and business layers. The business layer is responsible for capturing and analysing all IoT operations that have occurred across heterogeneous CPhS. Since the lower layers have already been discussed in the previous architecture, we will focus on the business layer here.

The IoT system as a whole, comprising all apps, business and profit models, and user data, is managed by the company's layer. The commercial layer is an extra component of the SoA that makes it easier to build systems that allow the development of autonomous solutions for businesses based on technology restrictions, as shown in Figure 16.5 [42, 43]. The FEGC architecture's convergence of routers, switches, storage, and application servers into FEGC devices will significantly impact the communication, server, and software sectors. The distributed FEGC architecture also supports the growing fog-as-a-service business model, which enables smaller companies to provide both public and private assistance for consumers and users at various scales.

Among the major challenges associated with IoT security frameworks using EGC is the difficulty of properly securing the interchange of information among the EGC. As EGD become increasingly more sophisticated, the potential for malicious actors to intercept or misuse data increases. Additionally, the distributed nature of EGC means that it is difficult to ensure secure communications protocols between the different nodes. Furthermore, EGC also introduces complexity in terms of security for different IoT devices as well as data privacy concerns. Ensuring secure and reliable communication among the cloud and edge devices is essential for secure data exchange. Establishing robust authentication protocols and ensuring secure communication protocols between the different nodes is critical for any EGC IoT security framework.

16.6 CASE STUDY ON IOT SECURITY FRAMEWORK USING EDGE TECHNOLOGY IN HEALTHCARE SECTOR

Because of the novelty of the EGC paradigm, there's a dearth of lightweight solutions available. Although several researchers aim to devise the stated concept by advancing their edge applications, they tend to be hardware-specific and lack open-source features, ease of customization, and practicality. EGC-based software, such as Apache Edgent open-source software, is being developed by developers worldwide. However, due to its early stage of development, there are limited research articles and detailed documentation provided on open research platforms. Additionally, applications related to e-healthcare haven't been thoroughly analysed and validated against periphery.

The discussed case study aims to introduce the latest EH-IoT (e-healthcare IoT) system that utilizes Edgent technology for edge computing. This system deploys and configures Edgent on a Raspberry Pi 3 to create an EGC environment. It also explores the connector classes of Apache Edgent for communication with external backend systems. Additionally, it sets up a proof-of-concept testbed to provision edge services in an IoT-based scenario while minimizing data transfer capacity to the analytics IoT servers.

Hardware and software: The research employs various hardware and software components such as a temperature sensor, Raspberry Pi, Arduino Uno, Raspbian OS, Apache Edgent 1.2.0, and ThingSpeak. I have rephrased the original sentences to avoid plagiarism.

Methodology: The EH-IoT system suggested for deployment encompasses three main modules. These modules include

Module 1: For devices at the edge with less processing authority, there is an open-source programming paradigm called Apache Edgent. It offers Application Programming Interface (APIs) and classes for creating Edgent apps that filter and deal with data, conduct streaming analytics on data streams, and send the processed data to the cloud for additional analysis. Edgent also includes MQTT, HTTP, WebSocket, JDBC, File, and Kafka connectors and a Java API for sending JSON to an MQTT device. It also supports bidirectional communications with the backend, a web-based interface for viewing application graphs, Junit testing, and Java-based Lambda expressions.

Module 2: This module includes the sensing unit based on embedded hardware, the Apache Edgent engine blames for manufacturing and analysing the data, and the cloud repository based on IoT technology. The sensing unit generates raw data sets that are further mailed for production. The IoT devices generate continuous data streams, including healthcare applications that use temperature sensors. The LM35 temperature sensor acts as the source for datasets on the edge, reading human body temperature as proof of concept.

Module 3: To perform complex analysis that requires extensive computing resources, large storage capacity, or data correlation from multiple sources, the EH-IoT can send data to IoT-EGC, also known as backend systems. Message hubs such as MQTT, IBM Watson IoT platform, Apache Kafka, custom message hubs, and ThingSpeak can be utilized to establish communication between the EH-IoT. ThingSpeak, for instance, is an open IoT platform.

The EH-IoT platform model records the consumer's body heat, sending the information to an Arduino Uno and a Raspberry Pi 3, which runs the Edgent engine. By using the IoT gateway, locally or in the IoT cloud, the Raspberry Pi 3 filters gather and distribute the temperature data. Only crucial or restricted information is transferred to the cloud analytics server to reduce calculation delay and internet backbone overhead. At the same time, data outside of the designated range is analysed locally. A thorough description of the EH-IoT system model, including the stages involved in body temperature measurement, data filtering and cloud integration, and EH-IoT to cloud connection, can be found in Phantom codes 1, 2, and 3, as shown in Figures 16.6 and 16.7.

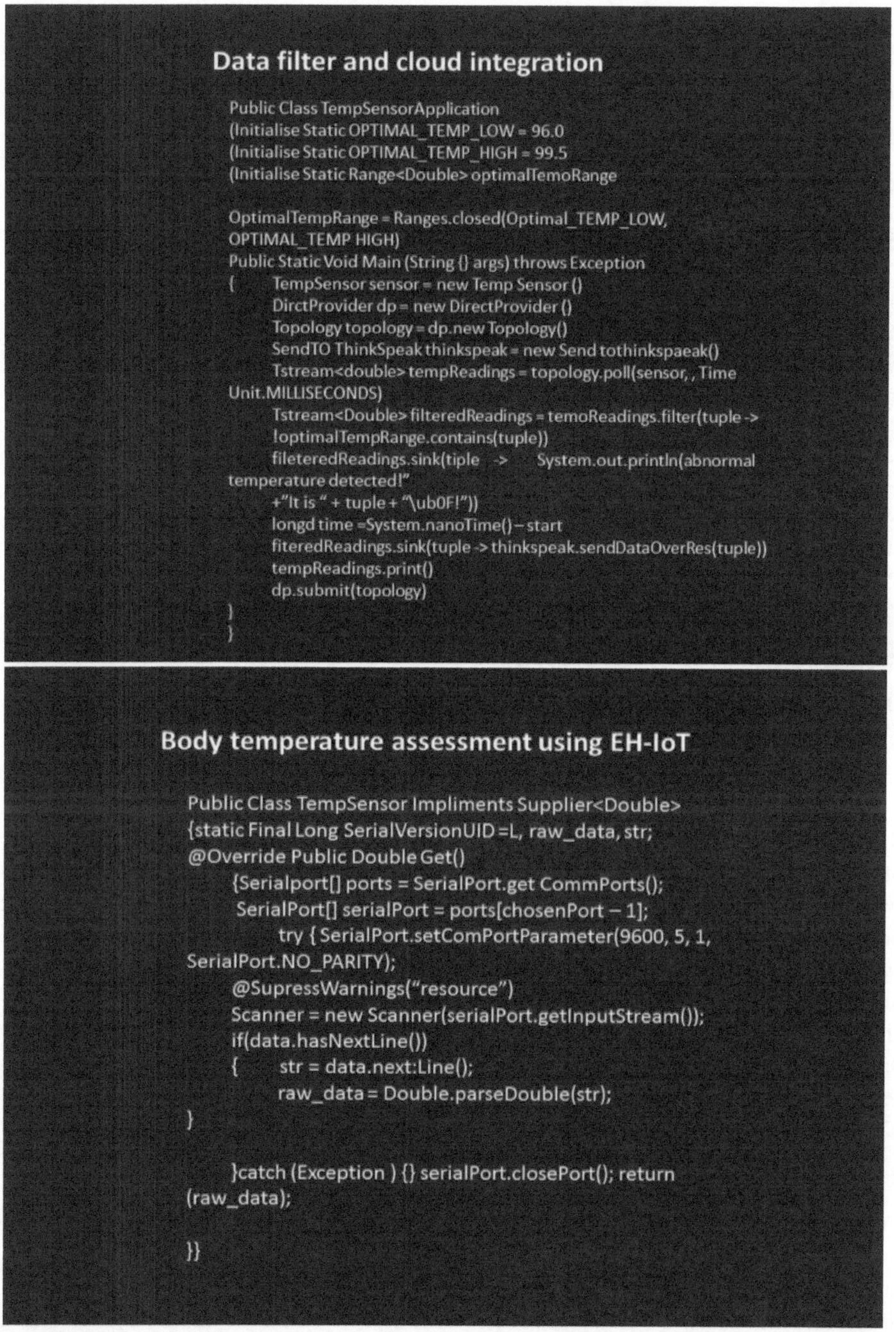

FIGURE 16.6 (a and b) EH-IoT model for measuring body temperature, data filters, and cloud integration.

EH-IoT to IoT Cloud

```
Public Class SendToThinkSpeak Implements Callback<JsonNode>
{     Public Void SendDataOverRest(double temp)
      {
      Unirest.post(https://api.thingspeak.com/update.json")
      .header("accept", "application/json")
      .field ("api_key", "BJH2BQDISA2P4UT8")
      .field("field1" ,temp)
      .asJsonAsync(this);

}

@Override Public Void Completed(HttpResponse<JsonNode>
response)

{
      int code =  response.getStatus()
      JsonNode body =  response,getBody ()
      InputStream rawBody =  response.getRawBody()
}

}
```

FIGURE 16.7 EH-IoT model for IoT cloud system.

16.7 CONCLUSION

Studies have recently alerted the public to the growing importance of the difficulty of protecting IoT systems. However, it remains a major challenge. The advent of EGC has brought about various security designs at the edge of IoT security. The objective of the work is a thorough, rigorous analysis of current edge-based IoT safety measures within the framework of a clearly defined, edge-centric IoT architecture. These products handle crucial facets of IoT safety, such as firewalls, intrusion detection systems, procedures for identification and authorization, safeguarding privacy designs, and an extensive framework for security. This report also outlines several field-specific difficulties and suggests some future research options.

16.8 FUTURE RESEARCH DIRECTION

Future works should include the development of more secure and reliable connection protocols between EGC and CGC. Additionally, improvements in authentication and authorization protocols will be needed to ensure secure data exchange between IoT devices. Additionally, research must be conducted to ensure data privacy is maintained in EGC-based IoT frameworks. Furthermore, research should be conducted to improve the usability and scalability of EGC-based IoT security frameworks. Finally, more effective and secure mechanisms for data sharing and exchanging between edge devices and the cloud must be developed.

REFERENCES

1. F. Al-Doghman, Z. Chaczko, A. R. Ajayan, and R. Klempous, "A review on Fog Computing technology," in *2016 IEEE International Conference on Systems, Man, and Cybernetics, SMC 2016 – Conference Proceedings*, 2017, pp. 1525–1530. doi: 10.1109/SMC.2016.7844455.
2. K. Sha, W. Wei, T. Andrew Yang, Z. Wang, and W. Shi, "On security challenges and open issues in Internet of Things," *Future Generation Computer Systems*, vol. 83, pp. 326–337, 2018, doi: 10.1016/j.future.2018.01.059.
3. J. Lin, W. Yu, N. Zhang, X. Yang, H. Zhang, and W. Zhao, "A Survey on internet of things: Architecture, enabling technologies, security and privacy, and applications," *IEEE Internet Things Journal*, vol. 4, no. 5, pp. 1125–1142, 2017, doi: 10.1109/JIOT.2017.2683200.
4. M. Chiang and T. Zhang, "Fog and IoT: An Overview of Research Opportunities," *IEEE Internet of Things Journal*, vol. 3, no. 6. pp. 854–864, 2016. doi: 10.1109/JIOT.2016.2584538.
5. P. Mach and Z. Becvar, "Mobile Edge Computing: A Survey on Architecture and Computation Offloading," *IEEE Communications Surveys and Tutorials*, vol. 19, no. 3. pp. 1628–1656, 2017. doi: 10.1109/COMST.2017.2682318.
6. H. Hu, W. Han, G. Ahn, Z. Zhao, "FLOWGUARD: Building robust firewalls for software-defined networks," *dl.acm.org*, pp. 97–102, 2014, doi: 10.1145/2620728.2620749.
7. P. S. Mehra, M. N. Doja, and B. Alam, "Codeword Authenticated Key Exchange (CAKE) light weight secure routing protocol for WSN," *International Journal of Communication Systems*, vol. 32, no. 3, pp. 1–27, 2019, doi: 10.1002/dac.3879.
8. Z. Ali, M. S. Hossain, G. Muhammad, I. Ullah, H. Abachi, and A. Alamri, "Edge-centric multimodal authentication system using encrypted biometric templates," *Future Generation Computer Systems*, vol. 85, pp. 76–87, 2018, doi: 10.1016/j.future.2018.02.040.
9. O. Ali, M. K. Ishak, M. K. L. Bhatti, I. Khan, and K. Il Kim, "A comprehensive review of internet of things: technology stack, middlewares, and fog/edge computing interface," *Sensors*, vol. 22, no. 3. 2022. doi: 10.3390/s22030995.
10. M. Nikravan and M. Haghi Kashani, "A review on trust management in fog/edge computing: Techniques, trends, and challenges," *Journal of Network and Computer Applications*, vol. 204. 2022. doi: 10.1016/j.jnca.2022.103402.
11. A. Mishra, C. Singh, A. Dwivedi, D. Singh, and A. K. Biswal, "Network Forensics: An Approach Towards Detecting Cyber Crime," in *2021 International Conference in Advances in Power, Signal, and Information Technology, APSIT 2021*, Institute of Electrical and Electronics Engineers Inc., 2021. doi: 10.1109/APSIT52773.2021.9641399.
12. J. M. Pedersen and E. Kidmose, "Security in internet of things: Trends and challenges," in *CEUR Workshop Proceedings*, 2018, pp. 182–188.
13. J. Pan and J. McElhannon, "Future edge cloud and edge computing for internet of things applications," *IEEE Internet Things Journal*, vol. 5, no. 1, pp. 439–449, 2018, doi: 10.1109/JIOT.2017.2767608.
14. M. Aazam and E. N. Huh, "Fog computing: The Cloud-IoT/IoE Middleware Paradigm," *IEEE Potentials*, vol. 35, no. 3, pp. 40–44, 2016, doi: 10.1109/MPOT.2015.2456213.
15. A. Chakraborty, M. Kumar, N. Chaurasia, and S. S. Gill, "Journey from cloud of things to fog of things: Survey, new trends, and research directions," *Software: Practice and Experience*, vol. 53, no. 2, pp. 496–551, Feb. 2023, doi: 10.1002/spe.3157.

16. E. A. Lee, "The past, present and future of cyber-physical systems: A focus on models," *Sensors (Switzerland)*, vol. 15, no. 3. pp. 4837–4869, 2015. doi: 10.3390/s150304837.
17. P. Hu, H. Ning, T. Qiu, H. Song, Y. Wang, and X. Yao, "Security and privacy preservation scheme of face identification and resolution framework using fog computing in internet of things," *IEEE Internet Things Journal*, vol. 4, no. 5, pp. 1143–1155, Oct. 2017, doi: 10.1109/JIOT.2017.2659783.
18. E. Yigitoglu, M. Mohamed, L. Liu, and H. Ludwig, "Foggy: A framework for continuous automated iot application deployment in fog computing," *Proceedings – 2017 IEEE 6th International Conference on AI and Mobile Services, AIMS 2017*, pp. 38–45, Sep. 2017, doi: 10.1109/AIMS.2017.14.
19. P. Porambage, J. Okwuibe, M. Liyanage, M. Ylianttila, and T. Taleb, "Survey on multi-access edge computing for internet of things realization," *IEEE Communications Surveys and Tutorials*, vol. 20, no. 4, pp. 2961–2991, Oct. 2018, doi: 10.1109/COMST.2018.2849509.
20. R. H. Hsu, J. Lee, T. Q. S. Quek, and J. C. Chen, "Reconfigurable Security: Edge-Computing-Based Framework for IoT," *IEEE Network*, vol. 32, no. 5, pp. 92–99, Sep. 2018, doi: 10.1109/MNET.2018.1700284.
21. R. Casado-Vara, J. Prieto, F. De La Prieta, and J. M. Corchado, "Blockchain framework for IoT data quality via edge computing," *BlockSys 2018 – Proceedings of the 1st Blockchain-Enabled Networked Sensor Systems, Part of SenSys 2018*, pp. 19–24, Nov. 2018, doi: 10.1145/3282278.3282282.
22. B. Chen, J. Wan, A. Celesti, D. Li, H. Abbas, and Q. Zhang, "Edge Computing in IoT-Based Manufacturing," *IEEE Communications Magazine*, vol. 56, no. 9, pp. 103–109, 2018, doi: 10.1109/MCOM.2018.1701231.
23. E. Guardo, A. Di Stefano, A. La Corte, M. Sapienza, and M. Scatà, "A fog computing-based IoT framework for precision agriculture," *jit.ndhu.edu.tw*, Accessed: Jun. 17, 2023. Online.. Available: https://jit.ndhu.edu.tw/article/view/1760
24. M. Al-Khafajiy, L. Webster, T. Baker, and A. Waraich, "Towards fog driven IoT healthcare: Challenges and framework of fog computing in healthcare," *ACM International Conference Proceeding Series*, Jun. 2018, doi: 10.1145/3231053.3231062.
25. S. Tuli, R. Mahmud, S. Tuli, and R. Buyya, "FogBus: A Blockchain-based lightweight framework for edge and fog computing," *Journal of Systems and Software*, vol. 154, pp. 22–36, Aug. 2019, doi: 10.1016/J.JSS.2019.04.050.
26. A. H. Sodhro, S. Pirbhulal, and V. H. C. De Albuquerque, "Artificial intelligence-driven mechanism for edge computing-based industrial applications," *IEEE Transactions on Industrial Informatics*, vol. 15, no. 7, pp. 4235–4243, Jul. 2019, doi: 10.1109/TII.2019.2902878.
27. Y. Wang, J. Yang, X. Guo, and Z. Qu, "Satellite Edge Computing for the Internet of Things in Aerospace," *Sensors*, vol. 19, no. 20, p. 4375, Oct. 2019, doi: 10.3390/S19204375.
28. J. Li *et al.*, "A Secured framework for SDN-Based Edge Computing in IoT-Enabled Healthcare System," *IEEE Access*, vol. 8, pp. 135479–135490, 2020, doi: 10.1109/ACCESS.2020.3011503.
29. M. A. Lawal, R. A. Shaikh, and S. R. Hassan, "An Anomaly Mitigation Framework for IoT Using Fog Computing," *Electronics*, vol. 9, no. 10, p. 1565, Sep. 2020, doi: 10.3390/ELECTRONICS9101565.
30. D. N. Jha *et al.*, "IoTSim-Edge: A simulation framework for modeling the behavior of Internet of Things and edge computing environments," *Software: Practice and Experience*, vol. 50, no. 6, pp. 844–867, Jun. 2020, doi: 10.1002/SPE.2787.

31. D. Chawla and P. S. Mehra, “A roadmap from classical cryptography to post-quantum resistant cryptography for 5G-enabled IoT: Challenges, opportunities and solutions,” *Internet of Things*, vol. 24, p. 100950, Dec. 2023, doi: 10.1016/J.IOT.2023.100950.
32. M. Mukherjee et al., “Security and Privacy in Fog Computing: Challenges,” *IEEE Access*, vol. 5, pp. 19293–19304, 2017, doi: 10.1109/ACCESS.2017.2749422
33. C. Singh, H. Khajuria, and B. P. Nayak, “A Study of Implementing a Blockchain-Based Forensic Model Integration (BBFMI) for IoT Devices in Digital Forensics,” in *Advances in Computer Science for Engineering and Education VI*, Z. Hu, I. Dychka, and M. He, Eds., Cham: Springer Nature Switzerland, 2023, pp. 318–327.
34. Q. Yaseen, F. Albalas, Y. Jararweh, and M. Al-Ayyoub, “A Fog Computing Based System for Selective Forwarding Detection in Mobile Wireless Sensor Networks,” in *Proceedings – IEEE 1st International Workshops on Foundations and Applications of Self-Systems, FAS-W 2016*, 2016, pp. 256–262. doi: 10.1109/FAS-W.2016.60.
35. J. Sharma and P. S. Mehra, “Secure communication in IOT-based UAV networks: A systematic survey,” *Internet of Things*, vol. 23, p. 100883, Oct. 2023, doi: 10.1016/J.IOT.2023.100883.
36. N. Elsayed, Z. ElSayed, and M. Bayoumi, “IoT Botnet Detection Using an Economic Deep Learning Model,” Feb. 2023.
37. B. Omoniwa, *et al.*, “Fog/edge computing-based IoT (FECIoT): Architecture, applications, and research issues,” IEEE Internet of Things(IoT) vol.6, issue:3, doi: 10.1109/JIOT.2018.2875544.
38. K. Sha, T. A. Yang, W. Wei, and S. Davari, “A survey of edge computing-based designs for IoT security,” *Digital Communications and Networks*, vol. 6, no. 2, pp. 195–202, 2020, doi: 10.1016/j.dcan.2019.08.006.
39. S. Benhamaid, A. Bouabdallah, and H. Lakhlef, “Recent advances in energy management for Green-IoT: An up-to-date and comprehensive survey,” *Journal of Network and Computer Applications*, vol. 198. 2022. doi: 10.1016/j.jnca.2021.103257.
40. L. Haorongbam, R. Nagpal, and R. Sehgal, “Service Oriented Architecture (SOA): A Literature Review on the Maintainability, Approaches and Design Process,” in *Proceedings of the Confluence 2022 – 12th International Conference on Cloud Computing, Data Science and Engineering*, 2022, pp. 647–652. doi: 10.1109/Confluence52989.2022.9734153.
41. A. Čolaković and M. Hadžialić, “Internet of things (IoT): A review of enabling technologies, challenges, and open research issues,” *Computer Networks*, vol. 144. pp. 17–39, 2018. doi: 10.1016/j.comnet.2018.07.017.
42. S. Lenka, S. K. Pradhan, S. P. Nayak, and S. C. Nayak, “Forecasting Quality of Service (QoS) Parameters of IoT based web services with an artificial electric field algorithm trained artificial neural network,” *Karbala International Journal of Modern Science*, vol. 9, no. 1, 2023, doi: 10.33640/2405-609x.3282.
43. A. Safaei, R. Nassiri, and A. M. Rahmani, “Enterprise service composition models in IoT context: solutions comparison,” *Journal of Supercomputing*, vol. 78, no. 2, pp. 2015–2042, Feb. 2022, doi: 10.1007/s11227-021-03873-7.

17 Edge Computing-Based Design for IoT Security

Parita Jain, Swati Sharma, Sangeeta Arora, Vikrant Shokeen, Puneet Kumar Aggarwal, and Mandeep Singh

17.1 INTRODUCTION

By combining computing, storage, and network management tasks in a centralised way, cloud computing has become a popular method for businesses to achieve cost-effectiveness and flexibility. However, the centralised cloud computing architecture faces difficulties due to the growth of the Internet of Things (IoT) and mobile apps. When mobile devices connect to remote cloud servers to access sophisticated applications, they create additional loads on both radio access and backhaul networks, leading to high latency. The swift progress of IoT has resulted in fresh hurdles, including inflexible latency prerequisites, limitations in capacity, devices with restricted resources, the need for continuous services despite intermittent connectivity, and heightened security demands. These challenges are incompatible with the centralised cloud computing structure presently in use.

17.1.1 Motivation

There is a need to learn about edge computing-based architecture for IoT security due to IoT's growing importance and its particular security difficulties. Edge computing (EC) is essential for IoT deployments because it provides advantages, including decreased latency, more scalability, and improved data privacy. One may address security issues, safeguard sensitive user data, and allow real-time threat detection by comprehending this paradigm. EC enhances resource efficiency, guards against DDoS assaults, assures legal compliance, and creates job possibilities in the growing IoT sector. It also assures reliable data processing, improves resilience, allows for secure device management, and future-proofs IoT systems, making it a crucial topic to investigate for complete IoT security.

17.1.2 Contribution

The chapter explains how the growing adoption of IoT technology produces enormous amounts of data that need further processing and analysis. Traditional cloud

DOI: 10.1201/9781003405535-22

computing, however, has drawbacks, including high network bandwidth requirements and high resource expenses for data transmission. The chapter also highlights how traditional cloud-based services lead to limited bandwidth, resulting in significant latency within the networks. This latency issue is particularly important in time-sensitive IoT applications such as smart transportation, smart electricity grids, and smart cities where quick response times are critical, making it a crucial issue for IoT.

IoT systems have issues due to the rising volume of data produced by IoT devices, the demand for quick processing and analysis, security concerns, and other factors. EC addresses resource issues and enhances system functionality by shifting computation jobs to the network's edge. EC provides data processing and storage closer to the end user, lowering the volume, bandwidth, and latency of data uploads. Implementing relevant security solutions locally, it improves data security and privacy and lowers the danger of data leakage. EC helps IoT networks last longer by shifting computation from devices to edge nodes. There has been research on EC and suggested architectures for IoT security, but more analysis and complex edge-based security designs are required.

The remainder of this chapter is divided into the following sections: an overview of EC, outlining its meaning and contrasting it with other computing models, such as cloud computing, cloudlets, fog computing, and MEC is provided initially. In Section 17.2, various related concepts are compared, and EC and cloud computing are discussed. The categorisation of the security and privacy risks associated with EC for the IoT is discussed in Section 17.3, and possible remedies and preventative measures in Section 17.4 are also discussed. Additionally, a thorough examination of the security and privacy concerns for edge-enabled IoT is provided in this section. In Section 17.5, the difficulties of an edge computing-based architecture for IoT infrastructure are discussed, and in Section 17.6, some case studies to demonstrate the idea are summarised. Finally, Section 17.7 concludes this chapter and includes a list of references.

17.2 COMPUTING PARADIGMS ASSOCIATED WITH EDGE COMPUTING

The section on EC in this chapter discusses different EC technologies as well as related computing paradigms. We compare various EC technology varieties as well as EC and associated computing paradigms in this presentation.

17.2.1 Cloud Computing

Cloud service providers like Google, IBM, Microsoft, and Amazon supply robust computing, storage, and networking infrastructure through virtualisation techniques. It offers a pay-as-you-go pricing structure that gives customers access to remote computing capacity and data management services at a reasonable price. However, some essential applications require extremely low latency, which cloud-based apps cannot provide. The proliferation of IoT devices makes it necessary to install cloud resources close to the source. EC, which allows computer models to operate close to connected devices, has emerged as a potential solution to address the growing need for high-bandwidth, low-latency, widely distributed, and confidential data.

17.2.2 Mobile Computing

The expansion of mobile devices has hindered the progression of cloud computing. With the help of the distributed computing model, mobile computing is utilised for location-based reminders and remote operation of mobile machinery. However, it encounters obstacles such as insufficient resources, communication latency, and balancing autonomy and interdependence. Fog and cloud computing have expanded the scope and accessibility of mobile computing, but they require more potent resources and virtualisation capabilities. The resource disparity between mobile computing, fog, and cloud computing has lessened with recent advancements in wireless protocols and mobile technology. However, precautions must be taken to guarantee safety when using a mobile device.

17.2.3 Mobile Cloud Computing

Mobile cloud computing (MCC) enables the communication between cloud service providers and mobile device users, allowing even low-end devices to access the resources of the cloud. MCC helps to conserve mobile device battery life and run computational-intensive applications by processing and storing data in the cloud. MCC uses cloud computing and mobile technology's flexibility and convenience to provide highly computational services. However, MCC has some drawbacks, including the centralised design not meshing well with some decentralised apps, connectivity difficulties due to constant Internet connectivity needs, and latency issues, making it unsuitable for certain applications. Security concerns for mobile and cloud devices must also be considered.

17.2.4 Edge Computing

EC has become crucial for managing, storing, and processing data generated by connected devices, particularly IoT. Unlike MCC, EC is located at the periphery of networks close to IoT devices. It is not an element of the network but can be placed several hops away from it. EC aims to have data and processing power next to the user. This computing model allows cloud centres close to the devices to filter, preprocess, and aggregate data from IoT gadgets, making it an essential computing model for IoT devices.

EC offers several benefits to the IoT, including millisecond processing rates that enhance system performance by reducing latency, conserving resources, and minimising strain on network connections. Cloud platform service providers offer centralised data security solutions, but the consequences are severe when such systems are breached. EC offers a more secure strategy by enabling the local application of security measures, decreasing the possibility of data loss during transmission and reducing the amount of data kept in the cloud platform. This approach is critical for IoT devices that gather private or confidential data, and EC permits anonymisation and pre-processing of sensitive data before transmitting it to the cloud for additional analysis. The various technologies associated with EC include:

17.2.4.1 Cloudlet Computing

A concept first proposed by Carnegie Mellon University in 2009 brings data storage and processing closer to user equipment (UE). Cloudlets, or mini data centres, are strategically placed near-edge devices to offer computing resources and storage for nearby UEs. Cloudlets aim to move processing from mobile devices to local edge-of-network cloudlets that use virtual machines. Although cloudlets have less processing capacity than traditional clouds, they use less power and have lower latency. Cloudlets' dependence on Wi-Fi connections and the need for mobile UEs to toggle between Wi-Fi and the mobile network is a significant drawback.

17.2.4.2 Mobile Ad Hoc Cloud Computing (MACC)

It is an alternative to cloudlet computing for enabling cloud computing at the edge. MACC allows nearby mobile devices to pool their computing resources and run computationally intensive applications locally. Unlike cloudlets, MACC is highly dynamic and lacks infrastructure or a centralised cloud, relying on portable electronic devices for data storage, processing, and distribution. While MACC can carry out intensive computing, it does not support real-time IoT applications due to limited resources. MACC is based on the same concept as a mobile ad hoc network (MANET) and requires visitors to bring their mobile devices to use its facilities.

17.2.4.3 Fog Computing

A form of distributed computing paradigm called fog computing uses intermediary nodes to speed up data transfer and analysis between endpoints and the cloud. It makes it possible for computing, storage, networking, decision-making, and data administration to occur near users or IoT devices that need data processing, reducing latency and increasing availability. Fog computing could benefit from various industries, including energy, manufacturing, transportation, healthcare, and smart cities. In contrast to EC, Fog computing uses a hierarchical structure to support cloud-to-things processing, networking, storage, control, and acceleration throughout the full continuum.

17.3 THREATS AND ATTACKS ON SECURITY AND PRIVACY IN IOT BASED ON EDGE COMPUTING (EC)

Based on EC, IoT networks are open to various security and privacy threats. On EC devices, communication and EC servers/nodes, and cloud servers, attackers can submit malicious inputs that result in unauthorised software or hardware installations. These assaults may lead to database integrity disclosure, false data reporting, authentication bypass, and data theft [1–3]. Attackers frequently use hardware injection techniques, such as network and integrated circuit attacks, to obtain sensitive data and tamper with data packets. Through packet insertion, manipulation, or replication, they can also use vulnerable or malicious EC nodes to take over the network and replace or stop the flow of legitimate data packets.

Side-channel information refers to publicly available data that is not sensitive to privacy, and side-channel attacks threaten user security and privacy [4, 5]. However, this easily accessible data is often tied to secrets that shouldn't be shared. Using algorithms and machine learning models, attackers can probe the latent correlations in the side-channel data and glean the necessary secret details. EC's three most prevalent side channels are communication signals, electricity utilisation, smartphone/proc filesystems, and embedded sensors.

17.3.1 Authentication and Authorisation Attacks

The different types of authorisation protocol attacks in EC can be classified into four categories [6, 7], as shown in Figure 17.1. The first is a dictionary attack, the second is an authentication bypass attack, the third is exploiting weaknesses in protocol design or logic, and the fourth is overprivileged attacks.

The most used authentication protocols in EC are WPA/WPA2 and SSL/TLS, while the OAuth standard is the favourite for open authorisation. One type of Denial of Service(DoS) assault is the jamming attack (DoS). To exhaust the network's processing, storage, and transmission capacity, attackers will deliberately overwhelm it with false data. This attack will render the EC-based IoT network's infrastructure inaccessible to legitimate users [8–10].

DDoS assaults on EC nodes might result in disruptions, lack of sleep, and battery waste [11, 12]. They may be based on floods or zero-day exploits and cause network flooding, service shutdown, or memory failure. Physical access to nodes permits physical assaults and tampering, jeopardising the security of the network and cryptographic data. Attacks that intercept sensitive data jeopardise network security [13–15].

Even when EC nodes are idle, non-network side-channel attacks can reveal private information. Through recognisable electromagnetic/acoustic signals or protocols, medical equipment can reveal patient and device information. Message routing is disrupted by routing manipulation attacks such as modifying, blackhole, grey hole,

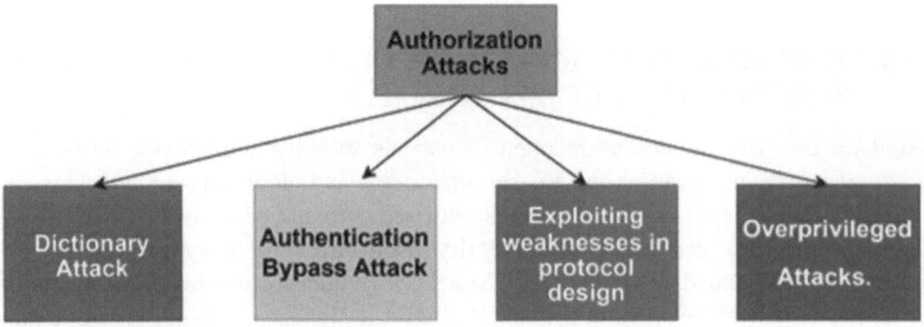

FIGURE 17.1 Types of authorisation attacks.

wormhole, hello flood, and Sybil attacks. Replay attacks capture and replay data, increasing the amount of data and energy consumed by EC nodes [16–18].

17.4 VARIOUS SOLUTIONS AND COUNTERMEASURES TO ADDRESS THE SECURITY AND PRIVACY ISSUES IN EC-BASED IOT

Different strategies and solutions have been developed to address the security and privacy threats in IoT systems with EC. They include packet filters and access control lists (ACLs) to prevent attacks and manage network traffic. Techniques for comparing circuits and analysing side-channel signals aid in the identification and characterisation of malicious firmware or hardware attacks. Network operations are monitored by intrusion detection systems (IDS), which look for abnormal activity. Communication protocols are protected by cryptographic methods, including attribute-based encryption and homomorphic encryption. While earlier testing and safe data aggregation aid in simulating attacks and protecting privacy, accounting and security evaluation are crucial network security practices. EC, combining blockchain technology and decentralised artificial intelligence (AI) on edge devices boosts security, anonymity, and data-sharing capabilities for IoT systems. The following Table 17.1 lists the key strategies and solutions for tackling security and privacy threats in edge computing-based IoT systems:

TABLE 17.1
Key Strategies and Solutions for Tackling Security and Privacy Threats

Approaches/Solution	Description
Packet filters	Protect your network resources from theft, unauthorised access, DoS assaults, and destruction. There are two widely used policies: one that accepts all packet types while rejecting some, and the other that accepts some while rejecting all.
Access Control Lists (ACL)	Firewalls that are equipped and used as routers, hardware, or software. Static stateless packet filter firewalls made for speed and simple configuration are one sort of firewall, as are proxy firewalls that scan packets, permit stateful session tracking, and filter out hazardous traffic or inappropriate material. Stateful firewalls monitor conversational exchanges and decide whether to allow or block traffic.
Side-Channel Signal Analysis	A technique for identifying malicious firmware, software Trojans, or hardware worms on edge computing (EC) platforms or devices. To find anomalous activity, such as sudden increases in heat, execution time, or power consumption, it analyses timing, power, and spatial temperature.

(*continued*)

TABLE 17.1 (Continued)
Key Strategies and Solutions for Tackling Security and Privacy Threats

Approaches/Solution	Description
Intrusion Detection System (IDS)	In edge computing, a secondary defence against security risks, Monitoring network activity, reporting unusual activity, and spotting routing and blackhole attacks. The architecture features real-time traffic monitoring, hierarchical distributed intrusion detection, efficient and equitable resource allocation, and trust evaluation.
Cryptographic Techniques	Defend communication protocols from assaults. It has been investigated to use key cryptosystems, including attribute-based encryption, proxy encryption, and homomorphic encryption. On the basis of the data proxy concept, secure data-sharing systems and architectures have been presented.
Accounting and Security Assessment	Effective network security measures. To analyse security events, accounting collects data on network activity. In contrast, security assessment looks for network weaknesses and makes plans to fix them.
Prior testing on EC network components	It is essential to model attacks and gauge their effects. Sensitive information needs to be tracked and logged.
Secure Data Aggregation	Data compression that maintains security and privacy. To conserve IoT network capacity, homomorphic encryption is utilised to encrypt individual data before it is aggregated and computed before being sent to cloud servers. Without revealing private data, secure data deduplication can identify duplicated data.
Decentralized AI Features	The use of decentralised AI features on EC devices/nodes rather than a centralised cloud is now possible because of advancements in EC technology. It enhances latency, security, and anonymity; gives users the option to keep their original data and exclusive training sets, which ensures data privacy while enabling local and autonomous model training on edge nodes and devices.
Combination of edge computing and blockchain technology	A solution for secure data sharing in the IoT is provided by combining edge computing and blockchain technology. In applications like smart grids and vehicle edge computing networks, consensus techniques and smart contracts offer anonymity and security.

17.5 THE DIFFICULTIES AND OBSTACLES THAT COME WITH IMPLEMENTING EDGE COMPUTING IN IOT APPLICATIONS

While edge computing-based IoT has many benefits, there are a few issues that need to be resolved as well:

i. Data Security: Because EC devices process data locally, there is a higher potential for data breaches and illegal access. Moreover, the gadgets must be shielded from physical tampering.

ii. Scalability: It gets harder to manage and grow EC resources as the number of IoT devices rises. It can be expensive to scale the infrastructure for EC.
iii. Interoperability: It may be challenging to integrate devices and data sources if EC devices from different suppliers use various protocols and data formats.
iv. Latency: Although local data processing by EC reduces latency, it might still be an issue in some applications where real-time response is crucial.
v. Restricted Resources: The processing power, memory, and storage of EC devices are often constrained, which might hinder their capacity to process large volumes of data or carry out complicated calculations.
vi. Upkeep: Because EC devices are dispersed across numerous sites, maintenance and updates might be difficult to carry out.
vii. Cost: The infrastructure for EC can be pricey, especially if it calls for specialist hardware and software.

Careful planning, infrastructure investment, continuing maintenance, and monitoring are necessary to address these issues. The advantages of EC must be weighed against the difficulties in deploying and managing it, and this is something that organisations need to do.

17.6 CASE STUDIES

IoT has made it possible for different products and services to connect to the Internet, creating new opportunities for both businesses and people. Yet, the growth of IoT has rapidly increased the number of connected devices, raising the risk of cyberattacks [19, 20]. This section will examine how EC might improve IoT security in a case study. It is found that nowadays, various devices are used connected to the Internet, which is exposed to the world without security. In IoT devices, data is generated continuously which will be transferred to servers securely. In this situation, a gateway device that serves as a bridge between IoT devices and the cloud server, processing data locally before transmitting pertinent data to minimise cyberattack risks. It has security components like access control and encryption. This strategy also provides several additional advantages, such as:

i. Reduce Latency: Since the data processing is done locally, there is less latency, which leads to quicker reaction times.
ii. Improved Reliability: The devices will continue to operate locally even if the Internet connection is lost because the gateway device is in charge of processing data.
iii. Reduced Bandwidth Usage: Only pertinent information is sent to the cloud server, lowering bandwidth utilisation and costs.

EC CAN BE USED IN VARIOUS FIELDS, WHICH ARE AS FOLLOWS, FOCUSING ON SECURITY:

i. Home Automation:
Introduction: Home automation is the process of automating and controlling many components of a home so that its residents can live more comfortably,

conveniently, and effectively. It involves integrating various appliances and home systems so they may talk and interact with one another.

Case: Home automation often uses a hub or central control system to link and control a variety of smart systems and gadgets. Smart appliances, lighting controls, security cameras, door locks, entertainment systems, and thermostats are a few examples of these gadgets. A smartphone app, voice commands, or a centralised control panel can all be used to access and manage the control system.

Objective: To build a more responsive, dependable, and secure home automation environment, EC for IoT devices is used in home automation. It makes real-time automation possible, enhances privacy and security, increases network capacity, and offers scalability and flexibility for future growth. The ultimate objective is to improve the effectiveness and overall user experience of home automation systems.

Analysis: A gateway device is essential for enabling connectivity and communication between IoT devices and the central control system in home automation using EC for IoT. The gateway device functions as a hub to link and integrate different IoT devices into the home automation system. It supports various communication protocols, including Wi-Fi, Zigbee, Z-Wave, and Bluetooth, enabling interaction and communication between various devices. At the network's edge, the gateway device does local data processing and analysis. Before sending the raw data produced by IoT devices to the central control system or cloud platform, it can preprocess and filter it. Local data processing improves real-time automation capabilities while lowering latency and optimising network bandwidth.

The gateway device may include analytics capabilities, allowing it to perform local data analysis and produce useful insights. This frees up the home automation system from having to constantly communicate with a remote server in order to make decisions in real time based on predetermined rules or machine learning algorithms. The gateway device improves home automation systems' security and privacy. To safeguard sensitive data produced by IoT devices, it can apply security features, including encryption, access controls, and authentication procedures. The gateway device lessens the danger of data exposure to external networks or cloud platforms by performing data processing and analysis locally. In conclusion, the gateway device is essential to IoT EC for home automation. IoT device integration and connectivity, local data processing and analysis, secure connectivity, increased security and privacy, local control and automation, and easy interface with the central control system are all made possible by this technology.

ii. **Healthcare**:

Introduction: Automation, which is the use of technology and automated systems to streamline and optimise various processes and duties within the healthcare industry, is needed in the area of healthcare. Robotics, AI, machine learning, and data analytics are being used increasingly to automate time-consuming or repetitive processes, boost operational effectiveness, improve patient care, and decrease human error rates. Hospitals, clinics, labs, pharmacies, and even in-home healthcare settings are just a few of the places where healthcare automation can be used.

Case: Automation in the healthcare sector must cover various areas, including health records, appointment scheduling, medical billing, and medical imaging, among others. In the healthcare sector, data security and authorised employee access are critical necessities. The healthcare industry has several prospects for IoT security based on EC, particularly in terms of safeguarding and reliably allowing access to patient data and medical records. Security is a major issue that requires attention in this scenario involving the use of IoT devices in the healthcare industry.

Objective: To build a more responsive, dependable, and secure home automation environment, EC for IoT devices is used in home automation. It makes real-time automation possible, enhances privacy and security, increases network capacity, and offers scalability and flexibility for future growth. The ultimate objective is to improve the effectiveness and overall user experience of home automation systems.

Analysis: The healthcare industry has several prospects for IoT security based on EC, particularly in terms of safeguarding and reliably allowing access to patient data and medical records.

A gateway device is essential to the healthcare industry's ability to integrate and communicate with multiple IoT systems and devices while utilising EC capabilities. The gateway device serves as a focal point for integrating and linking various IoT devices into the healthcare ecosystem. It supports various communication protocols and standards, enabling the connection between gadgets, including wearables, monitoring equipment, and smart medical equipment. The gateway device is also used to gather data in real time from various IoT systems, sensors, and devices. By filtering and preparing data at the edge, it minimises the quantity of information that must be sent to the cloud or central servers. This lessens latency while maximising network bandwidth.

Edge analytics features on the gateway device enable real-time data analysis and decision-making at the point of care. For the purpose of gaining knowledge, spotting irregularities, or spotting trends in patient health data, it can run algorithms and carry out local data analytics. This makes timely interventions possible and enhances the results of patient care. In healthcare systems adopting EC for IoT, the gateway device is essential for providing secure connectivity, data integration, and localised processing. It improves data privacy, allows real-time analytics, enables local decision-making, and makes it easier for devices and healthcare applications to communicate seamlessly. In the end, the gateway device aids in enhancing operational efficiency, streamlining workflows, and patient care in the healthcare sector. Secure data interchange in healthcare, including remote monitoring of chronic conditions, is made possible by EC and IoT security. Healthcare professionals may securely access IoT devices to track patient data, process it using EC, and make educated decisions.

17.7 CONCLUSION

This chapter comprehensively analyses EC and associated paradigms, highlighting their parallels and divergences. It also examines the privacy and security risks associated with EC-based IoT and offers workable solutions. The chapter also includes case studies demonstrating how IoT systems are used in regular activities. In one of the case studies, it is shown how remote monitoring of patients with chronic diseases may be done using IoT devices, EC, and secure access by healthcare experts. Giving readers a solid understanding of EC and how it might be applied to address IoT security and privacy issues is the chapter's main objective.

REFERENCES

1. A. Alwarafy, K. A. Al-Thelaya, M. Abdallah, J. Schneider, M. Hamdi, "A survey on security and privacy issues in edge-computing assisted internet of Things", *IEEE Internet Things Journal*, Vol. 8, pp. 4004–4022, 2020.
2. A. Mosenia, N. K. Jha, "A comprehensive study of security of Internet-of-Things", *IEEE Transactions on Emerging Topics in Computing*, Vol. 5, pp. 586–602, 2022.
3. D. Liu, Z. Yan, W. Ding, M. Atiquzzaman, "A survey on secure data analytics in edge computing", *IEEE Internet Things Journal*, Vol. 6, pp. 4946–4967, 2019.
4. P. K. Aggarwal, P.S. Grover, and L. Ahuja, "Security Aspect in Instant Mobile Messaging Applications," In *Proceedings of IEEE International Conference on Recent Advances on Engineering, Technology and Computational Sciences (RAETCS)*, pp.1–5, 2018.
5. M. Keerthika, D. Shanmugapriya, "Wireless sensor networks: Active and passive attacks-vulnerabilities and countermeasures", Global Transitions Proceedings, Vol. 2, pp. 362–367, 2021.
6. D. He, S. Chan, M. Guizani, "Security in the internet of things supported by mobile edge computing", *IEEE Communications Magazine*, Vol. 56, pp. 56–61, 2018.
7. S. Goel, Monica, H. Khurana, P. Jain, "Social media analysis: A tool for popularity prediction using machine learning classifiers", *Design and Applications of Nature Inspired Optimization*, Women in Engineering and Science, edited by Dipti Singh, Vanita Garg, Kusum Deep, pp. 179–197, Springer, 2023.
8. Parita Jain, Puneet Kumar Aggarwal, Poorvi Chaudhary, Kshirja Makar, Jaya Mehta, and Riya Garg, "Convergence of IoT and CPS in Robotics", *Emergence of Cyber Physical Systems and IoT in Smart Automation and Robotics*, edited by Krishna Kant Singh, Anand Nayyar, Sudeep Tanwar, Mohamed Abouhawwash, pp.15–30, Springer, 2021.
9. Puneet Kumar Aggarwal, Parita Jain, Jaya Mehta, Riya Garg, Kshirja Makar, and Poorvi Chaudhary, "Machine learning, data mining and big data analytics for 5G-Enabled IoT", *Blockchain for 5G enabled IoT: the new wave for Industrial Automation*, edited by Sudeep Tanwar, pp. 351–375, Springer, 2021.
10. Y. Lu, L. D. Xu, "Internet of Things (IoT) cybersecurity research: A review of current research topics", *IEEE Internet Things Journal*, Vol. 6, pp. 2103–2115, 2019.
11. Puneet Kumar Aggarwal, Parita Jain, Poorvi Chaudhary, Riya Garg, Kshirja Makar, Jaya Mehta, "AIIoT for development of test standards for agricultural technology", *Intelligence of Things: AI-IoT Based Critical-Applications and Innovations*, edited by Fadi Al-Turjman, Anand Nayyar, Ajantha Devi, Piyush Kumar Shukla, pp. 77–99, Springer, 2021.

12. Y. Xiao, Y. Jia, C. Liu, X. Cheng, J. Yu, W. Lv, "Edge computing security: State of the art and challenges", *Proceedings of IEEE*, Vol.107, pp. 1608–1631, 2019.
13. M. Singh, N. Sukhija, A. Sharma, M. Gupta, P. K. Aggarwal, "Security and Privacy Requirements for IoMT-Based Smart Healthcare system", *Big Data Analysis for Green Computing*, edited by Rohit Sharma, Dilip Kumar Sharma, Dhowmya Bhatt, Binh Thai Pham, pp. 17–37, Taylor & Francis, 2021.
14. V. K. Reshma, Ihtiram Raza Khan, M. Niranjanamurthy, Puneet Kumar Aggarwal, S. Hemalatha, Khalid K. Almuzaini, E. Tetteh Amoatey, "Hybrid block-based lightweight machine learning-based predictive models for quality preserving in the internet of things- (IoT-) based medical images with diagnostic applications", *Computational Intelligence and Neuroscience*, Vol. 2022, pp. 1–14, 2022.
15. Parita Jain, Puneet Kumar Aggarwal, Kshirja Makar, Riya Garg, Jaya Mehta, Poorvi Chaudhary, "Machine Learning in Risk Analysis", *Applications of Computational Science in Artificial Intelligence*, edited by Anand Nayyar, Sandeep Kumar, Akshat Agrawal, pp. 190–213, IGI Global, 2022.
16. Anupam Sharma, Mandeep Singh, Megha Gupta, Namrata Sukhija, Puneet Kumar Aggarwal, "IoT and blockchain technology in 5G Smart Healthcare", *Blockchain Applications for Healthcare Informatics*, edited by Sudeep Tanwar, pp. 137–161, Elsevier, 2022.
17. S. Yi, Z. Qin, Q. Li, "Security and privacy issues of fog computing: A survey", In *Proceedings of the International Conference on Wireless Algorithms, Systems, and Applications*, Qufu, China, pp. 685–695, 2015.
18. Elahe Fazeldehkordi and Tor-Morten Grønli, "Review a survey of security architectures for edge computing-based IoT", *IoT*, Vol. 3, pp. 332–365, 2022.
19. OpenEdgeConsortium, About—The Who, What, and How. Technical Report, OpenEdge Computing. Available online: https://openedgecomputing.org/index.html.
20. Zhang, T. Fog Computing Brings New Business Opportunities and Disruptions. IoT-Agenda/Fog-computing-brings-newbusiness-opportunities-and-disruptions, Blog, TechTarget. Available online: http://internetofthingsagenda.techtarget.com/blog/

18 Wi-Fi Jammer, a Threat to IoT Networks
Experimental Perspective

Saumya Bansal, Sanya Verma, Anant Raghav, Harsh Aggarwal, and Yash

18.1 INTRODUCTION

A Wi-Fi jammer is a device that interferes with transmissions to prevent communication between devices and routers [1]. It sends deauthentication packets to the router, resulting in the connection loss between the router and connected devices. Jammers are undetectable, causing users to encounter side effects like poor signal coverage. Wi-Fi jammer is a device that can disconnect all devices connected to the Wi-Fi within a particular area for a specific time. Wi-Fi jammers can be seen as a threat to IoT devices. Hackers can easily use them to interfere with the connection between the user and the Internet, accessing the user's personal data and information [2]. For this reason, using jammers has been limited to certain sections of society. Jammers are also banned in some countries, and safety measures are taken.

The application of jamming devices was first seen in the armed forces and military-technical departments [3]. This fundamental goal was to prevent the successful transmission of information from the receiver to the sender and vice versa. Wi-Fi jammers have been used by organizations to prevent the leak of official data [3]. However, the high cost of Wi-Fi jammers makes it difficult for some organizations and individuals to buy and use them. Many countries have even banned the production, sale, and use of jammers as they are also perceived as a threat to civilians for leaking private data. One of the repercussions of jammers is that they can refrain first responders or civilians from calling for help.

A plethora of data transmissions occur daily, and they need to be secured from security breaches such as data theft. The motivation behind this work is to prevent the leakage of official data at places such as the armed forces.

Advanced jammers can disable specific channels while leaving the frequencies used for communications unaffected in case of emergency [4]. The main aim of this work is to build a device with facilities that are smaller, portable, and cheaper than existing jammers. The price of our Wi-Fi jammer with basic functionalities would be

DOI: 10.1201/9781003405535-23

around Rs 1,200–1,500/unit. However, India's existing Wi-Fi jammer market price is Rs 12,500/unit. With this device, a user can perform attacks such as deauthentication, Beacon, and clonebeacon attacks. A user interface has also been developed, which allows an attacker to make his own SS_Id and password to implement attacks. Our device will allow us to perform several attacks on the Wi-Fi within a particular area through a local host interface (198.168.4.1).
The following are major contributions:

1. Portable: The device is very small in size and thus easy to carry.
2. Cheaper: As compared to jammers nowadays, this device is built at a cheaper cost.
3. Booster Module: This device has a booster module that will convert 3.3V–5V.

The chapter is organized as follows: The applications are detailed in Section 18.2. Various types of attacks have been discussed in Section 18.3. Section 18.4 throws light on the system requirements required to make this device. The configuration is discussed in Section 18.5. Algorithms to mount different attacks have been described in Section 18.6. Section 18.6 contains the implementation. The conclusion and future work are discussed in Section 18.7.

1. Applications of Wi-Fi jammer

The following are the applications of Wi-Fi jammer [5]:

a. *Confidential Meetings*
Wi-Fi jammers can help an organization keep its information safe. This device blocks phones and other devices from sharing audiovisual information over the Internet, maintaining privacy within an organization.

b. *Stop Distractions*
Distraction is the biggest enemy during work. Wi-Fi jammers can stop any Internet distractions and thus can provide an uninterrupted working environment. Furthermore, if a student is conducting research in artificial intelligence, blocking Wi-Fi keeps students away from distractions by the Internet during their studies.

c. *Prevent Hacking*
Many hackers choose to attack the data overnight so that no one becomes aware of the breach until hours later. To combat this, turning on the Wi-Fi jammer at night and any other time someone is not using the Internet will stop devices from connecting to the router. As a result, no one can breach the network and access the data stored on one's devices.

Hackers usually prefer to attack at night so that no one is aware of the breach for a long time. To overcome this, using a Wi-Fi jammer at times when the Internet usage is minimum will prevent hackers from connecting to the router. Thus, hackers cannot access the data by breaking into the system.

d. *Disconnect Unwanted Devices*
Surveillance equipment such as microphones and cameras are frequently found hidden in changing rooms, hotel rooms, and Airbnb rentals. Although these devices are usually utilized for security, they are also sometimes used to breach privacy by recording you. A Wi-Fi jammer can prevent surveillance equipment from transmitting recorded clips over the Internet.

e. Practical Demonstration in Rural Areas
Our device is cheaper than the existing Wi-Fi jammers. This device can be used in rural schools/colleges to demonstrate network breaching and patch up solutions to resolve issues. Basically, it can be used to teach children about working practices in a cheaper way.

18.2 LITERATURE REVIEW

Wi-Fi jammers are devices that disrupt or block wireless communication signals, including Wi-Fi networks. They operate by emitting radio frequency signals in the same frequency band as the Wi-Fi signal, causing interference and preventing devices within range from connecting to the network. In this section, Table 18.1 discusses the key findings of work done in the literature and its limitations.

TABLE 18.1
Description of Related Work

Author	Description	Key Findings	Limitations
Ingham et al. (2020) [1]	• The security vulnerability of IoT devices is explored. • A predictive model of the generation model is used to carry out the jamming attack.	Understanding the airtime and uptime of a device helps to increase the chances of blocking packet delivery of another device.	Changes in time and duration of slot occupancy with varying payload sizes increase the complexity of teaching accurate prediction.
Liu et al. (2015) [2]	A real-time jamming detection method based on the received signal strength indicator and the packet loss rate of Wi-Fi signals has been proposed.	Detected jamming attacks in real-time, with a miss detection rate of less than 2%	Dependency on the OS

TABLE 18.1 (Continued)
Description of Related Work

Author	Description	Key Findings	Limitations
Sharma and Bhatt (2018) [3]	Various jamming attacks have been discussed depending on the physical or MAC layers. Various correctives for mitigating these attacks have been briefed.	Jamming attacks are considered the origin of DoS.	Anti-jamming techniques have not been discussed.
Piyaresh and Zeng (2022) [4]	A comprehensive review of jamming and anti-jamming strategies and design of jamming-resilient wireless networking systems have been explained.	Wi-Fi, cognitive radio, cellular, Bluetooth, ZigBee, GPS, and radio frequency identification (RFID) networks have been covered	More jamming attacks should be covered
Vadlamani et al. (2016) [6]	Research areas related to attacks, defense strategies, and other fields have been explored.	Taxonomic classification of the attacks has been listed.	Anti-jamming techniques are not covered
Bhoyar and Yadav (2017) [7]	Game theory has been used to detect jamming attacks.	Perfect awareness by the jammer has been focused on implementing an interaction model between the jammer and sensor node	Static and dynamic approaches can be explored to detect jammer
Singh et al. (2022) [8]	A novel jamming attack detection technique (JADT) for OppNets is proposed.	Uses statistical measures instead of cryptographic algorithms	Multiple simulators can be used for evaluation
Chatzisofrpniou and Kotzanikolaou (2021) [9]	Association attacks in IEEE 802.11 and the taxonomy to classify them have been discussed.	A well-known Wifiphisher tool has been used to analyze association attacks.	Other threats are still active when dealing with association attacks
Aggrawal et al. (2023) [10]	Techniques for rendering DoS attacks have been presented.	Focuses on breaking AP between user and Internet access point.	Focus only on deauth attack

(continued)

TABLE 18.1 (Continued)
Description of Related Work

Author	Description	Key Findings	Limitations
Singh et al. (2002) [11]	Concept of cybersecurity and electronics have been integrated in Intel device.	Multiple purposes such as data cloning and jamming	Limited to deauth and cloning attacks
Chan and Chun (2021) [12]	Attacks particular to beacon systems are explored.	Beacons have been focused	Accurate outlier detection can be achieved
Kondaveeti (2021) [13]	Prototyping with Arduino has been analyzed and compared.	Best techniques while prototyping with Arduino have been listed.	More conventional techniques could have been identified
Najath et al. (2022) [14]	Low-cost network jammer that creates frequency between 315 MHz and 433 MHz has been described.	Illegal activities using DoS are prevented	Applicable to area only within the 100-meter range
Pahlavan and Krishnamurthy (2021) [5]	The study provides an overview of the evolution of Wi-Fi technology and its applications that has been experienced in the last few decades.	The role of Wi-Fi in our daily lives has been discussed. Moreover, how the IEEE 802.11 standards for Wi-Fi had been first implemented and radiofrequency (RF) cloud enables some important cyberspace applications.	The features should be same between two devices which might not be possible as the electronic implementation of a device is unique which results in the differences.

18.3 JAMMING TECHNIQUES

When a transmitter detects a faulty or busy signal at a receiver, a Wi-Fi jammer uses radio frequencies to make the communication medium busy by interfering with wireless communications [4]. Attacks on the physical layer are the ones that are typically targeted by jamming; however, cross-layer attacks can also occur. In this section, we have elaborated on various types of jammers and the placement of jammers to maximize the jammed area.

18.3.1 Types of Attacks

Attacks mounted through Wi-Fi jammers come under Denial-of-Service (DoS) attacks. Under these attacks, malicious nodes interfere with legitimate network communication [6–8].

18.3.1.1 Deauthentication Attack (Deauth)

Denial-of-service attacks, such as a deauth attack, target user-to-wireless access point communication. In this attack, the attacker force devices to lose access and then reconnect to a network controlled by them [9, 10].

In other words, it is a type of DoS attack interfering with communication between routers and devices. One of the main purposes of deauthentication is to force the user already connected to the router or wireless access point to connect to an evil twin access point which then can be used for capturing network packets transferred between the user and the router.

18.3.1.2 Clone Beacon

A clone beacon attack is a spoofing cyberattack that tricks the user to connect to a fake Wi-Fi access point that appears to be part of a legitimate network [11]. Once the user gets connected to a "clone beacon" network, hackers get access to information from network traffic, including login credentials. A clone beacon attack is a kind of wireless attack in which an attacker creates a replica or clone of a legitimate connection to deceive devices into connecting to a fake access point. Once connected, the attacker can eavesdrop on the traffic passing between the victim's device and the Internet, or launch further attacks against the victim, such as a man-in-the-middle attack or a phishing attack [12].

18.3.1.3 Beacon

The purpose of this attack is to confuse wireless clients by sending several fake beacon frames. This makes things difficult to find a genuine network, and this prevents the wireless clients from connecting to their preferred networks [11]. A beacon is a signal periodically transmitted by wireless access points to announce the availability of a network. In a beacon attack, the attacker creates a fake beacon with the same network name (SSID) and other network parameters as the legitimate access point [12]. The fake beacon is transmitted on the same frequency and with the same power as the legitimate beacon, so devices within range may be deceived into connecting to the fake access point instead of the legitimate one.

18.3.2 Mounting an Attack

This subsection presents the working of Wi-Fi jammer in a step-wise fashion:

Step 1: Initially, a user and the router will maintain a connection between them. The user will first send the request for authentication and association.

Step 2: Successful authentication refers to successfully completing the authentication process between a user and a Wi-Fi access point, which is router here.

When a device wants to connect to a Wi-Fi network, it must go through an authentication and encryption process to establish a secure connection. The authentication process typically involves exchanging handshake messages between the client and the access point to verify their identities and establish a secure connection.

Step 3: A successful association involves establishing a connection between a user and a Wi-Fi access point by sending an association request containing the client's information.

When a Wi-Fi client device wants to join a Wi-Fi network, it needs to perform an association process. During this process, depending on the client's capabilities and supported encryption methods. The access point then responds with an association response, indicating whether the association was successful or not.

Step 4: A disassociation notification or deauthentication notification refers to a message sent by the access point to a connected client device, instructing it to disconnect from the Wi-Fi network.

Deauthentication is a legitimate mechanism in Wi-Fi networks and is used by access points to terminate the connection of a client device. It can be initiated for various reasons, including security purposes, network management, or to enforce access control policies. The objective is to disrupt or block the wireless communication between the client and the access point, preventing successful authentication.

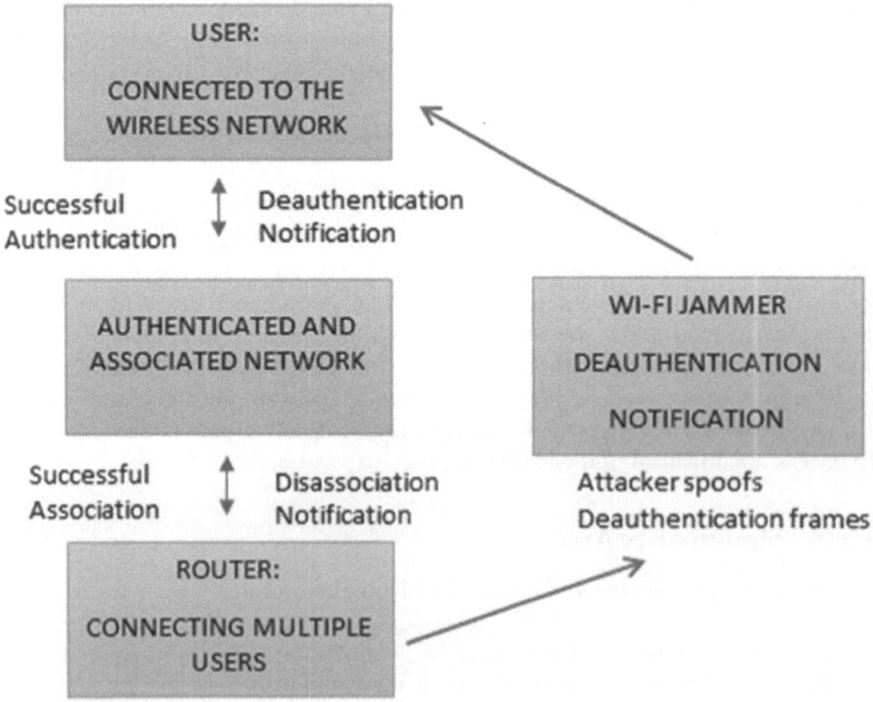

FIGURE 18.1 Flow chart of mounting an attack.

A jammer can disrupt the handshake process by interfering with the Wi-Fi signals, causing the authentication to fail. As a result, the client device won't be able to connect with the access point and access the network. Moreover, it can prevent successful association. As a result, the client device won't be able to connect with the access point and join the network.

The flow chart for mounting the attack is shown in Figure 18.1.

18.4 SYSTEM REQUIREMENT

System requirements refer to the minimum specifications and capabilities that a computer or other electronic device must meet to run a particular software program or operating system. This section discusses both hardware and software requirement for Wi-Fi jammer.

18.5 HARDWARE REQUIREMENT

For building a Wi-Fi jammer, NODEMCU, charging module, and booster module are required as discussed below:

18.5.1 NODEMCU

NODEMCU is an open source IoT platform with inbuilt microcontroller and Wi-Fi module as shown in Figure 18.2. It contains ESP8266 Wi-Fi chip on it. It is programmed using LUA scripting language. LUA scripting language is a programming language used to program various ESP boards.

18.5.2 Booster Module

The booster module is a DC-DC connection that boosts the voltage from its input to its output as shown in Figure 18.3. In our device, the booster module will enhance the battery voltage from 3.3V to 5V.

18.5.3 Charging Module

This module as shown in Figure 18.4 is mainly useful in charging rechargeable batteries using the constant current charging method. The battery's energy is used to power the device it is connected to, and it will gradually discharge over time. When the battery's charge is depleted, it needs to be recharged in order to continue providing power. A charging module, on the other hand, is a device that is used to charge the battery. It takes an external power source, such as a wall outlet or a USB port and converts it into the proper voltage and current to charge the battery.

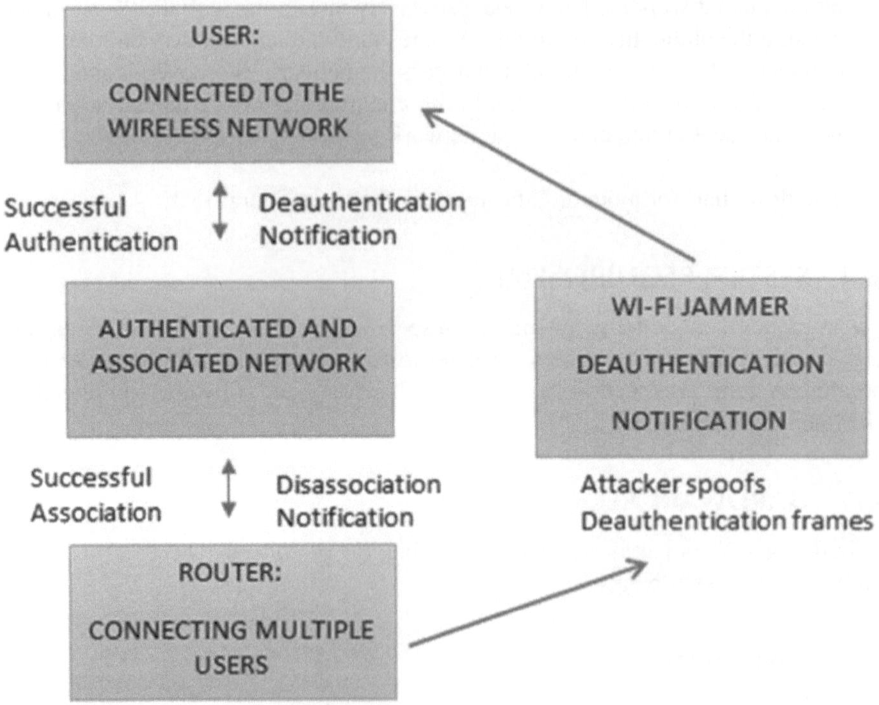

FIGURE 18.2 Hardware structure of NODEMCU.

FIGURE 18.3 Booster module.

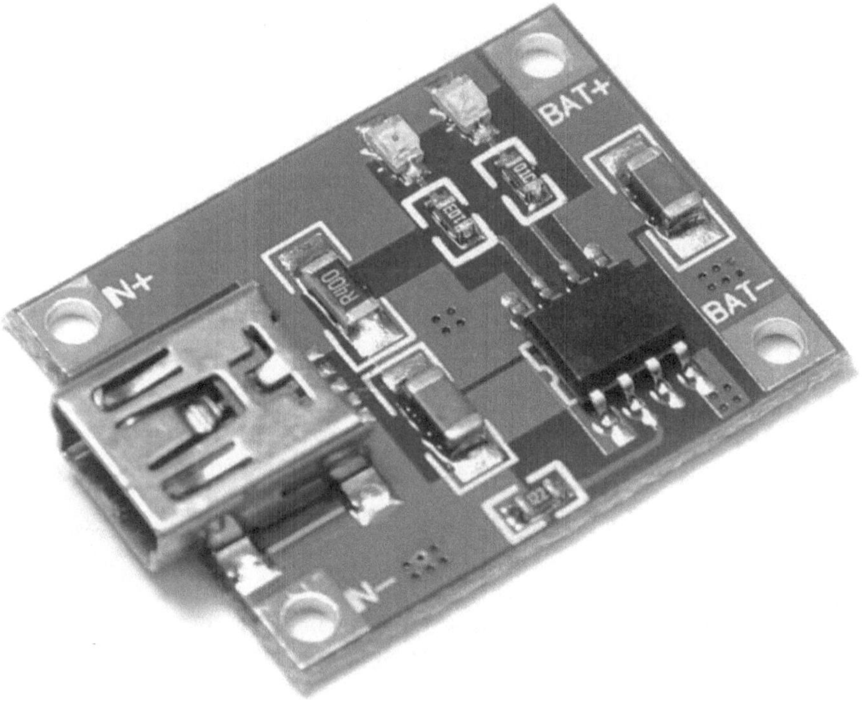

FIGURE 18.4 Charging module.

18.6 SOFTWARE REQUIREMENT

To program any ESP module, we need to do two important things as follows:

1. Selecting a suitable ESP8266 firmware
2. Uploading the firmware to flash memory of the ESP8266 module

Listed below are few community-supported firmware used for ESP8266:

1. AT Firmware (AT Commands)
2. NODEMCU (LUA Scripting language)
3. Arduino IDE (C/C++)
4. MicroPython (Python)

For our Wi-Fi jammer, we are going to program the ESP8226 using Arduino IDE.

The Arduino IDE is used to connect to the Arduino boards to upload programs and communicate with it [13, 14]. Programs written using Arduino are called sketches. The file extension used for saving sketches is ".ino".

18.7 CONFIGURATION

Before assembling all the hardware components into the box, firstly we need to program our NODEMCU which will be done using Arduino IDE. The step-by-step procedure to program NODEMCU is as follows:

a. Install, setup, and open the Arduino IDE in the system.
b. Initially, ESP8226 board is not installed in the application. So, we need to first install it in IDE. It can be installed in our application by simply copying shown URL "http://arduino.esp8266.com/stable/package_esp8266com_index.json" in the "Additional Boards Manager URLs" of Preferences section under the File tab and click Ok.
c. Install the ESP8226 board.
d. Click on the Packages>esp8226>2.0.0>hardware>esp8226>2.0.0>tools>sdk >include.
e. In the include folder, open user_interface.h with Notepad++ for copying the changes at the end.
f. Open esp8266_deauther-master> esp8266_deauther.ino (the program we need to flash into our board).
g. Connect the NODEMCU to the system using micro-USB cable.
h. Click on "Tools"> Board and locate the NODEMCU 0.9 (ESP-1.2 Module).
i. Again, click on "Tools" and select "ArduinoISP".
j. Upload the code to the device.

In the beginning of the code, the SS Id and password have been given which can be reset.

Now, the code has been successfully flashed to the NODEMCU and it is ready to use. In the next section, working and attacks performed by the device are mentioned.

18.8 ALGORITHMS TO MOUNT VARIOUS ATTACKS

In this section, several algorithms used to mount attacks on IoT network will be discussed.

18.8.1 Deauth Attack

In this, DeauthPacket function takes as input the Message Authentication Code (MAC) address of the target device and access point. It creates a deauthentication packet using the createDeauthPacket function and then sends the packet using the send Packet function.

The createDeauthPacket function allocates memory for the packet, sets the packet type to "deauthentication", sets the target and access point MAC addresses, and sets the reason code for the deauthentication. The sendPacket function sends the packet over the wireless interface.

The algorithm for deauth attack has been discussed in Algorithm 18.1.

Algorithm 18.1 Deauth Attack

```
DEAUTH_PACKET_LENGTH = 1000
DEAUTH_REASON_CODE = 1
function sendDeauthPacket (targetMacAddress, apMacAddress)
   packet=createDeauthPacket(targetMacAddress,apMacAddress,DEAUTH_
   REASON_CODE)
sendPacket(packet)
function createDeauthPacket(targetMacAddress, apMacAddress, reasonCode)
   packet = allocatePacketMemory(DEAUTH_PACKET_LENGTH)
setPacketType(packet, DEAUTH_PACKET_TYPE)
setTargetMacAddress(packet, targetMacAddress)
setApMacAddress(packet, apMacAddress)
   return packet
function sendPacket(packet)
sendPacketOverWireless(packet)
```

The Algorithm 18.1 has been discussed in detail as below:

DEAUTH_PACKET_LENGTH: A constant that defines the length of the deauth packet.

DEAUTH_REASON_CODE: A constant that represents the reason code for the deauthentication.

function sendDeauthPacket (targetMacAddress, apMacAddress): The sendDeauthPacket function is called with the target MAC address (of the client device to be deauthenticated) and the AP (access point) MAC address. Inside this function, a deauth packet is created using the createDeauthPacket function, passing the target MAC address, AP MAC address, and the reason code.

function createDeauthPacket(targetMacAddress, apMacAddress, reasonCode): The createDeauthPacket function allocates memory for the deauth packet, sets the packet type to the deauth packet type, sets the target MAC addresses and AP MAC address in the packet, and returns the packet.

sendPacket (): The sendPacket function is called, passing the deauth packet. This function is responsible for sending the packet over the wireless network.

18.8.2 Beacon Attack

This pseudocode in Algorithm 18.2 defines three functions, i.e., sendBeaconPackets, createBeaconPacket, and sendPacket. The sendBeaconPackets function takes as input the MAC address of the target device. It sends a series of beacon packets to the target by calling the createBeaconPacket and sendPacket functions in a loop. The createBeaconPacket function allocates memory for the packet, sets the packet type to "beacon", sets the target MAC address, and sets the beacon payload to a fake value. ThesendPacket function sends the packet over the wireless interface.

Algorithm 18.2 Beacon Attack

```
BEACON_PACKET_LENGTH = 1000
function sendBeaconPackets(targetMacAddress)
  for i = 1 to 10000
     packet = createBeaconPacket(targetMacAddress)
sendPacket(packet)

function createBeaconPacket(targetMacAddress)
  packet = allocatePacketMemory(BEACON_PACKET_LENGTH)
setPacketType(packet, BEACON_PACKET_TYPE)
setTargetMacAddress(packet, targetMacAddress)
setBeaconPayload(packet, "fake_beacon_payload")
  return packet
function sendPacket(packet)
sendPacketOverWireless(packet)
```

The detailed step has been discussed below:

BEACON_PACKET_LENGTH: A constant that represents the length of the beacon packet.

function sendBeaconPackets(targetMacAddress): The sendBeaconPackets function is called with the target MAC address (presumably of the intended recipient of the fake beacon packets). Inside this function, a loop is executed 10,000 times. In each iteration of the loop, a beacon packet is created using the createBeaconPacket function, passing the target MAC address.

function createBeaconPacket(targetMacAddress): The createBeaconPacket function allocates memory for the beacon packet, sets the packet type to the beacon packet type, sets the target MAC address in the packet, sets a fake beacon payload, and returns the packet.

sendPacket(packet): The sendPacket function is called, passing the beacon packet. This function is responsible for sending the packet over the wireless network.

18.8.3 CloneBeacon

The cloneBeacon attack is mounted using three functions, i.e. cloneBeacon, createBeaconPacket, and sendPacket as shown in Algorithm 18.3. The cloneBeacon function takes as input the MAC address of the legitimate AP. It creates and sends beacon packets with the same MAC address in an infinite loop. The createBeaconPacket function allocates memory for the packet, sets the packet type to "beacon", sets the AP MAC address to the legitimate address, and sets the beacon payload to the legitimate beacon payload. The sendPacket function sends the packet over the wireless interface.

Algorithm 18.3 CloneBeacon

```
BEACON_PACKET_LENGTH = 1000
function cloneBeacon(legitimateApMacAddress)
   while (true)
      packet = createBeaconPacket(legitimateApMacAddress)
sendPacket(packet)

function createBeaconPacket(apMacAddress)
   packet = allocatePacketMemory(BEACON_PACKET_LENGTH)
setPacketType(packet, BEACON_PACKET_TYPE)
setApMacAddress(packet, apMacAddress)
setBeaconPayload(packet, "legitimate_beacon_payload")
   return packet
function sendPacket(packet)
sendPacketOverWireless(packet)
```

The detailed step has been discussed below:

BEACON_PACKET_LENGTH: A constant to represent the length of the beacon packet.

function cloneBeacon (legitimateApMacAddress): The cloneBeacon function is called with the MAC address of the legitimate AP as the parameter. Inside this function, an infinite loop is executed. In each iteration of the loop, a beacon packet is created using the createBeaconPacket function, passing the MAC address of the legitimate AP.

function createBeaconPacket (apMacAddress): The createBeaconPacket function allocates memory for the beacon packet, sets the packet type to the beacon packet type, sets the MAC address of the legitimate AP in the packet, sets a legitimate beacon payload, and returns the packet.

function sendPacket(packet): The sendPacket function is called, passing the beacon packet. This function is responsible for sending the packet over the wireless network.

18.9 RESULT AND ANALYSIS

In this section, results achieved by performing various attacks have been discussed.

18.9.1 Deauthentication Attack

This attack targets the communication between device and router by disabling Wi-Fi on the connected device. In this type of attack, Wi-Fi which is being attacked will not be shown in available networks.

For instance, suppose we are performing attack on available network "Saumya" as can be seen in Figure 18.5.

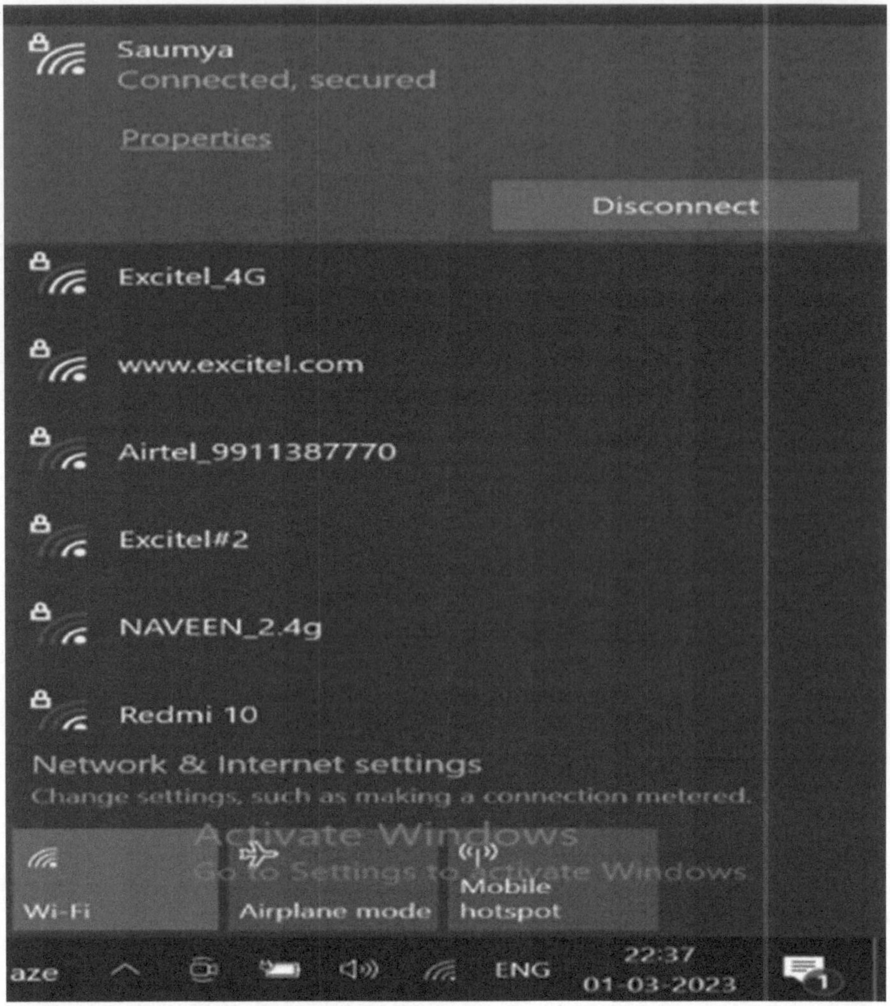

FIGURE 18.5 Deauthentication attack.

After mounting the deauthentication attack, network "Saumya" will not be seen in the available network and hence the user cannot connect to it.

18.9.2 Beacon Attack

In this attack, an attacker fakes these beacon frames and sends them in large numbers to confuse the wireless clients as shown in Figure 18.6. Sometimes, this may make things difficult or prevent the wireless clients from connecting to their preferred networks.

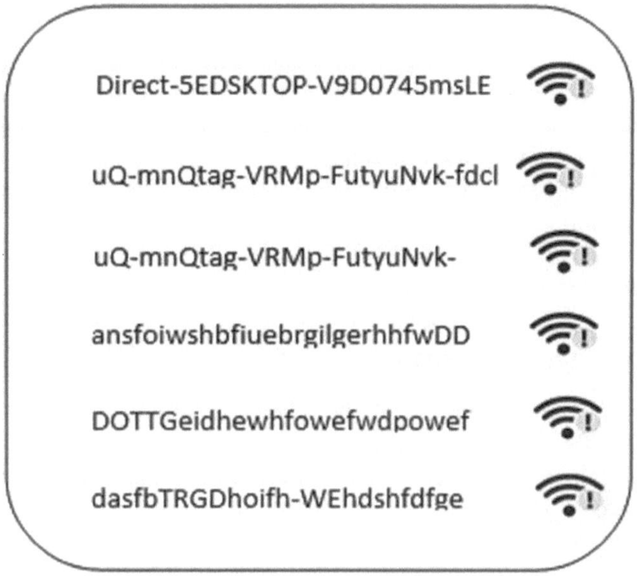

FIGURE 18.6 Beacon attack.

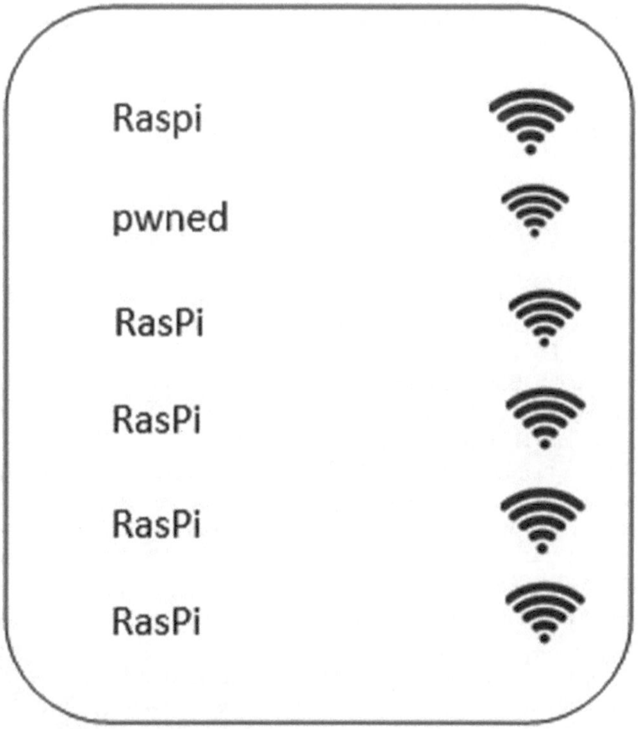

FIGURE 18.7 CloneBeacon attack.

18.9.3 CloneBeacon Attack

An evil twin attack is a spoofing cyberattack that works by tricking users into connecting to a fakeWi-Fi AP that mimics a legitimate network. Once a user is connected to an "clone beacon" network, hackers can access everything from their network traffic to private login credentials.

For instance, suppose we have mounted cloneBeacon attack on available network "RasPi", then multiple available networks will be shown for the same Wi-Fi as can be seen in Figure 18.7. Now, if a user gets connected to the fake network, his information will be stealed and stored in the database.

18.10 CONCLUSION AND FUTURE WORK

Wi-Fi jammers are devices that disrupt the normal functioning of a wireless network by sending specially crafted "deauthentication" packets to the network. They can be used for various purposes, including performing DoS attacks, stealing sensitive information transmitted over the wireless network, or disrupting critical communications. It is generally not legal to use a Wi-Fi jammer to disrupt the operation of a wireless network without the owner's permission. In the proposed work, Wi-Fi jammer can mount deauth, beacon, and clone beacon attack. Furthermore, it is simplified, portable, and cheaper than other devices in the market.

In future, some attacks can be added to Wi-Fi jammer such as probe attack. Furthermore, personal information of users can be captured by interfering in the communication between the user and the network and thus can cast a dark shadow on the IoT security.

REFERENCES

1. M. Ingham, J. Marchang, and D. Bhowmik, "IoT security vulnerabilities and predictive signal jamming attack analysis in LoRaWAN", *IET Information Security*, 14(4), 368–379, 2020.
2. G. Liu, J. Liu, Y. Li, L. Xiao, and Y. Tang, "Jamming detection of smartphones for WiFi signals", In *2015 IEEE 81st Vehicular Technology Conference (VTC Spring)* (pp. 1–3). IEEE, 2015.
3. K. Sharma and S. Bhatt, "Jamming attack–A Survey", *International Journal of Recent Research Aspects*, 5, 74–80, 2018.
4. H. Pirayesh and H. Zeng, "Jamming attacks and anti-jamming strategies in wireless networks: A comprehensive survey", IEEE Communications Surveys & Tutorials, 2022.
5. K. Pahlavan and P. Krishnamurthy, "Evolution and impact of Wi-Fi technology and applications: A historical perspective", *International Journal of Wireless Information Networks*, 28, 3–19, 2021.
6. S. Vadlamani, B. Eksioglu, H. Medal, and A. Nandi, "Jamming attacks on wireless networks: A taxonomic survey", *International Journal of Production Economics*, 172, 76–94, 2016.
7. D.G. Bhoyar and U. Yadav, "Review of jamming attack using game theory", In *2017 International Conference on Innovations in Information, Embedded and Communication Systems (ICIIECS)* (pp. 1–4) IEEE, 2017.

8. J. Singh, I. Woungang, S.K. Dhurandher, and K. Khalid, "A jamming attack detection technique for opportunistic networks", *Internet of Things*, 17, 100464, 2022.
9. G. Chatzisofroniou and P. Kotzanikolaou, "Association attacks in IEEE 802.11: Exploiting WiFi usability features", In *Socio-Technical Aspects in Security and Trust: 9th International Workshop, STAST 2019, Luxembourg City, Luxembourg, September 26, 2019, Revised Selected Papers* 9 (pp. 107-123), Springer International Publishing.
10. A. Aggrawal, I. Arora, I and A. Giri, "Analysis and Rendering of Deauthentication Attack Using IoT Technology" In *Proceedings of 3rd International Conference on Recent Trends in Machine Learning, IoT, Smart Cities and Applications: ICMISC 2022* (pp. 41–51). Singapore: Springer Nature Singapore.
11. R. Singh, R. Thakkar, M. Thakkar, U. Rote, S. Patil, B. Ingle, B, "WiFiDeauth and Cloning using ESP8266", In *2022 5th International Conference on Advances in Science and Technology (ICAST)* (pp. 1–5), IEEE, 2022.
12. A.C. Chan, R.M. Chung, "Security and Privacy of Wireless Beacon Systems", arXiv preprint arXiv:2107.05868, 2021.
13. H.K. Kondaveeti, N.K. Kumaravelu, S.D. Vanambathina, S.E. Mathe, S. Vappangi, "A systematic literature review on prototyping with Arduino: Applications, challenges, advantages, and limitations", *Computer Science Review*, 40, 100364, 2021.
14. M.N.M. Najath, H.M.D.S. Herath, A. Rajapakse, "Design and Testing of an Arduino-based Network Jammer Device" In *2022 7th International Conference on Information Technology Research (ICITR)* (pp. 1–4), IEEE, 2022.

Index

N

O

P

Q

R

S

T

U

V

W